W9-ABI-169

Wastes in Marine Environments

Wastes in Marine Environments

Office of Technology Assessment Task Force

Howard Levenson, Project Director
William D. Barnard, Senior Analyst

 SCIENCE INFORMATION RESOURCE CENTER

⊙ HEMISPHERE PUBLISHING CORPORATION
Cambridge   New York   Philadelphia   San Francisco
London   Mexico City   São Paulo   Singapore   Sydney

46.04

1-34-90

WASTES IN MARINE ENVIRONMENTS

Publishers' Note: This permanent edition contains the complete text of the Office of Technology Assessment Special Report, *Wastes in Marine Environments,* prepared by a distinguished Project Staff and Advisory Panel.

1 2 3 4 5 6 7 8 9 0   B C B C   8 9 8 7

Library of Congress Cataloging-in-Publication Data

Wastes in marine environments.

    "Howard Levenson, project director [and] William
D. Barnard, senior analyst."
    Reprint. Originally published: Washington, DC :
Congress of the U.S., Office of Technology Assessment,
1987.
    Bibliography: p.
    Includes index.
    1. Waste disposal in the ground—Environmental
aspects—United States. 2. Waste disposal in the
ocean—Environmental aspects—United States. 4. Environmental
impact analysis—United States. I. United States.
Congress. Office of Technology Assessment.
II. Science Information Resource Center (Philadelphia,
Pa.)
TD897.7.W38   1988   363.7'28   87-25211
ISBN 0-89116-793-5

Contents

Advisory Panel

Thomas Clingan, *Chair*
School of Law, University of Miami

Michael G. Norton, *Co-Chair*
Warren Spring Laboratory, United Kingdom
(formerly First Secretary, British Embassy)

Walter Barber
Waste Management, Inc.

Willard Bascom
Scripps Institute of Oceanography
(formerly with Southern California
   Coastal Water Research Project)

Rita Colwell
Department of Microbiology
University of Maryland

A. Myrick Freeman III
Department of Economics
Bowdoin College

John Gosdin
Governor's Office
State of Texas

Jon Hinck
Greenpeace, USA

Kenneth Kamlet
A.T. Kearney, Inc.
(formerly with National Wildlife
   Federation)

John A. Knauss
Graduate School of Oceanography
University of Rhode Island

George Lutzic
New York City Department of
   Environmental Protection

William J. Marrazzo
City of Philadelphia

Joseph T. McGough, Jr.
Parsons Brinckerhoff
(formerly with New York City
   Department of Environmental
   Protection)

Jerry Schubel
Marine Sciences Research Center
State University of New York at Stony
   Brook

Richard F. Schwer
Engineering Department
E.I. du Pont de Nemours & Co.

NOTE: OTA appreciates and is grateful for the valuable assistance and thoughtful critiques provided by the advisory panel members. The panel does not, however, necessarily approve, disapprove, or endorse this report. OTA assumes full responsibility for the report and the accuracy of its contents.

OTA Project Staff

John Andelin, *Assistant Director, OTA*
Science, Information, and Natural Resources Division

Robert M. Niblock, *Oceans and Environment Program Manager*

Howard Levenson, *Project Director*

William D. Barnard, *Senior Analyst**

Richard A. Denison, *Analyst***

Gretchen E. Hund, *Analyst*

Nicholas A. Sundt, *Analyst*

Kathryn D. Wagner, *Congressional Fellow*

Patricia A. Catherwood, *Analyst*

Dale T. Brown, *Analyst*

Administrative Staff

Kathleen A. Beil

Jim Brewer, Jr.

Brenda B. Miller

---

* Project Director through November 1985.

** Principal Author of OTA's report, ''Ocean Incineration: Its Role in Managing Hazardous Waste.''

Foreword

"Where can we dispose of our wastes?" is a question being faced by virtually every community and industry in the country. For years, one common answer for communities and industries located in coastal areas has been to intentionally dispose of large amounts of waste materials in the Nation's marine environments—estuaries, coastal waters, and the open ocean.

Two congressional committees—the House Committees on Merchant Marine and Fisheries and on Public Works and Transportation—requested OTA to undertake a broad assessment of waste disposal in marine environments. In addition, the Senate Committee on Commerce, Science, and Transportation endorsed the assessment. As part of the assessment, OTA issued a report on *Ocean Incineration: Its Role in Managing Hazardous Waste* (August 1986), and a staff paper on *Subseabed Disposal of High-Level Radioactive Waste* (May 1986).

This final report addresses two fundamental questions: what is the general condition of different marine environments and their resources, and what role can and should marine environments play in overall waste management? OTA's principal findings are that estuaries and coastal waters are in deep trouble around the Nation, and that more coordinated waste management efforts are needed in many areas. As a Nation, we have been only partially successful in protecting these waters—and their ecologically, commercially, and esthetically valuable resources—from degradation. Policies to maintain or improve their quality, however, must be implemented in conjunction with policies about waste management strategies in general, including disposal in the open ocean and on land.

OTA is particularly grateful for the considerable effort devoted to this undertaking by our advisory panel, numerous contacts in Federal and State agencies, and numerous individuals in industry, academia, and public interest and environmental groups. These individuals helped OTA examine the enormous amount of available information from a number of important perspectives. We greatly appreciate this help.

JOHN H. GIBBONS
Director

Chapter 1
Findings and Options

CONTENTS

Figure

Boxes

Findings and Options

OVERVIEW

The marine waters of the United States—estuaries, coastal waters, and the open ocean[1]—are used extensively for the disposal of various types of waste. Much public concern and debate has focused on the form of disposal known as dumping, which occurs when wastes such as sewage sludge, industrial wastes, and dredged material are transported by ships or barges to designated marine sites and dropped overboard. Relatively less attention has been given to other marine disposal activities such as the discharge of industrial and municipal effluents from numerous pipelines and to nonpoint pollution from agricultural and urban runoff. Pipeline discharges and runoff, however, are at least as important as dumping in causing impacts on marine resources.[2]

OTA believes the most productive way to look at the disposal of wastes in the Nation's marine environments is to understand two fundamental issues: first, the general condition of each of the types of marine waters that are used for disposal; and second, the nature and extent of the role that these waters can and should play in waste management. This study's major findings about the first issue point to several policy options that could be instituted to maintain or improve the condition of these waters. The study also explores the policy implications of these options within the broad context of the second issue—the role of marine waters in waste management.

OTA developed three major findings concerning the health of the Nation's marine environments. Although discussed later in this chapter and throughout the report, summarized briefly they conclude the following:

- **Estuaries and coastal waters around the country receive the vast majority of pollutants introduced into marine environments. As a result, many of these waters have exhibited a variety of adverse impacts, and their overall health is declining or threatened.**
- **In the absence of additional measures, new or continued degradation will occur in many estuaries and some coastal waters around the country during the next few decades (even in some areas that exhibited improvements in the past).**
- **In contrast, the health of the open ocean generally appears to be better than that of estuaries and coastal waters.** Relatively few impacts from waste disposal in the open ocean have been documented, in part because relatively little waste disposal has taken place there and because wastes disposed of there usually are extensively dispersed and diluted. Uncertainty exists, however, about the ability to discern impacts in the open ocean.

Managing Estuaries and Coastal Waters

Several Federal "pollutant control" programs have been established under the Clean Water Act (CWA) and the Marine Protection, Research, and Sanctuaries Act (MPRSA) to regulate the disposal (via both discharge and dumping) of wastes into marine waters and to control the levels of pollutants in these wastes.[3] The cornerstone of these programs has been the promulgation of *uniform* national regulations applicable to point sources of wastes or pollutants. Using this approach, some significant reductions in the quantities of pollutants entering marine waters have been and will probably continue to be achieved.

---

[1]These terms are defined in box A.

[2]These terms are described in box B. OTA analyzed the ocean incineration of hazardous wastes in a companion report, *Ocean Incineration: Its Role in Managing Hazardous Waste* (586) and the potential disposal of high-level radioactive waste under the seabed in a staff paper, *Subseabed Disposal of High-Level Radioactive Waste* (585). Box B lists other sources of pollution that are not covered in this assessment.

[3]These statutes are discussed in box A and in ch. 7. The term pollutant is defined and types of pollutants are described in box B.

Box A.—Marine Environments and Relevant Federal Statutes

Types of Marine Environments*

Marine environments are classified in this study into three general categories: estuaries, coastal waters, and the open ocean.

Estuaries refer to semi-enclosed bodies of water that have some connection to the open ocean and an input of freshwater that mixes with saltwater. Most often estuaries exist at the lower reaches of rivers, but estuarine waters also can include lagoons and tidal marshes. Well-known examples include the Chesapeake Bay and the San Francisco Bay.

Coastal waters are generally less enclosed and more influenced by oceanic processes than are estuaries; they generally lie over the continental shelf, within the territorial sea.** Examples include the Southern California Bight and the New York Bight. The New York Bight extends beyond the territorial sea but is considered to be a coastal water because it is located on the inner portion of the continental shelf and is influenced by freshwater input from rivers.

The *open ocean* refers both to deep waters beyond the continental shelf and to waters over the outer continental shelf that are not measurably affected by freshwater input from rivers. The Deepwater Disposal Sites, which are used for the dumping of sewage sludge and some industrial wastes, are located on the continental shelf break in the open ocean.

Major Statutes Regulating Waste Disposal in Marine Environments

The Marine Protection, Research, and Sanctuaries Act (MPRSA) regulates the transportation and dumping of wastes in waters seaward of the baseline (inner boundary) of the territorial sea (see figure 1). The waters subject to MPRSA can be either open ocean (e.g., the Deepwater Disposal Sites) or coastal waters (e.g., the dumping site in the New York Bight); estuarine waters are excluded because they lie landward of the baseline (45 FR 65944, Oct. 3, 1980).***

The Federal Water Pollution Control Act, known as the Clean Water Act (CWA), regulates all discharges into navigable waters of the United States, including the territorial sea (Sec. 502). Thus, CWA coverage generally extends to pipeline discharges and the dumping of waste from vessels in estuaries and some coastal waters. CWA also covers discharges from point sources other than vessels in waters beyond the territorial sea (e.g., in the New York Bight or the open ocean). CWA would thus apply to long outfalls from land-based facilities and to discharges from stationary drilling platforms, but not to the dumping of waste from vessels in waters beyond the territorial sea (the latter is regulated under MPRSA).

These two Acts overlap in their coverage of dumping from vessels within the territorial sea. Currently, such activity is restricted to dredged material disposal. In such cases, however, MPRSA preempts CWA (by the authority of MPRSA Sec. 106(a)). Thus, dumping that occurs in coastal waters or the open ocean is subject to the requirements of MPRSA; dumping in estuaries is subject to CWA.

---

*This assessment excludes consideration of the Great Lakes and wetlands. For a discussion of wetlands issues, see ref. 582.

**The *territorial sea* extends seaward for 3 miles from what is called the baseline. The *baseline* is defined in Sec. 502(a) of the Clean Water Act as the "belt of the seas measured from the line of ordinary low water along that portion of the coast which is in direct contact with the open sea and the line marking the seaward limit of inland waters."

***For domestic vessels, MPRSA applies to dumping in all waters seaward of the baseline; for international vessels, it applies only to dumping in waters within the contiguous zone.

Figure 1.—Jurisdictional Boundaries of Environmental Laws Affecting Marine Disposal

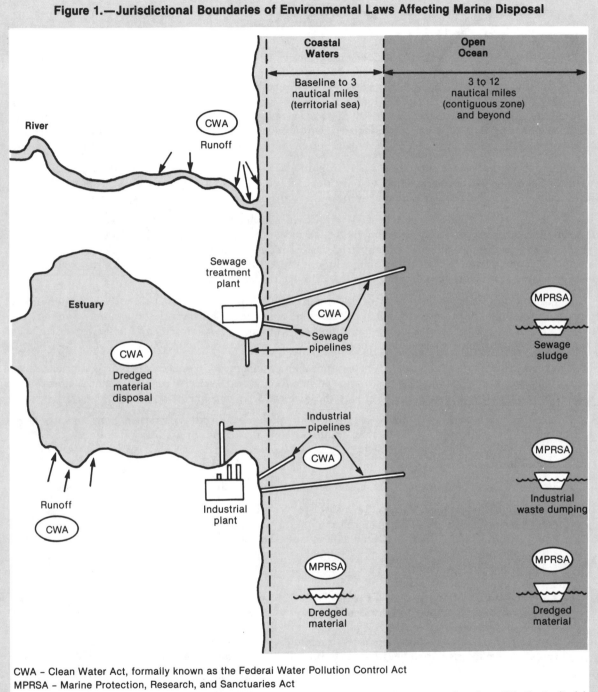

CWA - Clean Water Act, formally known as the Federal Water Pollution Control Act

MPRSA - Marine Protection, Research, and Sanctuaries Act

Dumping beyond the inner boundary of the territorial sea is covered by MPRSA (CWA covers dumping within the territorial sea in principle, but is preempted by MPRSA (see box A)). Estuarine dumping falls under CWA.

Pipelines (wherever they are located) are covered by CWA.

SOURCE: Office of Technology Assessment, 1987; adapted from National Advisory Committee on Oceans and Atmosphere, *The Role of the Ocean in a Waste Management Strategy* (Washington, DC: U.S. Government Printing Office, 1981).

Box B.—Sources and Types of Pollutants Entering Marine Environments

Sources of Pollutants

The term *pollutant* is used in this assessment to refer to a substance or organism present in a waste, or to a property of a waste such as acidity, that *has the potential to affect* marine resources or humans. This definition is broader than that used by some observers, who limit the term to substances that are demonstrated to *have actually caused an adverse impact*. For consistency, and because of acknowledged limitations in the ability to detect impacts associated with marine waste disposal, the broader definition is used throughout this assessment. Although impacts associated with a pollutant are generally adverse, in certain settings the same pollutant may exert a beneficial impact, or cause an impact whose character is difficult to assess.

Pollutants can enter fresh or marine waters from intentional *waste disposal* (also known as point sources) or from what are called *nonpoint sources*. While this study is focused primarily on waste disposal, it places this in context with some major nonpoint sources that introduce pollutants into marine waters.*

Waste disposal encompasses a variety of discrete activities, including the *dumping* of sewage sludge, dredged material, and other wastes (e.g., acid-iron wastes, seafood processing wastes, low-level radioactive wastes) and the *discharge* through pipelines of industrial and municipal effluents. Upon disposal, wastes can exert impacts as a consequence of two factors: the pollutants present in the waste, or the physical character of the waste itself (e.g., as in the burial of organisms by dredged material).

Major nonpoint sources of pollutants include runoff caused by precipitation on land surfaces, underground transport through aquifers, and atmospheric deposition. Rural (or nonurban) runoff can occur on agricultural, range, mining, and forest lands. Urban runoff generally reaches waterbodies by flowing over surfaces such as streets, lawns, and construction sites.

The distinction between intentional waste disposal and nonpoint sources becomes blurred, for example, when urban runoff (a nonpoint source) is channeled into "combined" sewers (which carry both stormwater runoff and sewage) that subsequently discharge the material from pipelines.

Both waste disposal and nonpoint sources contribute pollutants directly to marine environments. They also can introduce pollutants into rivers that eventually flow into estuaries and coastal waters.

Types of Pollutants

Pollutants are classified in the Clean Water Act into three groups** for purposes of regulation:

1. *Conventional pollutants* (listed in Sec. 304(a)(4)) consist of:
   - total suspended solids (TSS; material remaining suspended in a waste after treatment);
   - biochemical oxygen demand (BOD; a measure of the amount of oxygen consumed during degradation of waste);
   - pH (a measure of the acidity of a waste);
   - fecal coliform bacteria; and
   - oil and grease (designated by EPA (40 CFR 401.16)).
2. *Toxic pollutants* (mandated under Sec. 307(a)(1) and listed in 40 CFR 401.15) include metals and organic chemicals.***
3. *Non-conventional pollutants* comprise a "catch-all" category that can include any additional substance that may require regulation (44 FR 44501, July 30, 1979). The current Federal list includes nutrients

---

*This assessment does not consider sources of pollutants to marine waters that are distinct from land-based or land-originating sources. These include atmospheric inputs, discharges resulting from the operation of vessels (e.g., bilge waters), accidental releases (e.g., oil spills), and marine mining operations other than oil and gas drilling. In addition, wastes or pollutants such as plastics are considered only briefly (see box J in ch. 3).

**These classes are specified primarily for the purpose of developing effluent guidelines for pipeline discharges, but are generally useful in discussing pollutants present in other wastes as well.

***EPA currently lists a total of 126 toxic "priority" pollutants (40 CFR 122, app. D), an expansion of the initial list of 65. Other metals and organic chemicals, however, can also cause adverse impacts. OTA restricts the term *toxic* to the list of 126 substances and uses *metals or organic chemicals* more generally to refer to **any** metal or organic chemical.

such as total nitrogen, nitrates, total phosphorus, and other substances such as chlorine, fluoride, and certain metals (40 CFR 122, app. D).****

Regulated Pollutants

In principle, any substance discharged or dumped into marine waters is subject to regulation as a pollutant under the Clean Water Act and the Marine Protection, Research, and Sanctuaries Act: "catch-all" provisions in both statutes (Secs. 301 and 101, respectively) require that disposal of any material that would impede achievement of the broad goals of these statutes be regulated. To facilitate the development of regulations and to provide some degree of consistency, specific lists of pollutants have been developed by the Federal Government. However, for a variety of economic, technical, and environmental reasons, standards that actually limit release have been developed for only a subset of these substances. Moreover, many additional pollutants that have been identified in wastes may be of concern with respect to environmental or human health; in some cases, State or local limits on such substances have been developed, but many remain entirely unregulated (see ch. 8).

In practice, then, the term *regulated pollutant* has a rather limited meaning, referring only to: 1) those substances specifically included on government lists, or 2) the subset of these (plus any additional) substances for which limits are actually specified in discharge or dumping permits or ordinances. In this assessment, the term refers to a substance that meets one or both of the above criteria, with the understanding that even for many regulated pollutants, actual limits governing their disposal have not been developed. The term *unregulated pollutants* is reserved for other potentially significant pollutants that do not meet either of the above criteria.

****The statutory basis for regulation of substances other than conventional or toxic pollutants is provided by Sec. 301(b)(2)(F) of the Clean Water Act.

These programs represent reasonable approaches to address the problem of pollution in marine environments. However, while relatively easy to conceptualize, they have proven far more difficult to fully implement. Only partial implementation has been achieved to date and numerous obstacles hinder them from becoming fully operative. **Indeed, the prospect of ever achieving full implementation and enforcement is unlikely: the Nation's past commitment of resources has been insufficient to accomplish all the essential activities of existing programs (e.g., monitoring and enforcement, municipal treatment plant construction) and even these resources are now declining (chs. 7, 8, and 9).**

Moreover, even if total compliance with today's regulations is achieved, existing programs will not be sufficient to achieve some goals of the CWA, in particular to maintain or improve the health of all estuaries and coastal waters. In the absence of additional measures to protect our marine waters, the next few decades will witness new or continued degradation in many estuaries and coastal waters around the country (including some that exhibited past improvements):

- current programs do not adequately address toxic pollutants or nonpoint source pollution;
- pipeline discharges and nonpoint source pollution (particularly urban runoff) will increase as population and industrial development expand in coastal areas;[4]
- Federal resources available for municipal sewage treatment are declining, and the ability of States or communities to fill the breach is highly uncertain; and
- in many cases, economic, technical, or social factors will make it difficult or impossible to shift disposal or dumping of certain wastes out of estuaries and coastal waters.

_____

[4]The number of people living in counties near marine waters increased more than 80 percent from 1950 to 1984; by 1984, 40 percent of the U.S. population lived within 50 miles of a marine coastline (including counties near but not necessarily adjacent to marine coastlines, but excluding Hawaii, Alaska, and areas around the Great Lakes) (566). Coastal populations are projected to continue to increase, and the intensity of recreation, development, and waste disposal that can impact marine waters will increase accordingly.

This projection of continued or increasing degradation is of great concern because estuaries and coastal waters are among the most important of all marine environments, with respect to their commercial resources, recreational uses, and ecological roles (chs. 2 and 4). Moreover, the ability to detect such deterioration and to understand its causes will be hampered if funding for monitoring and basic research continues to decline.

The nature and extent of impacts, and their causes, show tremendous variation from one estuary or coastal water to another. This diversity suggests that any additional management efforts should be *site-specific*—i.e., tailored specifically to the needs and problems of individual waterbodies—regardless of whether such efforts are conducted by Federal, State, or local agencies. "Waterbody management" programs have been established for a few marine water bodies (e.g., the Chesapeake Bay Program), but many other estuaries and coastal waters need additional management.

If the Nation desires to maintain or improve the health of its estuaries and coastal waters, a two-tiered approach toward managing these waters will be needed. First, implementation of the present system of uniform national controls should be continued and enhanced to provide a consistent, minimum level of protection. Second, additional waterbody management that provides sufficient flexibility to address site-specific problems, while probably difficult to develop and implement, will be needed in many areas to supplement current programs. OTA's analysis of policy options for estuaries and coastal waters reflects this two-tiered approach.

Managing Open Ocean Waters

The health of the open ocean generally appears to be better than that of estuaries and coastal waters. Relatively few impacts from waste disposal have been observed, partly because the open ocean has been subject to relatively little waste disposal and because wastes are typically dispersed and diluted. Considerable uncertainty still exists, however, about the ability to discern impacts, particularly long-term ones, that may have occurred in the open ocean.

MPRSA has been relatively successful in managing dumping and providing some degree of protection for the open ocean. Nevertheless, the potential for harm to some valuable resources exists (e.g., from toxic chemicals such as polychlorinated biphenyls (PCBs), which have been detected in open-ocean fish).

Policy options for the open ocean discuss the implications of increasing, maintaining, or easing the current restrictions on open ocean disposal. If wastes are disposed of in the open ocean, it seems prudent to ensure that they contain low levels of toxic pollutants. Few long-term adverse consequences would be expected if relatively uncontaminated sewage sludge and dredged material were to be dumped in the open ocean under dispersive conditions.

Viewing Marine Waste Disposal in Broad Context

The environmental legislation passed in the 1960s and 1970s and the continued popularity of the environmental movement are clear expressions of society's desire to protect the environment, including marine waters. The expected degradation in many estuaries and coastal waters and the relatively greater degree of protection afforded the open ocean, however, in some respects reflect a lack of *comprehensive* waste management (ch. 2). Current programs established to manage wastes focus primarily on one waste source or on disposal in one environment. Such narrowly focused programs were reasonable steps in approaching pollution problems. Attempts to control one problem, however, sometimes generate other problems, and pollutants often have been merely transferred among environments or wastestreams without any significant overall reduction in associated risks.[5]

Some problems might be alleviated if policy choices about the role of marine waters in waste disposal were made within the context of a hierarchy of preferred waste management strategies (262, 377,586). These strategies include:

---

[5]For example, the processes used to remove conventional pollutants from municipal wastewater result in increased production of sewage sludge. Moreover, most sludge is contaminated with toxic pollutants from industrial and other discharges into municipal sewers. While disposal of uncontaminated sludge faces obstacles, the disposal of contaminated sludge is even more severely constrained: it often cannot be applied on land and may not be amenable to incineration, landfill disposal, or ocean dumping (ch. 9).

- reducing the generation of wastes;
- when possible, beneficially using or recycling wastes; and
- when beneficial uses are not possible, choosing treatment or disposal options that cause the least damage to the environment and human health and that are acceptable to society at large.[6]

None of these options eliminates risks entirely, and in some cases new risks can be created. **Moreover, not all waste generation can be eliminated.** Once wastes are generated, some type of "multimedia assessment" that compares the risks of different treatment and disposal options can help determine a preferred strategy in a given situation. Even then, a critical component will be public acceptability of the strategy itself and of the decision-making process (ch. 2).

To the extent that waste generation can be reduced, the need for disposal in different environments, including marine waters, can also be reduced. **It is evident, however, that large amounts of wastes requiring disposal (e.g., municipal effluents and sludge, industrial effluents, and dredged material) will continue to be produced, although the levels of specific pollutants in these wastes could be lowered.** At the same time, there is a strong desire for waste disposal to be inexpensive and to occur in remote locations. Several factors will increase pressure to use marine environments for waste disposal:

- the proximity of marine waters to major and/or growing urban areas that generate large amounts of wastes requiring disposal;
- the frequently lower costs of marine disposal;
- limits on the economic feasibility of land-based disposal for some highly voluminous wastes (e.g., municipal effluents); and
- limits on the availability of land-based disposal options for some wastes (e.g., sewage sludge,

dredged material, and some industrial wastes) because of increased public opposition and State or local regulatory restrictions.[7]

Policy Choices for Marine Waste Disposal

As indicated by OTA's analysis, the degradation of marine waters is most threatening in many estuaries and coastal waters. The open ocean, in contrast, exhibits relatively better health and has received a greater degree of protection. Thus, different policy choices are applicable to estuaries and coastal waters and to the open ocean.

Estuaries and Coastal Waters

With regard to impacts caused by waste disposal activities and runoff, the only policy choice available to maintain and improve the health of estuaries and coastal waters is to minimize pollutant inputs to these waters. One option to minimize inputs is to shift some disposal activities to the open ocean (depending on policies regarding open ocean disposal, discussed below), for example, by extending pipelines or moving the dumping of dredged material.[8] For a variety of technical, logistical, and economic reasons, however, it appears unlikely that a significant number of pipelines now located in estuaries and coastal waters could be extended much further offshore. Similarly, at least some dumping of dredged material in estuaries and coastal waters will be necessary. Some disposal activities might be moved to land, but the availability of some land-based options is becoming more restricted.

For these reasons, several other, more feasible options for minimizing waste disposal and pollutant inputs in estuaries and coastal waters deserve attention. These options are organized by OTA within a two-tiered approach:

1. **Maintain or consider expanding the current system of pollutant controls,** as exem-

---

[6]In this assessment, waste reduction includes those activities at the generating source that reduce the degree of **risk** associated with waste byproducts. OTA has analyzed the potential for, and obstacles to, achieving greater waste reduction (587). Reduction and reuse options may be applicable to some extent even to wastes commonly considered to be difficult to reduce or reuse. In some parts of the country, for example, municipal effluents are reclaimed for use in irrigation or groundwater recharge. Water conservation efforts (e.g., use of waterless toilets) could reduce the quantity of wastewater requiring disposal.

[7]Restrictions on land-based disposal, mandated by the 1984 Hazardous and Solid Wastes Amendments to the Resource Conservation and Recovery Act (RCRA) (see ch. 7), also could increase pressure to dispose of some hazardous wastes in the ocean (241,263).

[8]Some shifting of sewage sludge dumping from coastal waters (at the 12-Mile Sewage Sludge Dump Site in the New York Bight) to the open ocean (at the Deepwater Municipal Sludge Site, 125 to 150 nautical miles southeast of New York harbor) is already underway.

plified by CWA's uniform technology-based controls and requirements. Ensuring maintenance would require some combination of continued Federal, State, and local investments in:
—the construction of municipal treatment plants;
—increased and sustained support for enforcement efforts; and
—increased and sustained support for monitoring and research, to aid enforcement and evaluate long-term trends.
Expanding the system would involve regulating more toxic pollutants, industrial sources, and pathogens.

2. **Establish additional, site-specific controls on waste disposal and nonpoint pollution where needed.**[9] This would require:
—identifying those areas where such controls are needed (i.e., where the first tier of controls is not sufficient);
—establishing measurable, site-specific goals toward which progress could be evaluated; and
—in some cases, initiating or expanding formal "waterbody" management plans such as those developed for the Chesapeake Bay and Puget Sound.

Both Congress and the Environmental Protection Agency (EPA) are well aware of the need to continue supporting existing efforts and to develop new initiatives like those listed above. Congressional awareness of these needs is reflected in some of the major provisions of the Water Quality Act of 1987, which amended the Clean Water Act (box C). For example, Congress expressed its intent to continue Federal funding, for a limited time, of municipal treatment plant construction (although at a level considerably below that estimated by EPA to be needed); to promote additional management of various estuaries; and to provide funding for States to develop nonpoint source pollution programs. EPA has been involved in developing several waterbody management programs (e.g., the Chesapeake Bay and National Estuary Programs; see ch. 7), and has begun several efforts to iden-

tify waterbodies needing additional management (246,670). **Many of these initiatives to provide additional, site-specific controls are in their infancy, however, and they will require much more direction, support, and oversight from Congress. Furthermore, these efforts currently are not part of a single, integrated strategy.**

Establishing additional, site-specific controls could be aided by increasing the emphasis given to the "water quality" approach. This approach consists of designating desired goals such as fishable waters for a waterbody, developing pollutant-specific numerical criteria that establish the quality of water needed to attain the goals, and implementing controls on wastes or pollutants from point and/or nonpoint sources to meet the criteria. The water quality approach, which has always been a component of CWA, was intended to supplement the uniform pollutant controls once they were well-established and thus provide an additional layer of controls when and where necessary. EPA has developed some water quality-based controls, but **in general the water quality approach has not been systematically applied to estuaries and coastal waters. Given OTA's findings about the declining health of many estuaries and coastal waters and the limitations on the effectiveness of pollutant control programs, it now seems appropriate that Congress and EPA begin developing a systematic framework to implement the water quality approach more extensively.**

Open Ocean Waters

Several distinct policy choices about the use of the open ocean for waste disposal are possible:

1. **maintain current restrictions on and allowances for open ocean disposal,**
2. **tighten these restrictions, or**
3. **ease them.**

Deciding which policy to choose is not clearcut and depends on factors such as the availability of disposal options on land and in estuaries and coastal waters, as well as on the character of the particular waste in question. For example, *uncontaminated* sewage sludge and dredged material might best be used beneficially on land or in certain aquatic settings (e.g., sludge could be used to fertilize forestland; dredged material could

---

[9]Although specific policy options for nonpoint pollution are not developed here, the relative importance of pollutants from disposal activities and nonpoint sources (particularly runoff) is evaluated in ch. 3.

Box C.—The Water Quality Act of 1987

The Water Quality Act of 1987 reauthorized and amended the Clean Water Act. Some of the major provisions of the act that are pertinent to waste disposal and pollutants in the Nation's marine waters include:

- **Construction Grants Program and State Revolving Loan Funds:** Authorizes $18 billion through 1994 for construction of sewage treatment facilities. Construction grants program authorized at $9.6 billion through fiscal year 1990, to be allocated to States for direct grants to local communities; from fiscal year 1989 to fiscal year 1994, $8.4 billion authorized for capitalization of State revolving funds, to be used for loans to local communities. Coastal States receive 55 percent of funds (for projects that affect fresh or marine waters). Some funds set aside for marine problems caused by combined sewer overflows and for National Estuary Program.
- **Ocean Dumping:** Bans sludge dumping in New York Bight Apex by December 15, 1987 (or date named by EPA, whichever first); prohibits dumpers other than those now using New York Bight Apex from dumping at Deepwater Municipal Sludge Site.
- **National Estuary Program:** Authorizes $60 million for program to address pollution in estuaries. EPA authorized to spend 10 percent on management conferences for individual estuaries and to provide up to $5 million annually to the National Oceanic and Atmospheric Administration for estuarine-related research. Priority given to 11 waterbodies.
- **Chesapeake Bay:** Authorizes $12 million for continuation of Federal/State Chesapeake Bay Program and $40 million for grants to States.
- **Boston Harbor:** Authorizes $100 million for grants to improve municipal sewage facilities and water quality in Boston Harbor.
- **Control of Toxic Pollutants:** States must identify "hot spots"—waters not expected to meet water quality standards because of toxic pollutants in discharges, even after dischargers meet permit requirements. States must develop control strategies.
- **Sewage Sludge:** Requires EPA to identify toxic pollutants of concern in sewage sludge, and to establish numerical limits for each pollutant and management practices to achieve limits.
- **Nonpoint Source Pollution:** Authorizes $400 million for grants to help States reduce nonpoint pollution. States are to identify waters not expected to meet water quality standards and submit to EPA a management program for nonpoint pollution. Federal grants for up to 60 percent of costs of implementing State programs available to States with EPA-approved reports and programs.
- **Penalties:** Increases penalties for civil and criminal violations of clean water laws; gives EPA new authority to assess administrative civil penalties for violations of effluent limits, permit conditions, or State-issued Section 404 dredge-or-fill permits.
- **Compliance Deadlines:** Extends deadline for industrial compliance with national standards to within 3 years after EPA issues standards, but not later than March 31, 1989.
- **Anti-backsliding:** Prohibits relaxation of requirements when a discharge permit is renewed or rewritten, except in certain special circumstances (147).

be used to replenish beaches or wetlands[10]). The feasibility of such uses can sometimes be limited, however, by economic constraints, land availability, public opposition, and local and State regulations (chs. 9,10). These wastes, as well as acid or alkaline industrial wastes, also can be dumped in the open ocean under certain conditions with little likelihood of causing significant long-term impacts.

Contaminated material, on the other hand, can rarely if ever be used beneficially and therefore generally requires some other form of management. In such cases, the full range of available options, including some forms of marine disposal, needs to be evaluated. For example, it might be determined that disposal of some types of contaminated dredged material is best accomplished by "capping" it with clean material in marine waters; in other cases, disposal on land may be preferable.

Pressure to use the open ocean for disposal of sewage sludge, dredged material, and some in-

---

[10]For example, some observers have suggested that uncontaminated dredged material could be used beneficially to replenish eroding marshes and islands along the southern Louisiana coast (K. Kamlet, A.T. Kearney, Inc., pers. comm., November 1986).

dustrial wastes will probably increase, especially if greater protection is provided for estuaries and coastal waters.

In response to growing pressure for such disposal, Congress could choose to allow increased disposal of some wastes in the open ocean, deciding that some types of marine disposal are environmentally acceptable.

In contrast, Congress could opt to maintain or even strengthen the current restrictive policy, either because of concerns about the long-term health of the open ocean or because allowing such disposal could be a disincentive to developing better waste management options. This course of action might, however, interfere with attempts to implement other measures designed to improve the health of estuaries and coastal waters (e.g., shifting some dumping further out to sea).

Therefore, maintaining or increasing the availability of alternative, land-based management options (e.g., waste reduction, treatment, and disposal) would be critical to the success of this strategy.

Whether or not increased disposal is allowed, Congress may wish to provide guidance and oversight by ensuring that:

- the level of pollutants in wastes is reduced prior to disposal;
- disposal sites and methods are chosen so that impacts are minimized;
- long-term monitoring and research is properly designed and coordinated; and
- disposal does not provide a disincentive to the development of beneficial uses for these wastes or to reduced waste generation wherever possible.

INFORMATION NEEDS

Many types of information are essential for developing policies about marine waste disposal, including information about the value of marine resources, ecological relationships, the quantity and fate of pollutant inputs from disposal activities, environmental and human impacts, and the ability of different disposal technologies to lessen impacts. Without such information, it is impossible to identify problems in specific areas, support enforcement activities, or effectively evaluate progress toward specific goals.

Programs for gathering and analyzing information are conducted by numerous Federal, State, and local agencies, as well as by industrial firms that must comply with regulatory requirements (ch. 7). The effectiveness of these programs has often been questioned. Some observers contend that: 1) monitoring is not sufficiently linked with basic research to facilitate the understanding of why certain impacts are occurring; and 2) too much responsibility for monitoring has been delegated from the Federal to the State and local levels, with a concomitant loss of proper design and quality control (84).

The responsible Federal agencies contend that the design, implementation, and success of such programs has improved in recent years. Agencies such as the National Oceanic and Atmospheric Administration (NOAA), EPA, and the Army Corps of Engineers (COE) have initiated many new programs during the 1980s that are better designed than their predecessors and that address issues on a more comprehensive basis (ch. 7). NOAA, for example, has several ongoing programs including an inventory of resources in the Nation's estuaries, a project to map living resources in the U.S. Exclusive Economic Zone, and a survey of outdoor marine recreation (611).

Nevertheless, information gaps still constrain analyses of marine waste disposal, partly because of a lack of information-gathering in some areas of the country, a lack of systematic analyses of gathered data, and ineffective dissemination of results. For example, high-quality, systematically analyzed information is not available about overall compliance with discharge permits, the types of pollutants present in many waterbodies, or the nature and extent of impacts in many waterbodies.[11] Moreover, different programs that address the same issue often are not well-coordinated.

Many information programs also suffer from inadequate funding. Relatively little is invested in programs that obtain and analyze information in comparison with other expenditures (e.g., capital investments in pollution control technology). The effectiveness of pollution controls is difficult to evaluate without such information, yet funding levels for monitoring and other information programs generally are declining.

Increased political and financial support will be needed to ensure the coordination and proper design of these programs. The need for coordination and long-term support has been emphasized in recent endeavors. NOAA has developed plans under the National Ocean Pollution Planning Act, with input from other Federal agencies, that recommend establishing a national network to better coordinate and synthesize existing programs. A re-

cent symposium focused on improving the design of monitoring programs and their utility in the decisionmaking process (332).

Aside from actual information-gathering programs, continued support of basic ecological research and applied technological research also is needed, both to understand how waste disposal affects marine resources and to improve disposal methods. Observers have suggested the need for additional Federal funding of numerous research topics including:

- improving the engineering and design of disposal technologies (e.g., ways to produce higher dilutions of wastewater);
- predicting how marine systems will respond to waste disposal (the experimental discharge of municipal sludge is discussed in ch. 9);
- enhancing, possibly through genetic engineering, the ability of microorganisms to degrade pollutants such as organic material or chemical pollutants in municipal and industrial wastes, both before and after disposal in marine waters; and
- increasing the use of biomonitoring tests (e.g., effluent toxicity tests) or indices of environmental degradation to identify areas likely to suffer or actually suffering some degradation (105,375,385,412,659).

---

[11]For instance, EPA's Permit Compliance System is an automated data system for tracking National Pollutant Discharge Elimination System (NPDES) discharge permits that is intended to fill this need. Until recently, it was only used by some States and EPA regional offices, and therefore has been far from complete; its use is now mandatory (ch. 8).

POLLUTANT INPUTS AND IMPACTS IN MARINE WATERS

Waste Disposal and Pollutant Inputs

Estuaries and Coastal Waters

Many municipal and industrial wastes are *discharged directly* into estuaries and coastal waters. More than 1,300 major industrial facilities and 500 municipal sewage treatment plants discharge wastewater effluents directly into estuaries, and an additional 70 municipal plants and about 15 major industrial facilities discharge into coastal waters; only a few pipelines are used to discharge wastewater into the open ocean (ch. 3). Some sewage sludge is discharged through pipelines in southern California and in Boston, although these discharges are scheduled to be terminated.

The large quantities of waste entering estuaries and coastal waters through discharges reflect: 1) the close proximity of population centers and industries to these waters; 2) cost savings to waste generators that use this option; and 3) a management approach that allows certain discharges, generally based more on technological treatment capabilities than on resulting water quality.[12] The net effect is a considerable degree of "acceptance" of this routine but environmentally significant activity, espe-

---

[12]Increasing efforts to focus on water quality are evident, however. For example, some States require that discharges into marine waters meet ambient water quality objectives established by the State (e.g., the California Ocean Plan; ref. 68). In addition, water quality considerations can be included in the design of disposal systems.

Marine waste disposal activities (pipeline discharges and dumping operations) are overwhelmingly concentrated in estuaries and coastal waters. Over 1,300 major industrial and almost 600 municipal facilities discharge directly into estuaries and coastal waters, and at most a few discharge into the open ocean.

cially when contrasted with the far greater attention focused on marine *dumping*.

Dumping also occurs in estuaries and coastal waters. The majority (80 to 90 percent by volume) of all waste material dumped in marine waters originates from dredging operations. About 180 million wet metric tons of dredged material are dumped annually in marine waters.[13] According to COE, most of this material is relatively uncon-

taminated and does not contribute significant quantities of pollutants to these waters (ch. 10).[14]

The quantity of municipal sewage sludge dumped in marine waters has increased over the last decade and now totals about 7 million wet metric tons annually. Sludge dumping now occurs primarily in coastal waters at the 12-Mile Sewage Sludge Dump Site in the New York Bight, although it is scheduled to be shifted entirely to the Deepwater

---

[13]Almost two-thirds of the material is dumped in estuaries, about one-sixth in coastal waters within the territorial boundary, and one-sixth beyond the territorial boundary. Most dumping beyond the territorial boundary is still within coastal waters.

[14]Dredged material considered by COE to be heavily contaminated is disposed of, for example, in upland sites or by placing it in pits under water and covering it with uncontaminated material. Clearly defined, *quantitative* criteria are lacking, however, for deciding whether such material is contaminated.

Municipal Sludge Site in open ocean waters during 1988.

Much smaller amounts (about 50,000 wet metric tons) of acid and alkaline industrial wastes are currently dumped in coastal waters each year at the Acid Waste Disposal Site in the New York Bight. Other wastes (e.g., seafood processing wastes or drilling fluids from offshore oil and gas operations) are also dumped or discharged into marine waters.

Relative Contribution of Pollutants From Waste Disposal and Nonpoint Sources.—The relative contribution of pollutants from discharges, dumping, and nonpoint sources[15] varies with the type of pollutant and the location (ch. 3). In most estuaries and coastal waters, little or no dumping occurs and therefore discharges and runoff contribute greater amounts of pollutants.[16] Where dumping does occur, however, it can sometimes be the major source of pollutants. The most extreme case probably occurs in the New York Bight, where dumping of sludge and dredged material accounts for one-half or more of the cadmium, chromium, copper, PCBs, total suspended solids, and phosphorus introduced to these waters.

Metals and organic chemicals enter marine waters from various disposal activities, and they primarily originate from industrial discharges. A portion of the pollutants discharged by industries to municipal sewers passes through municipal treatment plants into receiving waters or contaminates sludge; thus, municipal plants can act as a conduit for industrial pollutants. Furthermore, the pollutants in industrial and municipal discharges can contaminate sediments that may later need to be dredged.

Pathogens enter marine waters through discharges of raw sewage (e.g., from septic systems

---

[15]Nonpoint pollution can arise from a wide variety of distinct sources (box B). Comprehensive data is available only for urban and nonurban runoff, however, so this section only discusses these sources.

[16]Quantifying nonpoint runoff is difficult because it tends to be diffuse and widespread, occurs along the shorelines of virtually all estuarine and coastal waters, and varies dramatically over time, but some data are available. In addition, the *availability* of pollutants in different wastes to organisms may differ somewhat. For example, many pollutants in dredged material tend to be bound to particles that are deposited on the bottom and then rapidly covered, processes that make these pollutants less likely to be taken up by organisms.

or combined sewer overflows) as well as treated effluent and sludge. Municipal treatment processes destroy most, but not all, bacteria, and they are less effective against viruses and parasites (chs. 6 and 9). (The shortcomings of current standards regarding pathogens are discussed in ch. 6).

Estuaries and coastal waters also receive large amounts of pollutants from upstream sources. Thousands of industrial and municipal plants discharge into rivers that subsequently flow into estuaries, and nonpoint runoff is a major contributor of pollutants to rivers. In some cases (e.g., the Mississippi River delta region), upstream sources are the major contributor of most pollutants; these pollutants may be highly diluted by the large flow of a river, however, so that their subsequent impact may be less than commensurate with their quantity.

Open Ocean Waters

In contrast with estuaries and coastal waters, relatively little dumping and discharge occurs in the open ocean. Some sewage sludge is now dumped at the Deepwater Municipal Sludge Site, and this site will eventually receive all of the sludge that is now dumped in the New York Bight as well as additional sludge from New York City's new treatment plants. About 30 million wet metric tons of dredged material (less than one-sixth of the material dredged from all estuaries and coastal waters) is dumped in the open ocean. Currently, about 150,000 metric tons of acid and alkaline wastes from two industrial facilities are dumped at the Deepwater Industrial Waste Site.

Over the past 10 to 15 years, industrial waste dumping has decreased dramatically, while dumping of sewage sludge has steadily increased; dumping of dredged material has fluctuated considerably during this period. Future pressure for dumping could take many forms. Some coastal municipalities (other than those already conducting such disposal) have expressed interest in renewing or initiating ocean dumping of sewage sludge if it were to be allowed (32,532), and certain large-volume, industrial wastes such as flue-gas desulfurization sludges and coal ash have been considered potential candidates for dumping.

Impacts on Marine Environments

General Nature of Impacts

Some conventional and nonconventional pollutants can contribute to excess nutrient levels (eutrophication) and low oxygen levels (hypoxia), particularly in estuaries and some coastal waters. Pathogenic organisms (e.g., certain bacteria, viruses, and parasites) contained in sewage or runoff can contaminate water and fish, resulting in direct risks to human health such as outbreaks of hepatitis and gastroenteritis. Their presence can also cause direct economic and recreational losses.

Many metals and organic chemicals can cause severe, short-term, acute impacts on marine organisms. Moreover, many organic chemicals and some forms of certain metals can dissolve and accumulate in the fatty tissues of these organisms. When these organisms are consumed by predators, some of these pollutants can increase in concentration (i.e., biomagnify). Because of their persistence and toxicity, they can cause long-term, chronic impacts on organisms, potentially including humans. The presence of metals and organic chemicals in sewage sludge and dredged material also greatly constrains the management of these wastes.

Because of the sheer physical volume of waste that is dumped in marine environments—particularly dredged material—the solid material in such waste can modify bottom sediments or bury bottom-dwelling organisms at disposal sites. Such impacts, however, are often transient or reversible once the activity is halted.

Evaluating the Relationship Between Pollutants and Impacts

The nature and severity of impacts vary greatly among waterbodies, reflecting differences in the physical characteristics of the waterbodies, the extent and types of disposal that take place, and the types and values of the marine resources present. The information available to OTA supports the conclusion that, even though the precise link between specific pollutants and impacts is often unclear, **many of the adverse impacts on marine waters and organisms are caused by the introduction of pollutants through the disposal of wastes.** The site-specific relationship between impacts and waste disposal is illustrated, for example, through selected examples (see below). Evidence suggests that losses in individual incidents attributable to waste disposal (e.g., closures of shellfish beds or restrictions on fishing) can amount to millions of dollars; nationwide, many millions of people can be affected directly or indirectly each year.

Several factors create some uncertainty about the absolute extent to which pollutants from individual waste disposal activities contribute to observed affects, but not about the general conclusion that they do indeed cause many impacts. Uncertainty exists, for instance, because:

- pollutants can originate from many sources;
- the significance of contamination to marine organisms and humans is often poorly understood;
- impacts can be caused by other factors (e.g., overharvesting of fisheries or natural reductions in oxygen levels);[17] and
- the information available is often incomplete.

For example, although good information exists about some areas, for many other areas little effort has been made to systematically collect and analyze needed data. This hinders attempts to rank estuaries and coastal waters according to their importance and extent of impacts, or even to confidently catalog all impacts that have occurred.[18]

Health of Estuaries and Coastal Waters

Estuaries and coastal waters are among the most ecologically and economically important of all aquatic environments (chs. 2 and 4). Many such waters around the country have suffered significant impacts, although the overall trend during the last 10 to 15 years has been mixed. Some areas that once exhibited severe impacts have improved, but noticeable deterioration continues to occur or is accelerating in many other areas. Much public at-

---

[17]Other activities (e.g., dredging and filling of wetlands, or hydrologic modifications such as channelization and regulation of freshwater flow) also can affect the quality of estuaries and coastal waters (582, 670).

[18]Some useful data on pollutant inputs are available from NOAA's National Coastal Pollution Discharge Inventory and from Resources for the Future (ch. 3), but few comparable data are available on actual impacts.

Photo credit: B. Sargent, The Coastlines Project

Many beaches have been closed because of contamination of water with fecal coliform bacteria, particularly from raw sewage in combined sewer overflows. Most closures are temporary, but some have been permanent.

tention has focused on well-documented problems in the Northeastern United States (including the Chesapeake Bay and the New York Bight), southern California, and Puget Sound. Serious impacts, however, have also occurred in the less-studied Gulf of Mexico and the Southeastern United States.

The extent of degradation varies greatly around the country—in type, spatial scale, duration, and commercial importance. Observed effects include:

- impacts on water quality (eutrophication, hypoxia, turbidity, elevated concentrations of pollutants);
- loss of submerged aquatic vegetation;
- impacts on fish and shellfish (bioaccumulation of toxic chemicals, disease and abnormalities, reproductive failure, mortality);

- impacts on entire marine communities (changes in diversity, abundance, and distribution as reflected, for example, in declines in commercial fisheries);
- closures of beaches and shellfish grounds because of microbial or chemical contamination;
- a rising incidence of reported human disease, from consuming contaminated shellfish or swimming in contaminated marine waters; and
- accumulation of toxic pollutants in sediments (in some cases, to levels that warrant classification as hazardous waste sites requiring cleanup under the Comprehensive Environmental Response, Compensation, and Liability Act, or Superfund).

Estuaries and coastal waters are susceptible to these problems for several reasons. First, many marine organisms use these waters during critical parts of the organisms' life cycles (e.g., for spawning or nursery habitat). Second, these waters (particularly estuaries) bear the brunt of marine disposal activities and nonpoint pollution. Third, the physical and chemical features of many estuaries (circulation patterns, semi-enclosed configuration, shallow depth, mixing of fresh and saltwater) cause pollutants to be flushed relatively slowly from these waters or to actually become trapped. Particles (and many metals and organic chemicals, which have a tendency to bind to particle surfaces) aggregate and settle to the bottom; in addition, metals dissolved in the water can become insoluble and also settle. In many estuaries, there is a net landward flow of these sediments, so that they are far less likely to be moved further out to sea by tides or currents.

Estuaries and coastal waters and their indigenous organisms can in some cases recover from certain impacts if the inputs of pollutants are reduced or terminated. For example, impacts on water quality such as low dissolved oxygen levels or eutrophication can be reversed, and areas where communities have been destroyed by physical burial can be recolonized.[19] In many cases, improvements can

---

[19]The terms "recovery" and "reversal" describe the degree to which a condition that existed prior to an impact is restored. This does not necessarily include restoration of other conditions that were affected by the original impact. For example, decreases in levels of dissolved oxygen could also lead to the decimation of fish populations. An area

result from better control of conventional pollutants and nutrients in municipal and industrial discharges or from halting the activity entirely. Other impacts, however, may require more time to be reversed or may in some cases be irreversible. For example, contamination of sediments with metals or persistent organic chemicals, or major changes in community structure (including ones caused by other, reversible impacts, such as loss of aquatic vegetation due to eutrophication) may be difficult, if not impossible, to correct.[20]

Health of Open Ocean Waters

Living resources in the open ocean also are commercially important, but they tend to be distributed unevenly (i.e., they can be concentrated in certain areas and relatively absent in others).[21] In general, the open ocean has exhibited few documented impacts that can be attributed to waste disposal activities, partly because fewer wastes have been disposed of directly in these waters. In addition, certain problems are less likely to occur there than in estuaries and coastal waters, because the physical character and processes of the open ocean (e.g., depth, currents, and wind) tend to dilute and disperse pollutants. For example, the open ocean is less susceptible to problems such as hypoxia or eutrophication, which generally occur only when certain conventional pollutants and nutrients are present in high concentrations, and to physical burial of organisms.

In contrast, metals, organic chemicals, and pathogens are of great concern, even though they also are dispersed, because: they can cause impacts at very low concentrations, many are persistent, some can accumulate in organisms, and some can increase in concentration as they are passed up marine food chains. Uncertainty exists about the ability to discern impacts from these pollutants, because detection of such impacts is generally difficult and the impacts may not be observed until long after the polluting incident is over. Some of these pollutants have been detected in significant concentrations, both in the water and in the tissues of various marine organisms including fish, seabirds, and marine mammals. The significance of such contamination is not always clear, however, because of gaps in our understanding of issues such as the nature of open ocean food chains, the concentrations of various chemicals likely to cause reproductive failure in marine organisms, or the likelihood of pollutants being transferred to humans.

could exhibit a rapid return to the higher levels needed to sustain aquatic life, but an equally rapid recovery in the fish population would not occur. In addition, restoration to original conditions might not be identical to conditions that would have existed had the unimpacted system continued to change naturally. For example, recolonization, which might require a period of several months to several years, could result in a species composition quite different from that of the original community.

[20]Some areas could continue to suffer impacts even if inputs of pollutants were halted; for example, the prior accumulation of toxic, persistent pollutants in sediments would remain a source of contamination for a long time. These pollutants could be buried under new, uncontaminated sediments, which might be considered a reversal of contamination because marine organisms would no longer be exposed to the pollutants. Later disturbance of the sediments from human activities or storms, however, could re-expose organisms to pollutants.

[21]Many open ocean organisms also spend a portion of their life cycle in estuaries and coastal waters.

Selected Examples

Selected examples help to illustrate the relationship between waste disposal activities and other sources of pollutants and impacts on marine resources and environments. The examples presented here were chosen to represent a range of problems and geographic areas. They describe the value of resources in different marine waters, illustrate the great variation in the nature and severity of impacts among these waters, and mention management programs developed for some of these areas (such programs are described in ch. 7). The seven examples are:

1. Puget Sound,
2. San Francisco Bay,
3. the southern Louisiana coast,
4. Mississippi Sound,
5. Chesapeake Bay,
6. New York Bight and adjacent coastal waters, and
7. the Deepwater Disposal Sites in the open ocean.

Puget Sound

Puget Sound, a large and relatively deep water-body in northwest Washington, contains numerous bays and inlets and is the receiving basin for many rivers. In contrast to other areas in the Northwestern United States, it is characterized by a high degree of urban and industrial development, dominated by Seattle. At the same time, however, numerous parks and wildlife refuges are scattered along its shores, and large adjacent areas are devoted to farming.

The Sound supports many commercially or ecologically important populations of fish and shellfish. The commercial and recreational harvest of fish and shellfish was valued at nearly $74 million in 1984. The Sound also is well-known for its diversity of birdlife and marine mammals —including harbor seals, gray whales, and killer whales (24). Its waters and shores annually attract 56 percent of the State's residents and support a large tourist industry (463).

The health of the Sound has become a major issue, in part stimulated by the designation of some areas as Superfund sites and by the publicity surrounding incidents like the unexplained death of eight gray whales in 1984. In general, problems in the Sound are most prevalent in embayments and inlets, which tend to trap and accumulate various pollutants. High concentrations of toxic metals and organic chemicals, for example, have made Commencement Bay one of the most contaminated sites in the country. Exposure of fish to toxic organic chemicals and metals has been linked to diseases and abnormalities—most notably, liver tumors in bottom-dwelling fish. As a result, many people are concerned that public health could be threatened by the consumption of contaminated fish. In addition, large parts of commercial shellfish beds have been closed because of fecal coliform bacteria contamination.

These impacts have arisen from past and current waste disposal practices and from nonpoint pollution. When conditions were worst, large volumes of untreated wastes were released, in some cases causing sediment contamination that still persists. Currently, over 400 municipal and industrial pipelines (both major and minor) discharge significant quantities of pollutants to the Sound (463). Runoff from agricultural and forest lands and urban surfaces is a major contributor of conventional pollutants.

Federal and State pollutant control programs have resulted in the reduction of some pollutants in municipal and industrial discharges. Further reductions are anticipated—e.g., Seattle is upgrading treatment levels at its municipal treatment plants. In an ambitious attempt to provide more coordinated management of all point and nonpoint sources, the State of Washington created and funded the Puget Sound Water Quality Authority in 1985. The Authority is developing a comprehensive plan for waste management in the Sound, and implementation of the plan will begin in 1987.

San Francisco Bay

San Francisco Bay actually is a large, shallow, enclosed estuary that receives water from rivers that drain California's Central Valley. Much of the Bay is ringed by intense urban development (including San Francisco, Oakland, and San Jose) and industrial development (including major petroleum refineries). Numerous parks and wildlife reserves, however, are located on the Bay's shores. Relatively little agricultural activity occurs directly adjacent to the Bay.

The Bay supports many valuable recreational and commercial fisheries, and serves as a magnet for a large tourist industry (510). It also serves many important ecological functions; for example, it provides very important habitat on the California coast for shorebirds as they migrate from Arctic breeding grounds (24).

A variety of activities—including filling of wetlands, discharges of municipal and industrial wastes, and diversion of freshwater inputs—have severely affected conditions in the Bay (395). At the turn of the century, for example, industrial wastes were linked to the collapse of the Bay's oyster industry. Conditions continued to deteriorate for many decades, into the 1960s, when oxygen deficiencies periodically resulted in massive fish kills. The shellfish industry by that time had been largely destroyed.

In some respects, the situation has improved markedly since the early 1960s, primarily because pollutant inputs from municipal discharges were reduced by building improved treatment facilities. Fish kills no longer occur regularly, and concentrations of bacteria have declined. In 1982, shellfish harvesting was sanctioned in parts of the Bay for the first time since 1930.

The Bay still receives substantial amounts of pollutants, however, from industrial and municipal discharges and from the rivers draining the heavily cultivated Central Valley. As a result, serious problems persist, especially in the shallower and more poorly flushed portions of the Bay. Eutrophication and low dissolved oxygen levels occur frequently in localized areas. Organisms inhabiting the Bay have been exposed to elevated concentrations of pollutants—e.g., pathogens, metals, PCBs, and DDT—and many impacts to bottom-dwelling organisms, fish, and birds have been documented (424,529,530). Physiological abnormalities and population declines in striped bass, for example, have been linked with pollutants from waste disposal activities. Selenium, which has been linked with severe problems in waterfowl at Kesterson National Wildlife Refuge in interior California, has been detected at high levels in two species of ducks in the Bay (201).

Various efforts to understand and mitigate the Bay's problems have been undertaken. Federal programs to control pipeline discharges are administered by the State's San Francisco Bay Regional Water Quality Control Board (70). The Aquatic Habitat Institute, a nonprofit organization formed in 1983, does not have a mandate to develop a comprehensive management plan, but is attempting to better integrate the efforts of parties involved with research on and management of pollution in the Bay.

The Southern Louisiana Coast

The coastal waters of southern Louisiana are dominated along their inner boundary by exten-

sive marshes and then extend outward along the gently descending continental shelf. The Mississippi and Atchafalaya Rivers, which together drain over 40 percent of the continental United States, flow directly into these coastal waters. This drainage serves much of the Nation's agricultural land, as well as many major cities and industrialized areas.

These coastal waters are extremely productive and valuable (614), and large numbers of fish and shellfish spend all or portions of their lives there. These populations support a large fishing industry, centered on menhaden and shrimp. About 1.7 billion pounds of fish and shellfish, valued at nearly $230 million, were landed at Louisiana ports in 1985—far larger than the weight of the second largest amount landed (in Alaska). In terms of the weight of annual harvest, three of the Nation's top four fishing ports are located in southern Louisiana.

This productivity is being undermined by extensive and periodic hypoxia (i.e., extremely low levels of dissolved oxygen in the water). The precise extent and duration of hypoxia change from year to year, but it is clear that large areas of the northern Gulf of Mexico have been affected. In the summer of 1985, hypoxia was evident in an area about four times larger than that seen recently in the Chesapeake Bay (610).

Few, if any, shrimp or bottom-dwelling fish are caught in such areas. In 1983, for example, the harvest of brown and white shrimp, as well as total fish biomass, declined along the Louisiana coast as levels of dissolved oxygen decreased (465,476). Hypoxic conditions during critical early periods of the shrimp life cycle also may have resulted in a more generalized decline in shrimp stocks.

The exact causes of the widespread hypoxia are not known. The waters of the Mississippi and Atchafalaya Rivers are laden with nutrients and conventional pollutants and appear to be contributors to the problem. At the same time, these nutrients are also one reason why these waters are so productive. The pollutants probably originate primarily upstream, from point source discharges and nonpoint runoff, but the relative importance of either source cannot be defined precisely.

While Federal and State pollutant control programs have reduced some municipal and industrial discharges, the hypoxia problem itself has not received major attention until recently. In 1985, the Louisiana Universities Marine Consortium initiated a research project on hypoxia in southern Louisiana, with an annual research budget of about $200,000, about half of which is provided by the Federal Government (D. Boesch, pers. comm., September 1986).

The Mississippi Sound

The Mississippi Sound is a long, shallow embayment stretching primarily along the southern end of Mississippi but also entering the coastal waters of Alabama and Louisiana. Its shoreward limits are characterized by marshes, bayous, rivers, and small bays; toward the more open coastal waters, it is bordered by a series of islands. The Sound is adjacent to several major urban and industrial centers, including Biloxi-Gulfport and Pascagoula-Moss Point. Among the area's industries are petroleum refineries, petrochemical plants, and pulp and paper plants.

The Sound is extremely productive and contains valuable nursery grounds for fisheries (23), as reflected by the area's large commercial fishing industry. Over 470 million pounds of fish—the fourth largest amount among all the States—were landed at Mississippi ports in 1985, with a value of over $40 million. Landings in the Pascagoula-Moss Point area, dominated by menhaden and shrimp, were among the largest for all Gulf of Mexico fishing ports (614). Recreational fishing in Mississippi has been valued at nearly $45 million, and coastal tourism generates an estimated $200 million each year (290).

A number of problems—eutrophication and hypoxia, and contamination by pathogens, metals, and organic chemicals—have been documented in localized areas of the Sound. For example, high concentrations of metals and organic chemicals occur in the Pascagoula River system and Biloxi Bay; the most troublesome of these pollutants are chlorinated hydrocarbons.

Significant quantities of pollutants have been entering the Sound for many years, from a variety of point and nonpoint sources. Metals and organic chemicals tend to concentrate in the sediments of rivers and bays near discharges or runoff sources, rather than in more open portions of the Sound (323). Because the area is undergoing continued rapid urban and industrial growth, marine resources in the area may be increasingly threatened (359).

At the State level, efforts to grapple with these problems are concentrated in the Bureau of Pollution Control and the Bureau of Marine Resources. The Gulf Coast Research Laboratory, an educational and research center founded in 1947, conducts research in the Sound; its funding is derived from the State and from grants and contracts.

Chesapeake Bay

The Chesapeake Bay—the Nation's largest estuarine waterbody—is located in Maryland and Virginia and is substantially sheltered from the forces of the open ocean. Numerous rivers flow into the Bay, draining parts of Maryland, Virginia, West Virginia, Pennsylvania, and Delaware. Its drainage includes extensive agricultural areas and several major cities; nearly 13 million people reside in the area (334).

The Bay is rich with marine life. Fish and shellfish are abundant and diverse, and many species use the Bay's waters as nursery areas (25). The Bay supports a fabled and valuable commercial fishing industry. The 1985 landings in Virginia and Maryland, most of which originated from the Chesapeake, totaled nearly 815 million pounds—about 13 percent of the total U.S. catch. These landings were worth nearly $124 million, over 5 percent of the total value of fish landed in U.S. ports (614). In addition, more than 1 million waterfowl—over 75 percent of the Atlantic flyway waterfowl population—winter in tidewater areas of the Bay. Submerged aquatic vegetation plays a pivotal ecological role for all of these organisms.

Several dramatic impacts have focused public attention on the Bay's overall health (334, 543,640)—in particular, declines in the striped bass population, contamination of the James River with the pesticide kepone, decreases in oxygen levels, and increases in eutrophication. Submerged aquatic vegetation has declined precipitously in recent years, in turn affecting valuable fisheries, bottom-dwelling organisms, and waterfowl populations. Numerous shellfish beds have been closed because of bacterial contamination. Contamination of sediments and organisms by metals and organic chemicals is severe in localized areas, in particular the Patapsco River near Baltimore and the James River and its tributaries.

Many of these problems—e.g., eutrophication and the decline in submerged aquatic vegetation—are caused by large inputs of nutrients and suspended solids into the Bay. Most of these pollutants originate from municipal sewage treatment plants and from agricultural runoff (both directly into the Bay and from upstream areas). Other pollutants such as metals and organic chemicals enter the Bay from industrial and municipal pipelines and from urban and agricultural runoff.

Because pollutants enter the Bay from many sources and States, coordinating efforts to deal with waste-related problems is complicated. Numerous local, State, and Federal agencies have jurisdiction over some aspect of the Bay and its resources. To address the complexity of problems and multiple responsibilities, Congress created the Chesapeake Bay Program. Coordinated by EPA, this program has pooled the energies of many local, State, and Federal agencies, as well as the general public and independent groups such as universities. Funding is derived from numerous sources, primarily the Federal Government, Maryland, and Virginia. To date, the program has evaluated the most important problems and potentially effective control mechanisms and has devised preliminary management plans.

The New York Bight and Adjacent Coastal Waters

The New York Bight is a large coastal waterbody located off the New York and New Jersey coastlines, adjacent to one of the world's largest urban and industrial concentrations. Its shorelines, bounded by Long Island and New Jersey, consist of open beaches or extensive bays and marshes protected by barrier islands. The Hudson River flows into Raritan Bay at its apex. The Bight extends seaward along the continental shelf to about 100 miles offshore.

The Bight contains important habitats, including crucial nursery and spawning areas, for various marine organisms (343). Shellfish and fish support sizable commercial and recreational fisheries. Birds migrating along the Atlantic flyway rely on the area as a stopover point or as wintering habitat. Recreational areas, including beaches and parks, and wildlife protection areas are scattered along the coastline (25).

The Bight receives large volumes of wastes from numerous sources, both directly and carried from upstream: industrial and municipal discharges; raw sewage; urban runoff; combined sewer overflows; some agricultural runoff; and dumping of sewage sludge, dredged material, industrial wastes, and construction debris. For example, several billion gallons of raw sewage, around 7 million wet metric tons of sewage sludge, and about 8 million wet metric tons of dredged material entered the Bight in 1985.

Pollutants from these sources have caused many problems, even though the Bight is large, relatively open, and more dispersive than enclosed waterbodies. Eutrophication and hypoxia have occurred in many areas. Pathogens, metals, and organic chemicals have contaminated sediments, the water column, and various organisms. These pollutants have been linked to diseases and population declines in marine organisms and to changes in community structure. High bacterial concentrations have resulted in widespread restrictions on shellfishing and permanent or temporary beach closures. High concentrations of PCBs have prompted restrictions on fishing and the sale of striped bass.

Public attention to these problems was galvanized in 1976, when reduced oxygen levels caused massive shellfish kills and when raw sewage washed ashore. These events were attributed by some of the public to sludge dumping, but other factors were equally or more important—natural events (seasonal water stratification), inputs of nutrients from other sources (including municipal discharges), and inputs of untreated sewage. Oxygen levels in 1985 were the lowest in some areas since the 1976 episode (610).

Determining the relative contribution of different sources to the Bight's problems has been difficult. Sludge dumping will soon be shifted out of the Bight to a site much further offshore, the Deepwater Municipal Sludge Site. The degree of improvement in water and resource quality in the Bight that will result from this shift is uncertain, however, given the magnitude of pollutant inputs from other sources. If an improvement does occur, it will only be detected if sufficient monitoring is conducted before and after the shift.

As might be expected, many Federal agencies (including EPA and the COE) and local and State agencies have some responsibility for different aspects of the Bight. Several public groups (e.g., the Coalition for the Bight and the New York Academy of Sciences) are bringing these agencies and the public together to discuss improving the quality of the Bight. Congress has considered amending MPRSA to establish a New York Bight Restoration Plan.

The Deepwater Disposal Sites

Two waste disposal sites are located in deep, open ocean waters above the continental slope, about 125 to 150 nautical miles southeast of the entrance to New York harbor—the Deepwater Municipal Sludge Site and the Deepwater Industrial Waste Site (49 FR 19005-19012, May 4, 1984; see ch. 3 for additional descriptions of

the sites).[22] Because the open ocean generally exhibits relatively low productivity, marine life at the sites is less abundant and of less commercial importance than in areas closer to shore or areas such as the Georges Bank. Nonetheless, some important species use the area as a migratory pathway, including commercial fish such as swordfish and tuna, endangered whales, and threatened sea turtles.

The two deepwater sites are located within the previously designated interim 106-Mile Ocean Waste Dump Site (or 106-Site). Dumping of various industrial wastes has occurred at the 106-Site since 1961. The amounts dumped peaked in the early 1970s at almost 5 million metric tons, and then declined to the present level of about 200,000 metric tons annually. Some very small quantities of sewage sludge also have been dumped at the site.

Monitoring has not detected any serious short-term impacts to marine resources, either at or near the 106-Site, associated with past dumping. This is not surprising, since the most easily detected impacts (e.g., eutrophication and hypoxia) are not likely to occur there for several reasons. First, the currents and depth of open ocean waters greatly disperse or dilute most of the dumped material. Second, industrial wastes tend to have relatively low concentrations of the pollutants (nutrients and suspended solids) which cause these impacts.

Although the 106-Site's health appears to be relatively good, some observers are concerned that the difficulty and expense of monitoring in these deep waters may preclude detection of long-term or cumulative impacts. Information about the nature and extent of impacts such as bioaccumulation of organic chemicals in migratory species, or persistence of difficult-to-detect pathogens, is currently scanty or uncertain.

The amount of industrial wastes dumped at the Deepwater Industrial Waste Site will prob-

[22]The discussion here is drawn largely from refs. 193, 374, 434, 589, 646.

ably remain stable during the next few years. The amount of municipal sewage sludge dumped at the Deepwater Municipal Sludge Site will increase substantially in the next few years; dumping of this material now takes place primarily in the New York Bight but is being shifted to the deepwater site. It is uncertain whether this increase will result in significant long-term impacts on marine resources; only a well-designed and supported monitoring program will be able to address this issue. In addition, some concern has been expressed about the ability of regulatory agencies to ensure that the wastes and disposal activities conform with legal requirements, since an expanded transport fleet will constantly be travelling to and from the deepwater site.

Three Federal agencies share responsibilities for the deepwater sites. EPA is responsible for reviewing and administering all ocean dumping permits, NOAA is involved with monitoring, and the Coast Guard is responsible for surveillance of dumpers and enforcement of maritime laws.

POLICY ISSUES AND OPTIONS FOR ESTUARIES AND COASTAL WATERS

Pressures to continue current disposal activities in estuaries and coastal waters will probably increase. Unless inputs of pollutants into estuaries and coastal waters are reduced, however, the extent and severity of impacts in these waters are likely to increase. The ideal strategy to protect most estuaries and coastal waters is to reduce waste generation or reuse wastes, thereby avoiding disposal. Even with extensive waste reduction efforts, however, large amounts of wastes now disposed of in marine waters will continue to require disposal for the foreseeable future.

Therefore, it will be essential to increase efforts to reduce the levels of pollutants in municipal and industrial discharges and to reduce nonpoint pollution where necessary, as well as to minimize waste disposal in estuaries and coastal waters wherever possible. The ability to minimize disposal in these waters, however, may be precluded by policy decisions made about disposal in the open ocean and on land. Reducing pollutant levels in discharges and reducing nonpoint pollution are likely to be more broadly applicable.

Any attempt to address impacts from disposal activities must therefore determine the ability of the current statutory and regulatory system to control pollutant inputs. Most Federal regulatory and management programs relevant to the control of discharges (as opposed to dumping) in estuaries and coastal waters fall under CWA.[23] Two basic types of regulatory programs have been established under the Act to address pollutant inputs into these waters: *pollutant control programs*, which regulate specific pollutants or sources; and *waterbody management programs*, which address the overall management of particular waterbodies.

The ability of these programs to achieve their stated goals is summarized below (based on chs. 7 and 8). Several options are described for improving the ability of pollutant control programs to reduce the inputs of pollutants and subsequent impacts. **For many waterbodies, however, improvements in pollutant control programs alone will not be sufficient. The policy question that must then be decided is whether additional efforts (and, if so, what types) should be undertaken to counter the onset, continuation, or increase in degradation in these waterbodies.** Options are presented below for providing additional waterbody management where necessary.

---

[23]Although CWA also addresses nonpoint pollution, the Federal Government has not been extensively involved in controlling this type of pollution. The Water Quality Act of 1987, however, authorized $400 million for grants to help States develop nonpoint management programs.

Issue 1: Current Pollutant Control Programs Will Not Protect All Estuaries and Coastal Waters

The two major pollutant control programs authorized by CWA are the National Pollutant Discharge Elimination System (NPDES) and the National Pretreatment Program. These programs establish effluent guidelines and other requirements to regulate the discharge of certain pollutants from municipal and industrial facilities. States or EPA Regions have primary responsibility for granting permits and setting standards that incorporate these requirements, and discharges that meet the standards specified in the permits are legal.[24]

These pollutant control programs have been responsible for important reductions in some pollutants. The construction or upgrading of sewage treatment plants has reduced the levels of conventional pollutants and nutrients in many municipal discharges and, as a result, the health of some estuaries and coastal waters has improved in some aspects. In southern California coastal waters, for example, kelp beds have partially recovered. Similarly, reducing the levels of pollutants in industrial discharges into sewers has improved the quality of municipal sludge in some cases, allowing it to be used beneficially as fertilizer on farmland and forests.

Such reductions, however, have been achieved at considerable expense. The Federal Government has spent over $44 billion since 1972 to build municipal treatment plants that meet requirements specified in CWA and implemented through NPDES (573). Industrial facilities also have made substantial investments in response to the regulations established under these programs.

If compliance with existing regulations is achieved, the levels of regulated pollutants in municipal and industrial discharges are likely to continue to decline. The extent of future reductions, however, is difficult to predict. Although compliance has improved during the last few years, the likelihood of achieving full implementation and en-

Photo credit: *Marine Resources Council, Melbourne, Florida*

Estuaries and coastal waters are among the most ecologically and economically important of all aquatic environments. Humans use them for a variety of activities, including recreation, waste disposal, commercial fishing, and development. Some previously undeveloped estuaries and coastal waters are being urbanized, and the pollutants from subsequent waste disposal activities and urban runoff can greatly affect the resources in these waters.

forcement is unclear because Federal funding of some critical activities has been inadequate and is declining. For example, proposed funding levels in the fiscal year 1987 budget for water quality enforcement and permitting and for municipal enforcement are lower than current levels. As a result, some municipal and industrial facilities will probably continue to discharge pollutants in amounts that exceed their permit limits.

Moreover, even if total compliance with existing regulations were achieved, these programs would not be sufficient to maintain or improve the health of all estuaries and coastal waters in the future because:

- **Pipeline discharges and nonpoint source pollution (particularly urban runoff) will**

---

[24]Some States also have developed plans for managing water quality that incorporate more stringent requirements (e.g., California; ref. 68).

Photo credit: *Southern California Coastal Water Research Project Authority*

Beds of giant kelp along the southern California coast provide important habitat for many valuable fish and shellfish and support a substantial kelp harvesting industry. Large acreages of these beds disappeared prior to the 1970s, in part because of pollutants discharged to nearby waters. Reductions in the discharges of some pollutants, accompanied by kelp restoration efforts, have helped reverse this trend and kelp bed acreage is now increasing.

increase as populations and industrial development expand in coastal areas.

• **Current pollutant control programs do not address all pollutants.** Standards have not been developed for some pollutants that are listed in CWA and present in wastestreams in large quantities (ch. 8) because control technology is not available or because EPA has determined that its use would impose unreasonable economic burdens on affected industries. Standards also have not been developed for other pollutants that can be important in some

situations but that are not listed in CWA (e.g., organic chemicals such as dibenzofurans and trichlorophenols; pathogens such as viruses).

• **Current pollutant control programs do not address all sources of pollution.** These programs already address the most easily controlled sources, in particular municipal and most industrial discharges, but they do not adequately regulate some additional important industrial sources of pollutants (e.g., textile mills and commercial laundries) or nonpoint sources.

• **Federal resources available for maintaining or improving current levels of municipal sewage treatment are declining, and the ability of States or communities to fill the breach is uncertain.** Federal funding for capital investments in new and improved municipal treatment plants is declining,[25] and the cost of maintaining operations at existing plants is likely to increase as the plants become older. In addition, some plants could be required to upgrade treatment to remove certain problem pollutants such as nitrogen and phosphorus.

• **Monitoring, research, and enforcement currently are inadequate, and funding levels for these activities are being reduced in some instances.**

• **The contamination of sediments with persistent toxic pollutants is not adequately addressed.** These sediments may be a source of contamination for long periods; many observers have proposed the need to develop sediment quality criteria analogous to those for water quality.

---

[25]In 1984, EPA estimated that about $110 billion would be required by the year 2000 for the Nation to meet its remaining municipal treatment needs (654). Prior to the Water Quality Act of 1987, about one-half of these needs (e.g., construction of secondary treatment plants and new "interceptor" sewers) would have been eligible for grants from the Federal Construction Grants Program; the Federal share would have been about $36 billion (569). EPA recently lowered its estimate of remaining municipal treatment needs to $76 billion by the year 2005 (676). Most of the reduction is attributed to changes in documentation requirements for responding States, and some State officials have criticized the estimate as not reflecting all water quality-related treatment needs (154). The Water Quality Act provided $18 billion for the Construction Grants Program and State revolving loan funds, with funding ending in 1994.

Options To Improve Current Pollutant Control Programs

Although total implementation of and compliance with existing pollutant control programs will not be sufficient to maintain or improve the health of all estuaries and coastal waters, these programs will continue to achieve important reductions in pollutant inputs and to provide the primary foundation for pollution control efforts. Therefore, maintaining and improving their capabilities is critical. Four sets of options for improving these capabilities are discussed below:

1. improving enforcement;
2. ensuring adequate funding by Federal, State, and/or local sectors of municipal treatment plant construction;
3. regulating ''important'' or additional pollutants and industrial sources; and
4. applying ocean discharge criteria to estuaries and coastal waters.

Option 1: Improving Enforcement

Enforcement of current regulations on point source dischargers is inadequate for many reasons (ch. 8). More rigorous enforcement would reduce pollutant discharges by the affected parties and would provide a greater deterrent to other facilities. Mechanisms for improving enforcement include the following:

- **Support continued or enhanced implementation and enforcement of the current NPDES and pretreatment programs, through oversight, financial support, and technical guidance.** Virtually any increase in financial resources for the implementation and enforcement activities of these programs should be helpful, although the cost of completely enforcing all regulations would greatly exceed current levels of funding committed to this activity.
- **Enhance EPA's enforcement authority by allowing administrative civil penalties in addition to court-imposed civil penalties.** Civil enforcement actions in court (fines or consent decrees) tend to be time-consuming. Many observers have suggested that the authority to levy administrative fines could improve the ability of EPA to pursue enforcement

in a timely and focused manner.[26] Congress also could consider the effectiveness of the provisions that encourage enforcement actions by private citizens.

- **Provide oversight to ensure that efforts to focus or target enforcement activities are based on consistently applied criteria.** EPA has implemented a policy to focus enforcement efforts first on major dischargers in ''significant'' noncompliance, then on major dischargers in less significant noncompliance and on minor dischargers. While attractive in theory, focusing enforcement could result in differential enforcement around the country, raising questions about equity. Some observers contend that selective enforcement makes more efficient use of available resources and is therefore justified; others question whether the deterrent effects of enforcement would be lessened for lower priority dischargers (502).

Option 2: Ensuring Funding of Municipal Treatment Plant Construction

Under the provisions of the 1987 Water Quality Act, only $18 billion in Federal funds will be provided for municipal treatment needs. EPA intends to continue requiring municipal treatment plants to comply with CWA's treatment requirements, however, whether or not Federal funding is available to help meet these requirements. A long-term capacity, therefore, still must be developed for funding new plant construction, replacing or repairing treatment plants as they deteriorate, and expanding capacity as needed. The large-scale feasibility of different non-Federal funding mechanisms such as State revolving loan funds, privatization, nondebt financing, and municipal bonds has been debated but remains uncertain (542,569).

Congress considered this problem and authorized: 1) a transition period until 1994 to allow States and localities to develop alternative funding mechanisms, and 2) about $8 billion (of the total authorization of $18 billion) to be used for the capitalization of State revolving funds. Congress could further

---

[26]In the Water Quality Act, Congress granted EPA new authority to assess administrative civil penalties. Several States which are authorized to administer pollutant control programs already have such authority.

support efforts to develop funding mechanisms, for example, by increasing direct incentives for their development; such incentives could include tax credits for privatization.

Option 3: Expanding Regulation of Important Pollutants and Sources

The coverage of the pretreatment and NPDES programs could be expanded by developing standards for additional individual pollutants and sources of pollutants, particularly industrial sources. The mechanisms to expand coverage generally are already available to EPA, but Congress could provide support for expansion initiatives through oversight and commitment of sufficient financial resources. This could entail several complementary approaches:

- **Promoting the development of effluent guidelines for pollutants that are listed in the Clean Water Act as priority pollutants but for which guidelines have never been developed.** Congress could increase its support of EPA efforts to identify unregulated pollutants present in large amounts in particular discharges and to develop feasible treatment technologies.

- **Supporting efforts to identify pollutants not on the CWA list but that can cause significant impacts, and supporting efforts under the pretreatment and NPDES programs to develop effluent guidelines for these pollutants.** EPA has undertaken preliminary efforts to develop screening processes and tests (e.g., effluent toxicity tests) to identify additional pollutants that are important in marine waters, but no new effluent guidelines have yet been developed. A screening effort that combined and augmented these efforts could expedite the identification of such pollutants and the development of effluent guidelines when necessary.[27] This would require more research on the potential impacts of unregulated pollut-

ants, and monitoring to search for specified pollutants in individual waterbodies.

- **Expanding pollutant control programs to improve coverage of important point sources that are not adequately regulated.** These include certain unregulated industrial categories (e.g., commercial laundries) and combined sewer overflows or stormwater outfalls.

- **Supporting EPA's ongoing effort to develop technical guidance on the quantities of toxic pollutants allowable for different municipal sludge disposal options.** Current sewage sludge regulations do not establish allowable levels of most pollutants for different disposal options. The development of comprehensive guidance or standards for sludge disposal would increase the ability of municipal treatment plants to require reduced industrial discharges of toxic pollutants into sewers.

- **Deciding how to best address the problem of hazardous waste discharges into municipal sewers.**[28] An exemption in the Resource Conservation and Recovery Act (RCRA) that allows such discharges could be abolished. If it is, problems in other environments could ensue (e.g., because of illegal dumping). If the exemption is retained, then improving the implementation and enforcement of the pretreatment program would become critical in ensuring adequate regulation and treatment of such discharges. This could include expansion of efforts by municipal treatment plants to develop local limits on such discharges.[29]

Option 4: Applying Ocean Discharge Criteria to Estuaries

The CWA Ocean Discharge Criteria (Sec. 403(c)) currently apply to discharges into coastal waters,

---

[27]Some observers argue that additional national effluent guidelines may not be necessary because permit writers can use "best professional judgment" to incorporate limits on any pollutant into individual discharge permits. Development of such limits, for example, could be part of a water quality approach (see Issue 2 below). On the other hand, this would not guarantee consistent development and application of limits, particularly for pollutants that are of significance in multiple industries or geographic regions.

[28]"Hazardous" refers to those substances or wastestreams specifically defined as such under RCRA.

[29]This problem is symptomatic of a larger issue, the role of municipal plants in the management of industrial wastes (ch. 9). Municipal wastes are often contaminated to some degree with metals and organic chemicals from industrial discharges, and some observers have suggested prohibiting industrial discharges into sewers. The near-term likelihood of a prohibition is low, although the practice could be partially restricted by prohibiting new industrial discharges into sewers. Water quality would then depend on the control (by NPDES) of direct industrial discharges and/or the implementation of other management options such as waste reduction, process substitution, recycling, and centralized treatment facilities.

but not to discharges into marine waters inside the baseline of the territorial sea (i.e., estuaries). These criteria specify additional factors that must be considered prior to the granting of a permit for discharging into non-estuarine marine waters, and in theory provide greater regulatory control.

Congress could consider applying the Ocean Discharge Criteria to discharges into estuaries, which would provide an additional means of control on such discharges. Whether this would increase actual protection would depend on the strength of the criteria. If necessary, the criteria could be strengthened by making the issuance of a discharge permit contingent on additional factors, such as:

* development of an acceptable monitoring protocol;
* specification and acceptance of conditions under which the discharge may be terminated or modified (e.g., if monitoring revealed severe impacts); and
* requiring that a need be demonstrated to discharge wastewater into estuarine waters.[30]

Issue 2: Some Estuaries and Coastal Waters Need More Comprehensive Management

More comprehensive planning and coordination of management efforts will be needed for several reasons if the Nation wishes to lessen, avoid, or reverse degradation of some estuaries and coastal waters. First, estuaries and coastal waters exhibit very *site-specific* characteristics with respect to physical nature, disposal activities and impacts, and economic importance. Second, these waters can encompass multiple political jurisdictions and fall under the authority of multiple agencies.

Third, the need to allocate available resources efficiently will become more critical, because financial resources for Federal and State pollution control efforts probably will not increase substantially in the future. In addition, because pollutants can be contributed by many sources, it is not always clear whether changes in current pollutant control efforts, such as regulating additional pollutants or achieving full compliance, would be sufficient to

achieve the desired improvements, or whether new efforts are needed.[31] Increasing the effectiveness of point source control programs might be sufficient in some areas, whereas in other areas efforts to control nonpoint source pollution may be critical.

These factors necessitate greater coordination and cooperation among responsible agencies to identify site-specific problems and allocate resources toward the most effective control efforts.

The need for comprehensive and coordinated management has led to the development of some "waterbody management" plans and programs by the Federal Government (e.g., Chesapeake Bay Program), the States (e.g., Puget Sound Water Quality Authority), and local authorities (e.g., Southern California Coastal Water Research Project). In most cases, numerous agencies from different levels of government share responsibilities for implementation. The Puget Sound program, for example, involves more than 10 governmental entities. Most programs address single waterbodies, although the National Estuary Program currently involves efforts in six areas.

Existing waterbody management programs vary greatly in their design. Some have the authority to set goals and establish plans (e.g., the Chesapeake Bay Program), while others are designed only to gather and share information about research needs or findings (e.g., Southern California Coastal Water Research Project). Most programs have the authority to perform only some of the following functions: address multiple disposal activities and pollutant sources, identify the most serious or tractable problems, allocate resources toward these problems, design and implement management plans, and coordinate various involved agencies.

In general, existing waterbody management programs are in the early stages of implementation and their effectiveness cannot yet be judged (ch. 7). The initial focus of many programs has been to characterize problems, identify sources of pollution, and develop pollution abatement strategies. The Ches-

---

[30]The first two of these additional criteria are currently not included among the Ocean Dumping Criteria either.

[31]It is difficult to discern in advance, for example, whether efforts to reduce nonpoint source pollution would be more cost-effective than requiring additional point source controls; the costs (as well as actual benefits to water quality) of implementing many individual nonpoint management practices must be compared to the costs and benefits of fewer, more expensive, point source controls.

apeake Bay Program, which is among the most advanced of existing programs, is currently entering the implementation stage.

While these efforts appear promising, programs have been established for only a few waterbodies to date; many of the estuaries and coastal waters in need of additional management are not covered by such programs. Furthermore, current programs generally have only limited authority and financial support, and may not be sufficient to ensure the health of the target waterbodies.

Options To Provide Additional Waterbody Management

Creating new waterbody management programs would probably be relatively straightforward, and in the Water Quality Act of 1987 Congress designated some specific waterbodies for which management efforts should be undertaken. **However, establishing a systematic approach for providing comprehensive and coordinated waterbody management will require additional, difficult policy decisions. The critical link that is lacking is a framework for making decisions about when and how to provide additional means of management, in particular, how to complement the current uniform national pollutant control programs to address situations that require additional, site-specific controls.**

A water quality-based approach could complement the system of primarily uniform, technology-based controls and provide a framework for addressing the site-specific needs of individual waterbodies.[32] Although the 1972 CWA Amend-

ments marked a shift away from this approach and toward the use of technology-based standards, the authority to institute water quality-based regulation was retained in the Act because Congress intended it to serve as an additional layer of pollution control, after the more uniform pollutant control programs were well-established. In general, however, it has not yet been used to provide comprehensive and coordinated management of estuaries and coastal waters,[33] although EPA has begun to develop a water quality approach to better control toxic pollutants in discharges (49 FR 9016-9019, Mar. 9, 1984).[34] EPA's Science Advisory Board recently recommended that the agency investigate applying water quality criteria to research conducted in marine waters (244,675).

Development of a framework that uses a water quality approach to provide additional waterbody management, where needed, now seems appropriate. Such a framework, which could build on existing mechanisms, would differ from or expand on current efforts by:

- **providing better means of evaluating progress in improving the quality of estuaries and coastal waters;**
- **identifying those waterbodies that will continue to be degraded, even after continued development and implementation of current pollutant control programs;**
- **developing new waterbody management programs for some of these waterbodies; and**
- **providing the guidance and flexibility needed for waterbody management programs to set site-specific goals and establish coordinated plans for achieving those goals.**[35]

---

[32]A water quality-based approach places controls on pollution sources, based on an assessment of the concentrations of pollutants in receiving waters below which unacceptable impacts will not occur. It relies on the development of water quality-based standards, which consist of designated uses (e.g., swimmable water) for defined segments of waterbodies and pollutant-specific numerical criteria designed to assure attainment of the uses. The States designate uses and set water quality standards, with Federal guidance. Individual dischargers generally have greater flexibility in choosing how they will comply with water quality standards than with technology-based standards (130). A water quality approach, however, requires enormous amounts of information, continuous monitoring, and the development of site-specific criteria for many pollutants. Furthermore, it can be difficult to ascertain the portion of the problem that is caused by disposal because, for example, ambient water quality is affected by episodic events (e.g., storms that cause excess runoff or low flow that causes salinity problems). Nevertheless, developing such an approach could provide the flexibility to address site-specific problems.

[33]For example, 9 of the 24 coastal States have not developed marine water quality standards for any priority pollutants (ch. 8). For the 8 coastal States that have marine standards for priority metals, standards have been developed for an average of 4.5 of the 14 metals. For the 15 coastal States that have such standards for priority organic chemicals, standards have been developed for an average of 6.8 of the 85 organic chemicals.

[34]In addition, several States have addressed seasonal variations or differences in the contribution of pollutants from different sources, while not relaxing technology-based standards, by using water quality-based techniques such as seasonal or variable permits (130).

[35]Some of the decisions that would be made within such a framework would be affected by the availability of land-based and open ocean disposal options. For example, if the relatively restrictive policy regarding waste disposal in the open ocean is maintained, then some methods for improving the quality of estuaries and coastal waters, such as shifting disposal further out to sea, will not be available.

Implementing these steps could take considerable time. In addition, the relative roles of Federal, State, and local governments in these activities is a central issue. Some observers advocate reliance mostly on State and local efforts, while others advocate a strong Federal role. This issue is addressed briefly below, but in general the question of primary responsibility will need to be addressed on a case-by-case basis.

Option 1: Establishing Measurable Goals and Evaluating Progress

CWA established as national goals the elimination of discharges of toxic pollutants in toxic amounts and the restoration or maintenance of fishable and swimmable waters. Clearly, such improvements will not, and were not expected to, occur overnight; many years may pass before the effects of changes in pollutant control programs or of now-developing waterbody management programs become apparent.

Moreover, it is often difficult to measure progress toward such broadly stated goals. Establishing goals toward which progress could more easily be measured could increase the ability to assess improvements in the health of estuaries and coastal waters, and, concomitantly, increase the ability to judge the need for additional controls. Congress might consider:

- **Refining the goals of CWA so that they apply explicitly to estuaries and coastal waters.** This could involve a statement of the intent to maintain the current quality of resources or to reverse any trends of degradation in these waters.
- **Supporting the further development and implementation of the water quality approach and the specification of site-specific, measurable goals toward which progress could be measured.** To be effective, such goals should be quantitative whenever possible and should be directly linked to tangible improvements in resources.[36] Examples of measurable goals include:

—avoiding specific impacts (e.g., no fish mortality in a specified area);
—achieving desired changes in ecological conditions (e.g., reestablishment of submerged aquatic vegetation in a specified area); and
—achieving desired changes in economic or recreational returns (e.g., the lifting of restrictions on harvesting shellfish).[37]

Progress toward such goals could be measured for individual waterbodies. If the waterbody met the goals, then no additional control efforts would be needed. If it did not meet one or more goals, then site-specific control efforts could be increased for those pollutants and waste disposal activities that most significantly impede attainment.[38]

Pollutant control programs would still need to be implemented and enforced, but they would only constitute a first step. Setting measurable, site-specific water quality goals would allow the effectiveness of these programs to be evaluated and judgments to be made about the need for more stringent controls on any pollutant sources.[39] Available resources could then be focused on the most viable or most cost-effective control efforts; additional permit limits on discharges, as well as the use of best management practices for controlling nonpoint pollution, could be required in site-specific situations.

Two additional uses of the water quality approach to waterbody management deserve mention:

1. **A water quality approach could be extended to address sediment quality.** The tendency for metals, organic chemicals, and pathogens to become concentrated in sediments suggests the need to develop sediment quality criteria

---

[36]For example, water quality criteria currently exist which specify a minimum level of dissolved oxygen for a particular waterbody. While these account for one condition that is necessary to protect the waterbody's living resources, such criteria need to be linked directly to a goal of improving the value or health of those resources (e.g., increase in fish population size or commercial yield).

[37]EPA has initiated a study to evaluate the potential economic benefits of improvements in the water quality of estuaries (K. Adler, U.S. EPA, pers. comm., December 1986).

[38]This concept is already a component of air pollution control plans required of individual States under the Clean Air Act. EPA recently indicated that extending this concept to marine pollution problems might provide an effective means of evaluating the effectiveness of management strategies (670).

[39]In theory, existing controls on point sources (i.e., NPDES permit limits) could also be relaxed or made less stringent if specified water quality standards in a waterbody were being met. Such "backsliding" was prohibited, except in some narrowly defined circumstances, in the Water Quality Act of 1987; the current national technology-based standards are unlikely to be modified to any great extent.

that would be analogous to those for water quality. Such criteria could be useful in determining, for example, if shellfish harvesting in a particular area should be restricted, or whether dredged material is sufficiently contaminated to pose undue risks to bottom-dwelling marine organisms. These criteria do not currently exist, although EPA is evaluating the feasibility of developing them for certain metals and organic chemicals (C. Zarba, EPA, pers. comm., November 1986).

2. **A water quality approach could be extended to account for pollutants that enter an estuary from upstream sources.** For example, many of the pollution problems in the Chesapeake Bay are aggravated by pollutants carried into it by the Susquehanna River. Although such an approach might be logistically difficult to implement, precedent exists for identifying important but distant sources and requiring that their pollutant inputs be reduced in order to achieve water quality goals in a particular waterbody.[40]

Some observers have argued that supplementing current technology-based controls with even stricter controls in response to water quality standards might impose unreasonable financial burdens on dischargers. Congress may wish to consider the use of financial incentives such as fees or taxes to ease this burden. According to some economists, this market-oriented approach to water quality management could be introduced in a manner that does not unduly compromise the technology-based approach (A.M. Freeman, Bowdoin College, pers. comm., July 1986; and refs. 130,305).

---

[40]A Federal court recently ruled in *Scott* v. *City of Hammond, Indiana, et al.* (741 F. 2d 992, 1984) that NPDES-permitted dischargers to Lake Michigan's tributaries must consider the effect of the discharges on the lake itself, not just on the tributaries. This water quality approach will involve difficult tasks such as assessing the capacity of Lake Michigan to accommodate wastes and allocating the rights to discharge certain amounts of wastes into tributaries. EPA Region V is developing a long-term toxic strategy to address this problem, including evaluation of the most cost-effective controls (whether they be best management practices for nonpoint pollution or controls on point sources). EPA considers this decision to be applicable nationwide, but some States disagree (L. Fink, U.S. EPA Region V, pers. comm., October 1986; and refs. 251, 674). In theory at least, the concept could be extended to estuaries and coastal waters impacted by pollutants from upstream sources.

Option 2: Identifying Waterbodies Needing Additional Management

Not all estuaries and coastal waters require additional management, so some mechanism would be needed to identify waterbodies likely to suffer degradation despite current pollution control efforts. **This need probably could be met by establishing a "screening" process to identify those waterbodies requiring additional management.**

Some States have developed criteria to identify such waterbodies, but uniform criteria probably should be used since waterbodies around the Nation are involved. Consistent criteria could be developed, for example, by the Federal Government and used by States to evaluate waterbodies within their boundaries. Some waterbodies, however, are bounded by and receive pollutants from several States, and multiple agencies could have responsibilities pertinent to waterbody management. In such cases, it may be appropriate to have the Federal Government conduct or coordinate the process. EPA's National Estuary Program and NOAA's National Estuarine Program are evaluating such an approach for some coastal waters, so a new program would not necessarily be required (611,670).

A screening process would need to precede other management decisions and thus should be relatively streamlined. Information would need to be collected for most or all of the Nation's estuaries and coastal waters, so decisions regarding what information will be needed should be made early in the process.

Option 3: Developing Management Plans

If a general goal of maintaining or improving the health of estuaries and coastal waters is to be pursued, additional programs will need to be developed for those waterbodies likely to suffer continued or new degradation. Several options exist for overseeing the development of management plans and defining the structure of individual waterbody management plans:

• **Decide whether a national program is necessary to coordinate and oversee the development of individual management plans.** While a strong Federal presence has not been necessary for the initiation of some programs (e.g., in Puget Sound), a national program could: 1) conduct any screening effort, 2) encourage States to develop plans for waterbodies entirely within their jurisdiction, and 3) initiate the development of programs for waterbodies encompassing multiple jurisdictions. A program could be newly developed or could build on ongoing efforts such as the National Estuary Program. If needed, a Federal program could provide incentives to the States, for example, by making grants contingent on the development of adequate plans.

- **Establish national guidelines for the development of individual plans.** Individual plans could be required to: 1) designate a lead agency to coordinate planning, 2) establish site-specific, measurable goals, 3) specify what efforts would be undertaken to achieve the goals, and 4) indicate how progress will be evaluated and reported. Existing planning mechanisms in CWA (e.g., Sec. 208 areawide plans or Sec. 303(e) water quality management plans) could provide the statutory authority for such plans.[41] Alternatively, existing plans, for example, those of the National Estuary Program, Chesapeake Bay Program, or the Great Lakes Program, could be used as models.[42] Regard-

less of the mechanism used, the requirement to define specific goals and evaluate progress could be structured in a manner analogous to State Implementation Plans under the Clean Air Act. Under this Act, planning agencies must determine whether air masses are in attainment with standards and establish plans describing how attainment will be achieved in nonattainment areas. For waterbody management plans, responsible planning agencies could be required to undertake similar planning.

- **Support the development of well-designed, long-term monitoring and data analysis programs whose results can be used to evaluate progress toward specific goals.** Such programs would have to be long-term because attainment of some goals is likely to require relatively long periods. If a monitoring or analytical program indicated insufficient progress toward attainment, then responsible agencies might need to shift planning or control efforts. Any planning and evaluation processes thus would have to be continuous and include mechanisms for modifying plans as needed.

Proposals to expand existing waterbody management programs or to develop new ones with multiple responsibilities might be dismissed for fear of large new expenditures. It is true that efforts to maintain or improve the health of estuaries and coastal waters will require new expenditures. If, however, management programs can identify site-specific problems and coordinate control efforts, the overall costs of such efforts could in some cases be less than the costs of separate, uncoordinated pollution control efforts.

---

[41]Numerous problems arose in the development and implementation of the 208 program (699). Nevertheless, the general concept of areawide planning seems viable.

[42]The Coastal Zone Management Act provides another possible vehicle for such plans, but it has generally focused more on development and land-use issues than on waste disposal activities. Some coordination of any waterbody management plan with State Coastal Zone Management plans, however, would still be essential.

POLICY ISSUES AND OPTIONS FOR OPEN OCEAN WATERS

Issues Regarding Waste Disposal in Open Ocean Waters

Waste disposal in the open ocean is generally limited to the dumping of acid or alkaline industrial wastes, sewage sludge, and dredged material.[43]

---

[43]The following discussion focuses on these wastes or on others that are possible candidates for dumping in the open ocean. OTA already analyzed the incineration of hazardous wastes at sea (586). An effective moratorium exists on the disposal of low-level radioactive wastes in the ocean.

The permitting system established under the Marine Protection, Research, and Sanctuaries Act has been relatively successful in managing such dumping: dumping of industrial wastes has declined dramatically, and the dumping of sewage sludge and dredged material is relatively well-controlled.

Some of the wastes currently dumped in the open ocean also can be used beneficially on land and in certain aquatic settings. When relatively uncontaminated, for example, dredged material can be used for beach or wetland replenishment projects, and sewage sludge can be used as a fertilizer or soil conditioner on farms and forests.

Most often, however, these wastes must be managed by other treatment or disposal options, and pressure to use the open ocean for dumping will probably increase. In light of these pressures as well as the 1981 court decision (*City of New York* v. *United States Environmental Protection Agency*) that required the balanced consideration of all available alternatives, a total ban on disposal in the open ocean seems unlikely.

It is essential, therefore, to consider whether there are conditions under which open ocean disposal might be environmentally acceptable. Some wastes (e.g., sewage sludge, dredged material, and acid and alkaline wastes) probably can be dumped in the open ocean, if levels of toxic pollutants in the wastes are low, without causing significant long-term impacts. Open ocean features and processes (e.g., large volume, well-mixed waters, high dispersal ability) reduce the likelihood of impacts such as hypoxia, eutrophication, and significant accumulations of suspended material. In addition, the open ocean is generally capable of quickly neutralizing acid or alkaline wastes because of its large natural buffering capacity. Some pipeline discharges might be environmentally acceptable for the same reasons.

However, some uncertainty is associated with these conclusions about the acceptability of open ocean waste disposal. For example, it is unclear whether pathogens and toxic chemicals, at concentrations likely to exist at disposal sites, can cause long-term impacts on open ocean organisms and populations, or whether the overall productivity of the open ocean would be affected by such impacts. In addition, since the productivity and corresponding biological activity of the open ocean is generally low, the degradation of wastes disposed of there could be slow relative to degradation in estuaries and coastal waters.

Other factors could constrain open ocean disposal of relatively uncontaminated wastes. In particular, most pipelines probably could not be sufficiently extended into open ocean waters (particularly on the Gulf and east coasts, where the distance to the open ocean generally is greater than on the west coast), and the shifting of dredged material dumping further out to sea may be seen as prohibitively costly in many cases.

In contrast, contaminated material can rarely if ever be used beneficially and therefore generally requires some form of disposal. In such cases, the full range of available options, including some forms of marine disposal, needs to be evaluated. Marine disposal that depends on containment rather than dispersion may sometimes be preferable to land disposal. For example, "capping" of some contaminated dredged material with clean material may cause fewer impacts than disposing the same material on land. Similarly, solidified coal ash potentially could be used in the construction of artificial reefs.

Options for Managing Waste Disposal in Open Ocean Waters

Two basic policy directions exist regarding waste disposal in the open ocean: maintain or strengthen the current restrictive policy, or allow increased disposal of some wastes under some conditions. Each choice involves some specific implications that are addressed by the options described below.

Option 1: Maintaining or Strengthening the Current Restrictive Policy

As currently implemented, MPRSA tends to restrict waste disposal in the open ocean. Maintaining or strengthening this policy could be justified, even though open ocean disposal is technically and economically feasible under certain conditions, because of concerns about the long-term health of the open ocean or because policymakers decide that allowing more ocean dumping might hinder the development of better management options. Thus Congress could strengthen this policy by amending the Act to specifically exclude particular wastes from eligibility for dumping, although total restriction of open ocean waste disposal appears implausible.

Choosing to maintain the current restricted policy, however, might preclude some measures for improving the health of estuaries and coastal waters (e.g., shifting disposal activities further out to sea).

Therefore, ensuring the availability of alternative options, such as beneficial use or land-based treatment and disposal options for wastes that could be disposed of in marine waters, would be a critical component of continuing this policy.

Option 2: Allowing Increased Disposal of Some Wastes

A choice also could be made to allow increased disposal of some wastes in the open ocean under certain conditions because of the environmental acceptability of and increased pressure for some disposal. If some increase in open ocean disposal is allowed, many of the necessary statutory and regulatory mechanisms to ensure sufficient control are already in place.

Whether or not open ocean waste disposal increases, Congress probably would need to support and oversee several important aspects by:

- **Ensuring that disposal sites and methods are chosen so that impacts are minimized.** MPRSA and its associated regulations define siting criteria for open ocean disposal, specifying that chosen disposal sites exhibit dispersive characteristics and contain few economically or ecologically important resources; sites also must exhibit a relative lack of pollutant inputs from other sources and a lack of use for other purposes. These criteria appear to be sufficient if rigorously implemented by EPA. It might be worth considering whether the use of several carefully selected sites is preferable to the use of only one or two dumpsites for a particular waste. Using several sites would reduce the input of pollutants (and the possibility of subsequent impacts) at any one site, but it also would require additional resources for monitoring and surveillance.
  Supporting efforts to reduce pollutants in wastes prior to disposal. The options for reducing pollutant levels in estuaries and coastal waters are equally applicable to waste disposal in the open ocean; they include, for example, greater implementation and enforcement of the pretreatment and NPDES programs and the development of comprehensive regulations for sludge disposal. In addition, Congress could require that stricter controls be imposed

on the composition of wastes that are to be dumped—in particular, to minimize the presence of toxic pollutants and pathogens.
- **Providing additional resources for properly designed and nationally coordinated monitoring and research programs, and ensuring that results are used in future policy decisions.** Greater support and coordination of monitoring and research is needed to ensure that significant impacts (including those that might become evident only after several years) are detected and that information on these impacts is effectively analyzed and disseminated. In addition, Congress could consider developing an explicit policy that allowed disposal to continue only if monitoring detected no significant impacts ("significant" or "unacceptable" impacts probably should be carefully defined prior to disposal). This could include specific provisions requiring that the disposal activity be phased out or modified if such impacts were detected. MPRSA currently appears to provide sufficient authority to phase out harmful disposal activities, as witnessed by the reduction of industrial waste dumping.
  Ensuring that open ocean disposal does not hinder the development and use of other options, such as land-based treatment or beneficial use. Existing provisions, if implemented consistently and rigorously, appear to provide a means of addressing this issue; for example, regulations (under Secs. 101 and 102 of MPRSA) for granting ocean dumping permits require:
  —initial analysis of all management and disposal options,
  —demonstration of a *need* for open ocean disposal, and
  —periodic reconsideration of other available management and disposal alternatives.
  Congress also could consider influencing market conditions to attain specific goals. For example, financial incentives such as fees, taxes, or tradable discharge permits (56,305) could be used to make the total cost of ocean dumping comparable to that of other options or to ensure that short-term economic factors alone do not drive decisions regarding dumping.

Photo credit: Florida Department of Commerce, Division of Tourism

Chapter 2
Understanding
Marine Waste Disposal:
The Broader Context

CONTENTS

Table

Boxes

Understanding Marine Disposal: The Broader Context

INTRODUCTION

Decisions about marine waste disposal are affected by many ecological, economic, and social factors that extend beyond purely technical considerations. These other important factors include the economic and aesthetic value of marine resources, philosophical perspectives on the use of marine environments for waste disposal, and the nature of public concerns over such disposal. Looking at this range of considerations can help define a broad context within which decisions can be made about using marine environments for the disposal of wastes.

Marine waters have enormous value. They are home to a tremendous diversity of marine organisms and play a critical role in nutrient and energy cycles. From an economic perspective, they harbor food species that provide sustenance to people, primarily from commercial and recreational fishing, but also from hunting coastal waterfowl, harvesting marine plants, and aquaculture and mariculture operations. Fishing supports numerous other commercial activities, such as shipbuilding, fish processing, and retailing. Marine resources are also important sources of products such as pharmaceutical chemicals and many common consumer goods (e.g., the base for toothpaste).

From an aesthetic perspective, marine waters afford the value of a relatively unspoiled and mysterious frontier, as well as numerous recreational opportunities. The sights, smells, and sounds of the sea and its life provide countless people with feelings of pleasure and well-being. Fishermen, mariners, poets, and beachcombers all recount the irresistable attractions of the sea. Commercial fishermen cling tenaciously to their way of life despite economic and physical hardships; recreational fishermen frequently continue to fish for pleasure even when advised not to consume fish due to high con-

tamination levels.[1] Although difficult to fully quantify, it is clear that marine resources are of substantial importance to a wide range of Americans.

Public interest in marine resources heightened in recent decades because of several marine pollution incidents. These include the detection of DDT residues and PCBs in parts of the deep ocean, closures of beaches and shellfish beds in the United States because of bacterial contamination, and a lethal incident in Japan that involved the consumption of mercury-contaminated fish. In addition, problems arising from waste disposal on land (e.g., the discovery of many toxic waste sites and increasing groundwater contamination) generally stimulated public concern about the impacts of wastes in all environments.

The images left after these and other pollution incidents have combined with aesthetic considerations to confer a special status on marine waters. The ocean is viewed by many in the general public as a unique and precious resource requiring careful stewardship because of its vital importance to the earth's ecosystem and the potential for rendering irreversible harm to it. In addition, in recent years the international community has begun to recognize marine waters in general as a global resource.

---

[1]A recent study of recreational fishermen in New Jersey found that 40 percent of the fishermen surveyed were aware that the fish they caught had unacceptable levels of contamination and refrained from eating them (31). A survey by the U.S. Department of the Interior in 1982-83 found that fishermen most frequently cite relaxation and enjoying nature as the reasons why fishing is a favored outdoor activity; the prospect of catching or consuming fish is cited much less frequently (626).

For these reasons, any discussion of the use of marine waters for waste disposal must consider not only the technical and economic feasibility of a disposal option, but also its political acceptability. Thus, it is increasingly important to consider marine disposal alternatives in the context of broader issues. In particular, the aesthetic and economic value of marine waters should be considered relative to land-based resources and marine waste disposal should be seen as one part of a more comprehensive strategy of waste management.

This chapter examines various perspectives toward marine waters, including their economic and social value from a recreational and commercial viewpoint, and the broad philosophical positions that affect use of these waters, ranging from a protectionist view to managerial stances. These differing philosophical perspectives are reflected in current statutes and could be obstacles to comprehensive marine waste management. Next, two elements of a more comprehensive waste management strategy are examined: a general waste management hierarchy and the use of "multi-media assessment." Finally, recognizing that credibility is crucial to the public acceptability of any waste management decisions, several specific public concerns about marine disposal are discussed: 1) questions of equity; 2) opportunities for public participation; and 3) risk acceptability.[2]

_____
[2]These public concerns were selected to illustrate the range of such concerns, but are only a sample of the various types of issues important to the public.

PERSPECTIVES ON MARINE WATERS

The Value of Marine Resources

It is impossible to accurately and meaningfully quantify the full value of all marine resources to all people. Thus, the following discussion focuses on marine resources that are directly important to large numbers of people or are economically significant, and in addition are especially vulnerable to changes induced by waste disposal. These resources include organisms dependent on marine waters, such as fish, birds, mammals and vegetation, and waters used for swimming and other recreational purposes.

These marine resources support commercial and recreational fishing, beach-going, and other activities generated by the tourist trade in coastal areas. The activities tend to be concentrated in estuaries and coastal areas, although a significant amount of fishing occurs in the open ocean. Marine resources are of substantial and direct importance to tens of millions of Americans and they generate annual expenditures of billions of dollars.

Fishing and beach-going are among the principal recreational uses. Almost 12 million Americans aged 16 or over fished recreationally in U.S. marine waters in 1980 and spent approximately $2.4 billion on food, lodging, transportation, equipment, licenses, tags, and permits (628). Approximately 30 percent of all U.S. finfish landings used for human food (as opposed to uses such as pet food or fish meal) in 1985 were caught by marine recreational fishermen (614).[3] About three-fourths of these fish were caught within 3 miles of shore (605,606).

Although the nationwide significance of beach-going has not been studied in detail, its importance is suggested by a study conducted in Florida (27). Over 13 million adults used the State's beaches in 1984, and direct and indirect beach-related sales amounted to $4.6 billion—nearly 3 percent of the State's gross sales. These sales generated about 180,000 jobs, with a payroll of about $1.1 billion, and over $164 million in revenues for the State.

The same study also attempted to quantify the social value of Florida's beaches. Based on extrapolations from a survey that asked people in Florida how much they would be willing to pay to use the beaches, the investigators estimated a social value ranging between $2 billion and $28 billion. This large range illustrates the uncertainty associated with such an estimate. Nevertheless, it draws attention to the enormous economic significance of recreational activities, and beach-going in particular, to some coastal economies.

_____
[3]This figure refers to fish landed in all U.S. marine ports, regardless of where they were caught.

Photo credit: Division of Tourism, Florida Department of Commerce

Almost 12 million Americans aged 16 or over fished for recreation in U.S. marine waters in 1980.

In addition to being drawn to marine waters for recreational fishing and bathing, people travel to or live near these waters for other recreational purposes, ranging from waterfowl hunting to whale and bird watching. The degree to which wildlife draws people to marine waters for these other activities is not known, but large numbers of people are involved. For example, National Park Service lands that include marine waters recorded more than 60 million recreational visits in 1985; over 22 million of these visits were recorded at National Seashores (627). A government survey found that wildlife *alone* attracted at least 5 million people to oceanside areas in 1980 (628).

Besides their recreational uses, wildlife resources are also of tremendous commercial value, primar- ily to commercial fishermen.[4] About 231,000 commercial fishermen were employed in the United States in 1984. Total commercial landings of fish and shellfish from all U.S. marine waters had a dockside value in 1985 of about $2.3 billion (table 1), and a retail value several times greater. About one-half of the total commercial value was generated by fish and shellfish harvested within 3 miles of shore. These figures do not include the value of support services, such as shipbuilding and fish processing. For example, nearly 110,000 people were seasonally employed in 1984 as processors and

[4]Other uses, while not discussed here, are locally important. These include activities such as the commercial harvesting of aquatic vegetation (e.g., kelp) and commercial exploitation of fur-bearing mammals.

Table 1.—Commercial Fish Landings in the United States, 1985

Coastal region	Million pounds	Million dollars
Northern Pacific	1,454	$ 730
California and Hawaii	380	155
Gulf of Mexico	2,412	597
Southern Atlantic	311	156
Northern Atlantic	1,556	644
Maryland, Virginia	(815)	(124)
Delaware, New Jersey, New York	(151)	(101)
New England States	(590)	(419)
Total	6,113	$2,282

SOURCE: U.S. Department of Commerce, National Oceanic and Atmospheric Administration, *Fisheries of the United States, 1985,* Current Fishery Statistics No. 8380 (Washington, DC: April 1986).

wholesalers for the commercial fishing industry (614).

Recreational and commercial activities have been affected by waste disposal activities in numerous instances. The effects are not always detrimental, and may in fact at times be beneficial. For example, wastes discharged from small fish-processing firms, if properly managed, can increase the food supply for local fish and improve nearby recreational fishing.

Unfortunately, in many cases the impacts are not advantageous. The nationwide magnitude of impacts certainly is very large, although its exact

Photo credit: U.S. National Oceanic and Atmospheric Administration

In 1985, marine fisheries supported well over 200,000 U.S. fishermen, and U.S. landings were valued at $2.3 billion. The single most important commercial marine species was menhaden, shown here being hauled aboard a fishing vessel. Some menhaden being caught along the Atlantic coast, from North Carolina south, exhibit skin ulcers that may be linked to pollutants, but a clear explanation for the affliction has yet to be found.

Box D.—Economic Economic Losses in the Atlantic Coast Striped Bass Fishers

Striped bass are large migratory fish which spawn in brackish waters, but spend most of their lives in ocean waters. The striped bass along the Atlantic coast (from North Carolina to Maine) are prized for their size (up to 100 pounds and almost 5 feet long) and combativeness. In 1980, they supported a commercial and recreational fishery that provided about 5,600 jobs and stimulated at least $90 million in direct expenditures and $200 million in related economic activity (403).

Over the last 15 years, however, the population of striped bass has fallen precipitously. The decline is believed to result from a combination of causes, including pollution, the destruction of critical habitats, and fishing. Of particular importance has been the degradation of the Chesapeake Bay, upon which 90 percent of east coast stripers once depended (625). Although the number of year-old striped bass in Chesapeake Bay increased during 1986, it is not known whether this presages a long-term increase or simply represents a short-term fluctuation.

In addition to declines in numbers, the fish in some areas have become highly contaminated with organic chemicals. For example, high levels of polychlorinated biphenyls (PCBs)—in excess of the Food and Drug Administration's allowable limits—have been found in striped bass in New York and New Jersey waters (406,518). The majority of the PCBs had been discharged into the Hudson River from 1950 to 1976 by two industrial plants.

As a result of these developments, the commercial and recreational exploitation of the striped bass has dropped sharply. The commercial catch fell from a record 14.7 million pounds in 1973, to 1.7 million pounds 10 years later, and catches had declined further by 1986. Between 1974 and 1980, declining catches resulted in the loss of 7,500 jobs and $220 million in economic activity (403). For many commercial fishermen, the decline brought an end to a way of life (278,342).

By late 1986, restrictions on commercial and recreational fishing were widespread. In many cases these measures were taken primarily to protect the fish themselves, as was the motivation behind a ban imposed in 1985 on commercial harvesting in the Chesapeake Bay. In some instances, however, as in New York and New Jersey, restrictions also reflected a concern for consumers. In 1986, New York prohibited the capture and sale of striped bass from all waters, both fresh and marine, after finding extremely high PCB concentrations in the fish over large areas of the State.

New York estimated that the closure would affect thousands of commercial and recreational fishermen and result in annual losses of up to $15 million to $20 million. In particular, the State estimated that employment for 1,400 commercial fishermen would be affected, with an annual income loss of $1 million. In turn, wholesalers and retailers would lose approximately $4.5 million to $6 million per year. Direct losses to recreational fishermen were estimated to be more than $4 million annually, and indirect losses of additional recreational expenditures could amount to $4 million to $8 million (393).

dimensions are unknown. The full significance of these impacts is often difficult to determine, in part because many other factors (such as overfishing or natural perturbations) contribute to the picture. Various examples, however, suggest that the losses associated with individual pollution incidents can often amount to millions of dollars annually and directly affect thousands of people. (See boxes D and E.)

The expected trend of degradation in many estuaries and coastal waters will affect many of these recreational and commercial values. The possibility of serious deterioration in U.S. waters could reduce the Nation's ability to provide an abundance of uncontaminated food even as demand for food and other products from the sea intensifies. Per capita consumption of seafood in the United States is high by world standards and growing extremely quickly.[5] Yet a growing gap has developed in our ability to meet this demand with products from domestic waters. Imports of seafood are at an all-time high: in 1985, net imports were estimated to be $5.6 billion (614), contributing to the total U.S. trade deficit of over $130 billion.

---

[5]Average U.S. per capita consumption of fish and shellfish in 1985 was 14.5 pounds of edible meat, the highest 3-year average ever recorded (614).

Box E.—Economic Consequences of PCB Contamination in New Bedford Harbor

New Bedford Harbor, located in southern Massachusetts, has been designated a Superfund site because its sediments are highly contaminated with polychlorinated biphenyls (PCBs). The PCBs were discharged primarily by two industrial operations, although significant quantities had also been released by New Bedford's municipal wastewater treatment plant (339,687). This contamination caused major economic consequences. The commercial lobster fishery, residential property values, and recreational activities all suffered damages (344,345,355).

After 28 square miles of commercial lobster grounds were closed because of PCB contamination, lobstermen either quit lobstering or moved their activities to other areas. Those lobstermen who shifted activities incurred a variety of additional costs that might reach nearly $3 million if the grounds are closed for over a century.

Not surprisingly, residential property values and recreational values also declined. In areas near the inner harbor, where contamination was most severe, the sales value of owner-occupied homes declined by approximately $20 million (in 1985 dollars), apparently as a result of their proximity to the contamination. Sale values declined an average of $7,300 per house, or 13 percent of the market value. It is estimated that total damages to recreational beach use and to recreational fishing will reach nearly $10 million by 2085.

The sum of these estimated damages is over $30 million. The figure is extremely rough, extending as it does a century into the future. It also should not be regarded as a complete estimate of total damages, because it is based only on the most easily quantifiable and conspicuous consequences of the PCB contamination. In all likelihood, the total damages would be considerably larger were all the impacts considered and somehow quantified.

Philosophical Positions

The social value placed on marine waters and resources depends largely on each individual's philosophical perspective. Indeed, many of the arguments related to marine waste disposal (or many other environmental issues) can be traced to broad, philosophical views about the most appropriate use of the ocean. Such views are influenced by factors such as an individual's ideology, educational and professional background, and possibly their employer's interests (322).

Three philosophical perspectives—protectionist, managerial, and exploitive—form the basis for a range of attitudes toward use of marine environments.[6] Briefly, the protectionist view of marine waters is that they should not be used for waste disposal or should only be used as a last resort. The management perspective recognizes that marine waters can be used for the disposal of some wastes under monitored conditions and tends to consider these waters on an equal basis with other potential disposal media. The exploitive perspective goes a step further and supports maximum use of marine waters as a disposal medium whenever it is economically attractive compared to alternative options.

Few individuals publicly advocate the extremes of complete exploitation or absolute protection of marine environments. Most views, as well as most policy debates over marine waters, involve an intermingling of the protectionist and managerial perspectives. Both of these perspectives are clearly reflected in existing statutes. The exploitive perspective is not explicitly incorporated into any existing policy and is considered here only as an extreme variant of the managerial perspective.

The Protectionist View

The protectionist position views the ocean as having unique properties and contends that it either should not be used for waste disposal or should be used only as a last resort. This perspective derives from an environmental ethic that dates back at least to the preservationism of Thoreau. From this per-

---

[6]Obviously, a wide range of attitudes are encompassed in these perspectives; finer distinctions are possible but are not necessary for this discussion. Conflicts between the protectionist and managerial perspectives have shaped many environmental policy debates since the Progressive era. See ref. 158 for a useful discussion of the key distinctions between these two perspectives, which are also referred to as "preservationist" and "conservationist" views.

spective, the ocean is considered a common resource that requires special protection to prevent its exploitation. The basic concern is to prevent a "tragedy of the commons," i.e., the overexploitation of the common resource of the oceans for individual gain (222). Special protection of marine waters by the government is considered justified on grounds similar to those used to argue for protection of national forests and other precious common resources.

The protectionist position argues further that the level of scientific uncertainty about marine environments requires extreme caution when considering their use for waste disposal. Given the potential irreversibility of negative impacts that might result from waste disposal activities or accidents, this view gives special weight to the effects of actions by today's society on future generations. As Jacques Cousteau explained:

> To fulfill a moral obligation that the legacy of the oceans be continued, our first concern must be directed to the future. Risks for our progeny must be weighed against anticipated short-term provincial benefits. Our responsibility toward them is overwhelming . . . Poisoning the sea will inevitably poison us. Let us act with wisdom, foresight, and prudence (117).

Most proponents of the protectionist perspective maintain that the first management priority is to reduce the generation of a waste at its source. When wastes must be disposed of, protectionists argue that marine waters should only be used as a last resort or at least not be considered equally along with other potential disposal media. One reason given for this strict stand is that whenever marine waters are considered as an option for waste disposal, they are chosen because marine disposal is often the least socially objectionable alternative (i.e., because it satisfies people's desire to have waste disposal occur at a distance). Marine disposal is also often the least costly disposal alternative for some wastes. For some coastal municipalities, for example, marine disposal of sewage sludge costs less than land-based treatment or disposal. Some observers, however, advocate changing this by having the costs associated with marine disposal (e.g., site selection and monitoring) be borne more directly by waste disposers rather than by the government.

An additional concern is that since "nothing in the sea is provincial," a global perspective must be maintained regarding marine waters. In particular, the United States has been considered a leader in environmental protection, so there is concern that if the United States increases its marine disposal activities, other nations will follow.

Even many strong proponents of protection, however, acknowledge that disposal in marine waters may be appropriate for certain wastes. For example, marine disposal of acid wastes might be considered acceptable in some instances, if properly managed and monitored. At the same time, there is general consensus that certain highly toxic wastes are probably never appropriate for such disposal. Most protectionists would argue that marine disposal should only be chosen after a comparison with land-based treatment and disposal methods (i.e., after conducting a multi-media assessment) (121).

The Managerial View

The managerial position contends that marine waters can be viewed in many ways, and "one of the uses of the oceans is that [of] a receptacle for wastes. If used properly, it should serve as a renewable resource" (189). This perspective is rooted in the conservation movement of the Progressive era; the movement emphasized wise and multiple use of natural resources. From the managerial perspective, many factors need to be balanced in deciding how to use marine waters. These factors include: environmental and human health considerations, technological feasibility, economic costs, and the availability of other disposal options. Depending on how these different factors are weighted, the managerial perspective can support a range of positions from strong protection to maximum use of marine waters.

A basic distinction between the protectionist and managerial perspectives is that the latter is a *human-centered* approach which views the environment, including marine waters, as a resource to be used for society's benefit. In contrast, the protectionist position places primary emphasis on the environment itself, treating anthropological concerns as peripheral in any policy decisions. It considers the environment for its own value, however difficult to quantify, rather than its value only in terms of its use for humans (558). Thus, the severity of the same impact can be interpreted differently.

When marine waters are viewed as a resource (i.e., from the managerial perspective), they are not necessarily used by society in the most beneficial or environmentally sound way. This problem is related to the "tragedy of the commons" argument: for any resource for which there are no individual property rights (i.e., a common resource), it is in each individual's interest to use the resource to his/her fullest advantage regardless of any long-term consequences. Ultimately, since all resources have the potential to be exhausted, disaster can result.

A managerial perspective can use several different approaches to encourage better management of such resources. The approach commonly used in the United States is a standard-setting regulatory approach, which delineates the allowable amount of resource degradation or use. Standard-setting can be based on environmental and human health considerations, as well as technological and economic factors; it is used extensively, for example, in the Clean Water Act. Another approach relies more on the use of economic mechanisms to control degradation. For example, a society could use fees or taxes to adjust the degree of resource use to a desired level, or it could use transferable property rights or tradable permits to allocate the rights to use the resource (172).

It is possible that an economic fee or charge approach and a standard-setting system might be integrated to better provide incentives for reduction and more efficient control of certain pollutants; the *combination* of a permit system with an economic charge system might be better than either system alone in providing the flexibility needed for responding adequately to changing circumstances (56).

Comparing the Marine Protection, Research, and Sanctuaries Act and the Clean Water Act

Currently, the two major statutes regulating marine waste disposal—the Clean Water Act (CWA) and the Marine Protection, Research, and Sanctuaries Act (MPRSA)—embody somewhat different expressions of these basic philosophical perspec-

tives. MPRSA includes protectionist provisions (e.g., establishment of marine sanctuaries), but allows managed use through a permit process for marine dumping. The permit system could also be used to increase or decrease protection. An assessment of all relevant factors, such as alternative options, potential effects, and economics is required before a dumping permit can be granted.

Under MPRSA, disposal of some wastes is absolutely prohibited (e.g., warfare substances and high-level radioactive waste). Other wastes such as some industrial wastes, sewage sludge, and dredged material can be disposed of under regulated conditions.

CWA is more consistently managerial in its orientation, and it stresses a "best available technology, economically achievable" and "best management practices" approach. Under CWA, permits include standards for allowable discharges, but do not require consideration of the full range of factors required for an MPRSA permit.

As a result, and because the two statutes also differ in their jurisdiction over marine environments (ch. 7), different marine environments have received varying degrees of protection from and use for waste disposal. In open ocean environments, dumping activities have generally been strictly controlled or reduced under the guidelines set forth by MPRSA. In estuaries and coastal waters, however, disposal activities regulated under CWA and MPRSA are much more frequent, and in general these waters have borne the brunt of marine disposal activities.

The basic orientation of a law, however, can evolve and change. In the case of MPRSA, Congress embodied a protective attitude in the law a decade ago by setting a 1981 deadline for terminating the disposal of sewage sludge which might "unreasonably degrade or endanger" human health, welfare, or the environment.[7] In a landmark case, *City of New York* v. *United States Environmental Protection Agency* (EPA) (543 F. Supp. at 1084, 1099 (S.D.N.Y. 1981)), the court held that not all dumping of sludge was necessarily prohibited by

---

[7]Congressional intent has been a source of confusion in the Environmental Protection Agency's implementation of this provision: Congress apparently imposed an absolute deadline, but also included language that can be interpreted to allow the dumping of "reasonable sludge" (12,291,531).

the 1981 deadline.[8] EPA decided not to appeal the case and currently interprets it to: 1) allow sludge dumping if it does not cause "unreasonable" harm, and 2) require development of criteria for comparing land-based and ocean alternatives to determine

---

[8]New York City had been dumping sludge under interim permits granted by EPA and was exploring alternative disposal options. It concluded that land-based alternatives would be more costly and potentially more environmentally harmful than marine disposal. EPA maintained that the 1981 deadline absolutely prohibited the dumping and denied the City's petition to continue ocean dumping. The City then sued, arguing that EPA was required to consider all of the statutory criteria listed in MPRSA (Sec. 102(a)) when evaluating permit applications. These criteria require EPA to take into account—beyond environmental criteria—such factors as the need for ocean dumping and the costs of land-based alternatives. The Court further held that MPRSA requires EPA to balance these statutory factors when evaluating permit applications (ch. 7). Some observers have raised the concern that this decision could reduce incentives to find land-based alternatives for sludge treatment or disposal, especially in light of the difficulties associated with siting land-based alternatives and their frequently higher cost (155,650).

when ocean dumping can be allowed. Several cities, including Philadelphia and Washington, have indicated that they would consider the ocean disposal option for sludge if it were to become available.

Although Congress has not officially removed the ban on dumping harmful sludge, it has allowed the 1981 deadline to pass. The law's originally protective attitude thus may be evolving into a more managerial approach to marine waste disposal, but the final policy direction is not yet clear. Specific deadlines for eliminating dumping at the 12-Mile Sewage Sludge Dump Site have been set by EPA and some dumping activity has already shifted to the Deepwater Municipal Sludge Site in open ocean waters. The House of Representatives has supported this shift (ch. 7). The extent to which sludge dumping activities should continue under MPRSA, however, has not yet been clarified by Congress.

COMPREHENSIVE WASTE MANAGEMENT

It is increasingly recognized that the existing suite of pollution control laws, each primarily focused on abating pollution in one particular medium (air, water, or land), has sometimes resulted in the shifting of wastes from one environmental medium to another and that long-term environmental and human health risks may not have been substantially reduced (110,111,263,378,382). As a result, there is a need for greater incentives to reduce or avoid the generation of wastes as the best means of reducing waste disposal-related risks (144,263,377, 586,587). As our understanding has grown, it has become clear that a highly protective policy toward the open ocean may be counterproductive and that, for particular wastes and situations, marine disposal should be carefully considered in context with land-based alternatives.

Thus, consensus is developing about the need for a more comprehensive waste management strategy in this country. Two key elements of such a strategy would be: 1) a hierarchical approach to waste management, and 2) multi-media assessment. A hierarchical approach ranks waste management methods according to their ability to reduce risk; for example, the highest tiers include methods that avoid the generation of waste (586). Multi-media assessment can be used as a tool to determine which

treatment or disposal method, in which environment, would most minimize risk.

Waste Management Hierarchy

The idea of a waste management hierarchy was developed originally for wastes classified as hazardous according to the legal definition in the Resource Conservation and Recovery Act (RCRA), but its principles are equally applicable to all wastes which can cause harm to the environment or human health (144,263). Tiers in the waste management hierarchy include:

- reduced generation of waste, with respect to both volume and toxicity (using techniques such as product or input substitution and process modification);
- recovery of waste for recycling or reuse of materials for energy (including the use of waste exchanges, shared central facilities, and third-party recyclers);
- destruction or treatment of wastes to reduce toxicity (using techniques such as land-based or ocean incineration);
- stabilization of waste through physical or chemical means (e.g., including neutralization and evaporation);

- isolation or containment (e.g., in surface impoundments or landfills); and
- dispersion in the environment (e.g., by dumping or discharge) (586).[9]

With regard to marine destruction and disposal methods, for example, methods such as ocean incineration, which has the potential to destroy 99 percent of certain hazardous wastes, occupy a middle position in such a waste hierarchy. Other methods such as sewage sludge dumping occupy a lower tier.[10]

Waste minimization was declared to be a national policy in the 1984 amendments to RCRA, but incentives to ensure its implementation are not yet sufficient (587). Strong incentives for reduction, recovery, and treatment of wastes prior to disposal, and for selecting the best available disposal options or improving disposal technology, could be made a more integral part of many environmental statutes (263,586,587). Some companies that have voluntarily implemented waste reduction strategies have found that they not only reduce the amount of waste generated, but also save money (279,490, 587).[11]

It is important to note, however, that even when waste reduction does occur large quantities of wastes may still result. Within a waste management hierarchy, the next objective would be to reduce the levels of toxic pollutants in the wastes, by recovering or recycling materials when possible, and then to select the best disposal option for any remaining wastes. The particular characteristics of a waste and the feasibility or availability of potential disposal media would determine the number and nature of options available, as well as their economy. In general, additional encouragement by Congress of a hierarchical approach for waste management could help facilitate the move toward more comprehensive environmental management.

Multi-Media Assessment

Multi-media assessment as an approach to waste management has gained considerable popularity. This procedure involves comparing the impacts of different treatment and disposal options, including impacts on environmental media other than the one directly used, and then selecting an option on the basis of the greatest reduction in overall environmental risk. Other social and economic factors can also be considered in the process.

Multi-media assessment can be difficult to implement, partly because the amount of information needed to perform such an analysis is large and expensive to obtain, and partly because estimating risks is difficult. Thus, this approach may be most useful as a qualitative gauging method for comparing options, rather than as a rigorous, exclusive, or formal basis for decisions.

The need for multimedia assessment arises in part because disposal in each environmental medium is generally regulated by separate statutes. Although possibilities exist for incorporating multi-media considerations into current statutes, to date waste management programs have operated quite independently. For municipal sludge, for example, MPRSA and CWA place strict limits on (and in some cases effectively prohibit) disposal of sludge in marine and surface waters generally. RCRA limits land-based disposal and land application options, and the Clean Air Act sets extensive technological requirements for sludge incineration operations. Each environmental medium may in theory be protected, but factors such as the cross-media transfer of wastes among the media and the effects of one regulatory program on another are not taken into account (49,283). Moreover, since the sludge must go somewhere, it commonly ends up in the least regulated medium. (EPA is developing comprehensive regulations for sewage sludge management to address these problems; see ch. 9.)

This lack of coordination arises in part because most major environmental statutes were developed

---

[9]Certain technologies may actually involve more than one tier of the hierarchy and more than one environmental medium. Ocean incineration of wastes entails the destruction of most of the wastes (and for this reason is considered to be in a middle tier of the hierarchy), but a small amount of the unburned wastes is dispersed into the air and the surface water of the ocean and any residues are contained in land-fills (586).

[10]Although in many cases the preferred strategy may be to eliminate or reduce the generation of a waste, it cannot automatically be assumed that this option will always best reduce overall risk. For example, process modifications can lead to a reduction in the *quantity* of a waste produced or change its composition, without necessarily reducing the *degree of hazard* of any remaining waste (586). Moreover, certain wastes may not be able to be reduced to any great extent (e.g., sewage sludge and dredged material).

[11]The 3M Co. estimates that it saved close to $300 million since 1975 as a result of its ''pollution prevention pays'' strategy (279).

independently, by different combinations of congressional committees and subcommittees, and are administered by different EPA offices and programs. Differences exist among the statutes with respect to their philosophy and intent, as well as their designation of management authority, and no mechanism provides for development of a comprehensive approach to waste management.

The use of multi-media assessment and a hierarchical approach to waste management could conceivably be integrated into current regulatory programs as a way to promote comprehensive waste management.[12] This would be an enormous under-

taking and almost surely would require Federal guidance. Yet, a general consensus is emerging both inside and outside of government that this is a necessary policy direction to ensure more efficient and effective environmental protection. It is increasingly essential that marine waste disposal options be viewed within the context of this general policy debate.

---

[12]A number of options are possible to further promote the use of multi-media assessment including: 1) using Sec. 304 of CWA, which

requires that guidelines on discharges into surface water include information on ''non-water quality environmental impacts,'' as a model for provisions in other statutes; 2) requiring the preparation of a multi-media impact statement for all disposal activities; and 3) requiring the development of more consistent criteria for assessing disposal options (where appropriate) among statutes, i.e., a common set of general criteria—perhaps focused on public health and environmental risk reduction—could be included in MPRSA, CWA, and RCRA to be used in comparisons of land-based and marine disposal options.

PUBLIC CONCERNS ABOUT THE USE
OF MARINE ENVIRONMENTS

A critical component of all decisions regarding waste disposal in marine waters is the public acceptability of disposal alternatives. A wide range of factors influence whether an individual or the public at large will accept a particular option, but this discussion only highlights several of the most important ones. One fundamental factor is the level of trust the public has for decisionmakers. While any government action is dependent on public trust for its legitimacy, such trust can be elusive.

EPA's credibility, for example, eroded in the early 1980s when several scandals involving the Agency and its dealings with some industries were uncovered (360). Efforts are underway to improve this situation, but rebuilding trust is a slow process. In addition, past violations of environmental regulations (that resulted, for example, in the creation of Superfund sites) have led industry and waste management companies to lose credibility. Given these problems, technical assessments about disposal options often hold little sway with the public, especially when such assessments are unclear about the risks of environmental degradation. Building credibility is closely linked to: 1) how equity issues are resolved during the decisionmaking proc-

ess, 2) how the public is allowed to participate; and 3) how risks and other public concerns are addressed.

Equity

Equity issues arise every time a waste disposal alternative is discussed because residents near a proposed facility or area of disposal fear that their health or property values will be disproportionately jeopardized (see box F). As a consequence, the public often believes that the generator of a waste not only should treat and dispose of it but also should be held liable for any impacts (see box G). One of the most frequently voiced objections at public hearings—whether about the siting of hazardous waste facilities or land disposal of sludge, or permitting for ocean incineration—is that one community should not have to bear the economic burden and potential health and environmental risks of another community's wastes. Although this is not an easy issue to resolve, several techniques to deal with these concerns are being attempted; when siting hazardous waste treatment facilities, for example, the use of risk-mitigating proposals (e.g., regular safety inspections by public officials and rep-

Box F.—The Siting Issue

One central public concern in all waste management policymaking is the actual siting of waste facilities or disposal operations. This is particularly true for land-based disposal, but siting is also an important issue when considering disposal in marine environments.

On land, siting decisions have been greatly constrained by property rights issues and what is commonly called the "NIMBY" ("Not in My Backyard") syndrome. The heart of the problem is that although society as a whole benefits from the proper disposal of wastes, the risks associated with disposal methods are localized, creating a basic problem of equity. Most communities are unwilling to accept the risks associated with disposing of other communities' wastes. As a result, no new hazardous waste management facilities were sited during the 1980s, despite the need for such facilities (361). It has been almost as difficult to select sites for land incineration or land application of sewage sludge.

The difficulties faced in siting land-based disposal or management activities can increase pressures to use marine environments for disposal of certain wastes (e.g., sewage sludge). Marine waste disposal, however, is not free of siting constraints, even though the concept of property rights does not extend to marine environments. Instead, public concern stems from the facts that: 1) these environments are common property that could be used for waste disposal in an "out of sight, out of mind" manner; and 2) most marine disposal sites continue to be used because of their past history of use, not because they have been selected on the basis of a technical evaluation.

For marine environments, the NIMBY syndrome associated with land-based sites has become translated into "not through my port" or "not off my shoreline." Public opposition to ocean incineration, for example, has been particularly intense in port and coastal communities near the proposed incineration sites.*

In general, decisionmaking processes have not effectively addressed several critical issues that contribute to this attitude: 1) credibility of the siting process, to ensure that scientific assessments and political judgments of a site's suitability are trusted by the public; 2) equity issues, to ensure that health and economic risks are not unfairly borne by the communities accepting wastes; and 3) public participation, to inform and involve the public in the siting decision. Studies of land-based siting decisions indicate that early and continual public participation, positive local-State rapport, and efforts to mitigate risks are crucial to winning public acceptance for siting waste management facilities (123,282,450).**

*See OTA's report, *Ocean Incineration: Its Role in Managing Hazardous Waste* (586).

**Mediation has been used in land-based siting disputes and has been suggested for marine-based siting. Thus far, however, it has not been successful in resolving the siting of any land-based hazardous waste facilities, partly because the interests involved were unable to compromise (114). If public participation is begun early in the siting process, however, it may be possible to avoid such impasses (236,264). A recent study provides a framework for selecting potential marine sites, including consideration of both socioeconomic and technical considerations (263).

resentatives) appears to be more attractive to communities than risk compensation proposals (e.g., lowering property taxes) (450).

Public Participation

The right of citizens to participate in decisions that directly affect their interests is a fundamental component of our form of government. Certainly one way to increase the credibility of a waste disposal decision is to involve the public early and throughout the decisionmaking process. The scientific and technical issues surrounding waste disposal options (e.g., the risks of siting a facility or disposing of waste in a certain location) are impor-

tant factors, but the way they are communicated to and discussed in the community is equally critical. For example, one of the most significant obstacles to the ocean incineration program proposed by EPA is public opposition, which stems in part from poor communication. In this case, the public was excluded from participating in decisions made early in the process (586,667).

Risk Acceptability

If information about a disposal option is highly uncertain, then risks associated with the option are likely to be perceived as high and its acceptance is less likely. Effective communication with the pub-

Box G.—The Liability Issue

The question of who is liable for effects on the environment or human health is a common issue in waste management today, and its implications extend to disposal activities in marine environments. Liability can be incurred not just from spectacular fatal incidents, such as the one that occurred in Bhopal, India, but also from less apparent hazards—for example, leaking sewers; industrial effluents that impair sewage treatment plants; discharges that, although permitted, cause human health or environmental damage; and current disposal methods that may not meet future standards.

In general, two major areas of concern exist with respect to current liability practices: 1) the availability and cost of liability insurance; and 2) the lack of compensation mechanisms for third parties (i.e., those people who have been injured in some way as a result of waste disposal practices). Two types of liability insurance cover environmental risks—general comprehensive liability (GCL) for sudden, accidental releases; and environmental impairment liability (EIL) for gradual, long-term contamination.

For land-based disposal activities, EIL insurance has been required for surface impoundments, landfills, or treatment facilities since 1982 (40 CFR Parts 264, 265). The demand for EIL insurance has increased since January 1986, when all new GCL insurance contracts were modified to contain a pollution exclusion, which virtually eliminated the availability of such insurance for waste handlers and generators (134). Only three companies currently sell EIL insurance for most firms (134). As a result of these factors, the price of liability insurance has increased dramatically during recent years, at the same time that its coverage has become more restricted.

Several statutory provisions address liability for activities in marine environments. Provisions applicable to Superfund sites (CERCLA Sec. 107) in marine waters establish a liability limit of $50 million plus cleanup response costs. Liability can also be incurred for damages from releases that occur during the transportation of hazardous wastes (CERCLA Sec. 306). Under the Marine Protection, Research, and Sanctuaries Act (MPRSA), an ocean vessel knowingly used while a violation occurs is liable for civil penalties or criminal fines (Sec. 105), but not for any damages. The Clean Water Act liability provision for oil and hazardous substances (Sec. 311) applies to vessels and to any onshore or offshore facilities; the provision establishes civil penalties up to $250,000, a liability limit of $150 per gross ton for oil pollution damages, and compensation for remedial actions. The American Shipowners Limitation and Liability Act also addresses pollution-related liability of vessel owners, as do provisions in other laws such as the Trans-Alaska Pipeline Authorization Act, the Outer-Continental Shelflands Act Amendments, and the Deep Water Ports Act.

In the past, it has been difficult for third parties to press liability claims. The Superfund Amendments and Reauthorization Act of 1986 overruled recent court decisions (e.g., *Illinois* v. *City of Milwaukee* (101 S.Ct. 1784); *Middlesex County Sewerage Authority* v. *National Sea Clammers Association* (453 U.S. 1)) and clarified that MPRSA does not preempt any person's right to: 1) seek damages or enforcement of any standard or limitation under State law, including State common law; or 2) seek damages resulting from noncompliance with any permit or requirement under MPRSA or under other Federal law, including maritime tort law.

lic regarding the nature of risks associated with a disposal option is an important aspect of building credibility, addressing equity issues, and encouraging effective public participation in helping to solve problems. Given that access to information influences an individual's perception of risk, efforts have been made recently to improve communication between government, industry, and citizens about risks (112). Public involvement has been encouraged by EPA, for example, to help make decisions about how to balance risks and other ethical, social, and economic considerations (652). Various States also are developing strategies for involving the public. The New Jersey Department of Environmental Protection, for example, is restructuring risk assessment activities so that information

Photo credit: Gilles Press

"In haul-seining, a net-filled dory is launched through the open surf, an enterprise that, on a rough Atlantic day, demands nerve and experience as well as skill. Without the striped bass, haul-seining is unlikely to survive, and the end of this fishery will mean the end of a surfboat tradition that began when the Atlantic coast was still the American frontier."
—Matthiesen, P., *Men's Lives: The Surfmen and Baymen of the South Fork* (New York: Random House, 1986).

about risks is communicated more effectively to the public.[13]

A number of factors influence risk acceptability (316). Two of the most crucial, especially with respect to marine disposal, are the controllability and irreversibility of potential risks. One of the primary reasons why the public has a protective attitude toward the oceans is a perception that any harm incurred as a result of waste disposal activities in marine waters may be irreversible. The public genuinely believes that the oceans are a resource requiring careful stewardship and that they should not be damaged perhaps irrevocably.[14] As one fisherman noted:

[13]One strategy is to involve citizen groups in the decisionmaking process *before* an issue becomes a news media event. For example, a New Jersey Department of Environmental Protection study recently found toxic contamination with dioxin in New York Bight lobsters; *before* any policy determinations were made, the agency invited representatives from the fishery cooperatives in New Jersey and the U.S. Army Corps of Engineers to meet with its staff. The Corps of Engineers was invited because lobstermen maintain that the marine disposal of dredged material is the primary source of the dioxin contamination. The involvement of both groups is intended to ensure that human health and economic issues will be adequately considered, that credibility in the process can be maintained, and that information will be effectively communicated to the broader public (29).

[14]Marine waters appear to be able to receive certain wastes (e.g., acids) in controlled and monitored quantities without suffering significant adverse impacts; water and sediment quality may be altered for a time, but long-term ecological change appears unlikely (see chs. 5 and 11). Certain land-based disposal methods also can lead to irreversible environmental effects (e.g., contamination of groundwater). There is a general perception, however, that these effects tend to be relatively localized compared to the more global contamination that might occur in the ocean. This perception is not always correct.

I've seen bluefish come and go in my lifetime, and striped bass, too. The bluefish is a wild fish and a hardy fish, and because he don't go up in them dirty rivers, he'll survive where the striped bass will go down. All the fish around here come and go in cycles, and years back, you could anticipate the cycles, but today, with the pollution the way it is, you can't be so sure that a fish that's gone will ever come back at all (342).

The logical conclusion drawn by most fishermen, then, is that some stewardship of marine resources is necessary (342).

This sentiment is generally shared by the public which—judging by its high level of recreational use of coastal areas—highly values marine environments. An observation frequently heard when discussing marine waste disposal options—that ''fish don't vote''—is literally true, but not completely accurate. Marine waters and organisms do have a constituency that attempts to represent their interests. A number of public interest groups are highly attentive to any decisions regarding the uses of marine waters and the potential impacts of waste disposal. At the same time, most groups would agree that decisions regarding the disposal of wastes in marine waters should be considered in the broader context of comprehensive waste management.

Chapter 3
Waste Disposal Activities
and Pollutant Inputs

CONTENTS

Tables

Figures

Boxes

Waste Disposal Activities and Pollutant Inputs

INTRODUCTION

To fully understand the potential for wastes to affect marine resources and ecosystems and to evaluate management options for reducing adverse impacts, it is important to have an understanding of the amounts of different pollutants entering marine waters from different sources. Marine waters currently receive a variety of wastes, including municipal and industrial effluents, sewage sludge, dredged material, and some industrial wastes. These wastes vary considerably in physical nature and in biological and chemical composition. In addition, many of the same pollutants can be carried directly into marine waters by nonpoint sources such as agricultural and urban runoff, and both disposal activities and nonpoint pollution can occur upstream in rivers that later flow into marine waters.

This chapter first discusses the quality of information available about pollutant inputs into U.S. marine waters, including the issue of unregulated but potentially significant pollutants. The chapter then reviews the extent and variability of waste disposal activities and nonpoint runoff, including a comparison of the relative contributions of pollutants from different sources, and describes the major sources of pollutants to marine waters.

AVAILABILITY AND QUALITY OF DATA

Ability To Make National Comparisons

Considerable information is available describing waste disposal and nonpoint sources and pollutant inputs from these sources. The quality of this information varies considerably, however, which creates some uncertainty in estimates of pollutant inputs on a national scale and places some constraints on our ability to make comparisons among different sources of pollutants. This section briefly describes the quality of available information and how well it can be used in making estimates and conducting comparisons.

For this report, data on dumping activities were obtained primarily from Environmental Protection Agency (EPA) reports and from the Corps of Engineers. Data on pipeline discharges and nonpoint source pollution were obtained primarily from analyses of various EPA computer databases such as the Industrial Facilities Data Base (139,503), and from databases provided by the National Oceanic and Atmospheric Administration (NOAA) and Resources for the Future (RFF).[1]

Information on the magnitude of major pipeline discharges and dumping operations that occur into or directly adjacent to marine waters is relatively complete and reliable. For example, the quantities of wastes dumped and the number of major industrial and municipal pipelines discharging into marine waters are relatively well-documented. This is because they occur in a limited area and involve a relatively small number of discrete events or continuous activities.[2] Such data can readily be used to compare the relative importance of these particular sources in different marine waters.

In contrast, the information that would be needed for an accurate national assessment of relative inputs of particular pollutants from all sources (discharges, dumping, nonpoint sources, and upstream activities) is often less complete and reliable, or is gathered and analyzed using differing methodologies and assumptions. Any comparison of information about different pollutant sources that

---

[1]NOAA's Ocean Assessments Division provided data from its National Coastal Pollutant Discharge Inventory, which will eventually contain estimates for all coastal regions. RFF provided data from its Environmental Data Inventory, which contains estimates for both coastal and inland areas. These inventories are referred to as the NOAA and RFF databases, respectively.

[2]Even then, however, extracting information about different types of discharges from existing databases can sometimes be difficult (139).

relies on different databases is constrained by the following:

- available databases rarely consider *all* significant sources using internally consistent methodologies;
- definitions of key parameters (e.g., the geographic boundary delineating an ''upstream'' source) can differ considerably among studies;
- information on the quantity and composition of different wastes (and variability in these parameters) often is not available, or is expressed in units that are difficult or impossible to compare;
- different studies often rely on different assumptions or models which are supported by varying degrees of field validation;
- available data for various sources may have been collected at different times, and may be out-of-date or unrepresentative of current circumstances; or
- aggregation of data in some studies can mask highly significant short-term fluctuations (e.g., even one day of low dissolved oxygen levels can cause a massive fish kill).

Nevertheless, some individual databases can be used to evaluate pollutant inputs from most (but not all) sources nationally. The NOAA and RFF databases used by OTA in preparing this report, for example, estimate pollutant inputs from discharges and runoff into all U.S. coastal waters. It will be essential to continue developing and refining national databases to provide a sound basis for assessing trends and evaluating policy and technical decisions regarding waste disposal in marine environments. However, several factors currently limit the usefulness of these databases (477,600). In particular, neither database includes readily comparable information on pollutant inputs from dumping activities. In addition, the NOAA database will not be completed until 1987 (D. Farrow, NOAA, pers. comm., Sept. 9, 1986).

Lack of Information on Unregulated Pollutants of Concern

The information now available about pollutant inputs to marine waters is largely restricted to the substances that are specifically regulated under the Clean Water Act (CWA) or Marine Protection, Re-

search, and Sanctuaries Act (MPRSA) because information programs and resources are generally focused on regulated pollutants.[3] Thousands of additional pollutants are present in the wastes disposed of in marine waters, however, and hundreds of these may have the potential to affect marine environments and human health. Most of these unregulated and potentially significant pollutants are either pathogens or organic chemicals. Little information is available about their presence in waste materials or marine environments.

These unregulated pollutants can be important. Hundreds of types of microorganisms—viruses, parasites, bacteria, fungi, and protozoa—can be present in waste discharges, sludge, or runoff, and many of these are capable of causing diseases. They can contaminate water and fish, and thus cause economic and recreational losses and direct risks to human health. Only one class of microorganisms—fecal coliform bacteria—is regulated as a CWA pollutant.[4] While not generally pathogenic, it is used to indicate the presence of sewage-derived material (and indirectly, pathogens). Recent studies have concluded, however, that the presence of fecal coliform bacteria is not a good indicator of the presence of these pathogens in marine waters (205).

Several hundred organic chemicals that are not on the CWA list of 126 toxic ''priority'' pollutants can also be present in waste material and sometimes in runoff. In one survey, EPA identified 385 organic chemicals (with hundreds of others unidentified for various technical reasons) in municipal and industrial wastestreams (644). The chemicals included xylenes, dibenzofurans, and trichlorophenols. In addition, the environmental degradation of chemicals can yield products that sometimes are as toxic or more toxic than the parent compounds. Since tens of thousands of organic chemicals are currently in commercial use and hundreds of new ones are produced annually (386), it is likely that many other chemicals are also present in waste

---

[3]**Regulated pollutants** are defined in box A of chapter 1. Information is not always available, however, even for regulated pollutants. For example, waste dischargers only report the quantities of those pollutants in their discharge for which some limitation has been specified in the discharge permit. Most discharge permits, however, include limitations on only a small fraction of listed toxic pollutants.

[4]EPA recently has developed a marine water quality-based standard for *Enterococcus* bacteria; however, it is restricted to recreational waters.

discharges and runoff (e.g., from pesticide application). An unknown portion of these may be potentially harmful and warrant regulation.

As coastal populations and developments increase, and as the land-based disposal of certain hazardous wastes is increasingly restricted, it is highly likely that the amounts of pollutants—both regulated and unregulated—entering marine environments will increase. This trend raises concerns over whether current regulations cover all of the ''important'' pollutants —those pathogens, metals, and organic chemicals that are most likely to cause impacts.

In marine environments, there is little disagreement that the conventional and nonconventional pollutants currently regulated under CWA should indeed be regulated.[5] This consensus is based on a long history of experience, research, and monitoring.

Substantial disagreement exists, however, about the need to regulate additional pathogens, organic chemicals, and metals. From an economic perspective, resources are not available to individually regulate the dozens of metals and hundreds of microorganisms and organic chemicals that have been detected in waste material. Moreover, our technical capabilities and scientific understanding are not sufficient to determine which of these substances are present in concentrations sufficient to cause impacts.

One way to evaluate and regulate the large number of potentially significant pollutants would be to develop better *pollutant screening* approaches to identify the unregulated pollutants that are of primary concern in marine environments. EPA has taken some initial steps to develop screening processes that, while broadly designed, could identify additional pollutants important in marine environments. In one effort, for example, EPA analyzed various industrial wastestreams and identified hundreds of unregulated organic chemicals (644). EPA identified six chemicals that were present in significant amounts, were not currently treatable, and which exhibited toxicity to humans or aquatic or-

ganisms: dibenzofuran, two trichlorophenols, carbazole, trichlorobenzene, and a form of dioxane. No standards have yet been developed for these six compounds, however. In a second effort, EPA is developing technical regulations for sewage sludge disposal options (including ocean dumping and various land-based options); the regulations will identify and focus on those pollutants that pose the greatest risks to humans and various environments.

These screening efforts have focused primarily on organic chemicals for several reasons. Many of these chemicals tend to persist in the environment for long periods and are acutely toxic to organisms. In addition, many are soluble in the fatty tissues of organisms and, once ingested from water or sediment, can bioaccumulate (i.e., concentrate) in these tissues. Some of these chemicals can also biomagnify (i.e., increase in concentration in higher levels of food chains) when the contaminated organisms are consumed by predators. Significant acute and long-term chronic impacts attributable to many organic chemicals have been documented in the laboratory and in the field. They are perhaps best exemplified by our experiences with DDT, the use of which has been banned since the early 1970s (54).

The continued development and use of screening procedures may help resolve existing uncertainties about which pollutants are of primary concern in marine environments (254). For example, a relatively simple test of the solubility of an organic chemical in certain organic solvents can serve as a measure of its potential to bioaccumulate in the tissues of marine animals (195). Similarly, the susceptibility of an organic chemical to degrade (e.g., by light energy or by organisms) or volatize can be used as a measure of its potential to be available to marine organisms or to cause impacts in marine environments.

For metals, additional factors such as the precise chemical form can be essential in determining bioavailability or toxicity. For example, organic mercury shows much higher toxicity and bioaccumulation potential than does inorganic mercury. Under conditions that generally prevail in marine environments, most metals bind strongly to particulate material, thus altering their environmental fates and impacts. Thus, screening efforts could focus on identifying those forms of metals that are

---

[5]Some concerns have been raised over the appropriateness of using standards for fecal coliform bacteria to control the level of microbiological pollutants in marine waters, as discussed previously.

actually toxic in marine environments, and those settings where toxic forms are likely to be present in levels sufficient to cause impacts to humans or marine organisms.

The presence in sewage sludge and other waste material of microorganisms that can cause diseases in humans often limits the availability of disposal options for these wastes. Monitoring for their presence is difficult because microorganisms are exceedingly difficult to detect in the field or characterize in the laboratory. Better culturing methods and indicators need to be developed before more extensive pathogen screening efforts can be undertaken. In addition, because many microorganisms are more likely to survive in sediments or in marine organisms than in the water column, monitoring programs must be designed to sample sediments and organisms.

Even for chemicals identified as being of potential significance through screening efforts, however, other factors must be evaluated in determining the need for, or the form of, regulation. For example,

in some areas an important source of polycyclic aromatic hydrocarbons (PAHs) to marine environments is the natural seepage of oil from the ocean floor (19). Similarly, certain known human pathogens are natural members of the bacterial communities in nearshore marine environments (ch. 6).

The screening approaches discussed here focus on identifying individual compounds that have the potential to cause significant impacts. Because numerous pollutants can be present in wastestreams and in marine waters, approaches that first consider the overall toxicity of a wastestream or the cumulative impacts of all pollutants in a waterbody could also be helpful. For example, as a first step biomonitoring procedures (including whole-effluent toxicity tests) (49 FR 9016-9019, Mar. 9, 1984; ref. 64) and environmental indices (414) could be used to identify an effluent or water quality condition that has the potential to cause or is actually causing impacts. Then more extensive screening, using the approaches discussed for individual pollutants, could be used to pinpoint particular pollutants.

THE EXTENT OF POLLUTANT INPUTS INTO U.S. MARINE WATERS

Waste disposal activities in marine environments are diverse and highly variable in type, frequency, volume, location, and potential to cause adverse effects. Despite this diversity, much of the debate about marine waste disposal has centered on two main issues: 1) the direct dumping of sewage sludge, industrial waste, and radioactive materials; and 2) the incineration of hazardous wastes at sea. Much less attention has been devoted to comparing the relative contributions of pollutants from other disposal activities or sources such as pipeline discharges and runoff.

While the available data about these two main sources exhibit serious deficiencies, some generalizations can be made about pollutant inputs in different marine environments. In addition, the data can be used to illustrate the complexity and site specificity of disposal activities and pollutant inputs in marine waters.

Pollutants From Pipeline Discharges and Dumping

Marine waste disposal activities (i.e., pipeline discharges and dumping operations) are overwhelmingly concentrated in estuaries and coastal waters (see tables 2 and 3). For example, over 1,300 major industrial and almost 600 municipal facilities discharge directly into estuaries and coastal waters, and at most a few discharge into the open ocean. The open ocean is used for the dumping of some dredged material, sewage sludge, and industrial wastes, but four-fifths of the marine-disposed dredged material and virtually all marine-disposed sewage sludge are dumped in estuaries and coastal waters.[6]

_____
[6]The New York Bight is included among coastal waters. However, the dumping of sewage sludge that now takes place in the New York Bight will soon be shifted to a site in the open ocean 106 miles from shore (see below and ch. 9). Current waste disposal sites are discussed below; both active and inactive sites are illustrated in (612).

Furthermore, most pipeline discharges and many dumping activities occur specifically in estuaries rather than in coastal waters. Almost 99 percent of industrial pipelines and 89 percent of municipal pipelines discharge directly into estuaries (table 2), and over half of all marine dumping of dredged material occurs there as well. The extent of these activities varies greatly around the country (tables 2 through 4). For example, over half of the major industrial and municipal pipelines are located in the Northern Atlantic region[7] and the western Gulf of Mexico; three-fourths of all municipal effluent is discharged from the Northern Atlantic States and California. The marine dumping of industrial wastes and sewage sludge is restricted to a few sites in the coastal and open ocean waters of the Northern Atlantic region.

---

[7]To facilitate discussion, OTA has grouped coastal States into various "regions"—northern Atlantic, southern Atlantic, Gulf of Mexico, California and Hawaii, and northern Pacific (see fig. 21 in ch. 5).

Table 2.—Number of Municipal and Major Industrial Facilities Discharging Directly Into Marine Waters

Coastal States	Number of dischargers[a]					
	Municipal		Major industrial		Total	
Northern Atlantic region:						
Maine	38	(3)[b]	35		73	(3)
New Hampshire	2		4		6	
Massachusetts	20	(1)	20		40	(1)
Rhode Island	8	(2)	24		32	(2)
Connecticut	22		75		97	
New York	47	(1)	29		76	(1)
New Jersey	48	(12)	129	(2)	177	(14)
Pennsylvania	9		33		42	
Delaware	4		30		34	
Maryland	34	(1)	120		154	(1)
Virginia	11	(4)	76		87	(4)
District of Columbia	1		1		2	
Total	244	(24)	576	(2)	820	(26)
Southern Atlantic region:						
North Carolina	10	(1)	41		51	(1)
South Carolina	11		22		33	
Georgia	4		26		30	
Florida (Atlantic)	34	(10)	24	(1)	58	(11)
Total	59	(11)	113	(1)	172	(12)
Gulf of Mexico region:						
Florida (Gulf)	22	(5)	17	(1)	39	(6)
Alabama	6		29		35	
Mississippi	6		30		36	
Louisiana	27	(1)	79		106	(1)
Texas	52		192	(1)	244	(1)
Total	113	(6)	347	(2)	460	(8)
California and Hawaii:						
California	50	(18)	112	(5)	162	(23)
Hawaii	13	(4)	?		13	(4)
Total	63	(22)	112	(5)	175	(27)
Northern Pacific region:						
Oregon	17	(1)	40	(5)	57	(6)
Washington	51		144		195	
Alaska	31	(5)	?		31	(5)
Total	99	(6)	184	(5)	283	(11)
Total United States	578	(69)	1,332	(15)	1,910	(84)

[a]Municipal category includes all municipal facilities. Industrial category includes those industrial facilities (excluding steam electric plants) discharging more than 10,000 gallons per day. The most recent available data pertain to dischargers as of 1982 or earlier.
[b]Numbers in parentheses indicate discharges directly into coastal waters. All remaining discharges are into estuarine waters.

SOURCES: Office of Technology Assessment, 1987; after EG&G Washington Analytical Services Center, *Industrial Waste Disposal in Marine Environments*, contract prepared for U.S. Congress, Office of Technology Assessment (Waltham, MA: 1986); Science Applications International Corp., *Overview of Sewage Sludge and Effluent Management*, contract prepared for U.S. Congress, Office of Technology Assessment (McLean, VA: 1986).

Table 3.—Quantities of Dredged Material Disposed of Annually in Marine Waters (mmt/yr)

Coastal region	Estuaries	0 to 3 miles offshore[b]	Over 3 miles offshore[b]	Total	
Northern Pacific................	5.4	10.0	0.3	15.7	(9)
Southern Pacific................	8.4	4.3	2.4	15.1	(8)
Gulf of Mexico	91.3	16.4	10.9	118.6	(66)
Southern Atlantic	6.1	1.5	10.6	18.2	(10)
Northern Atlantic	4.0	2.2	6.3	12.5	(7)
Total	115.2 (64)	34.4 (19)	30.5 (17)	180.1	(100)

The header spans "Average quantities disposed of annually".

[a]Data were obtained from each U.S. Army Corps of Engineers District Office in the form of an annual average; data were not obtained for individual years. The period over which the data are averaged varies from one district to the next, but generally includes most of the 1970s and early 1980s. Units are millions of metric tons per year (mmt/yr); numbers in parentheses are the percent of the total.

[b]The distinction between "0 to 3 miles offshore" and "over 3 miles offshore" was used by the Corps to classify its data, based on the statutory definition of the territorial sea. This division does not, however, correspond exactly to the division between coastal and open ocean waters used by OTA: some open ocean waters may be included in the "0 to 3 miles offshore" category, and some coastal waters may be included in the "over 3 miles offshore" category (see box A in ch. 1).

SOURCE: Office of Technology Assessment, 1987; compiled from data obtained through a 1985-86 survey of District Offices of the U.S. Army Corps of Engineers.

Table 4.—Relative Contribution of Pollutants (in percent) by Major Sources in Coastal Hydrologic Units,[a] Circa 1977-81

Region and source	BOD	TSS	TKN	TP	CD	CR	CU	PB	AS	FE	HG	ZN	OIL	CHL HCS	FEC COL
Northern Pacific:															
Industrial[b]	34	<1	11	4	98	7	35	46	96	3	86	40	29	99	0
Municipal	27	<1	27	23	<1	1	1	2	2	2	0	4	16	<1	4
Nonpoint	40	99	62	73	2	92	63	53	2	95	14	57	55	<1	96
Southern Pacific:															
Industrial	3	<1	3	<1	81	2	<1	6	99	6	88	16	10	90	0
Municipal	55	1	31	21	3	8	7	7	<1	4	9	6	58	9	34
Nonpoint	43	99	67	78	16	90	91	87	<1	90	3	78	33	<1	66
Gulf of Mexico:															
Industrial	34	<1	31	32	95	39	45	45	100	18	93	47	53	97	0
Municipal	26	<1	32	34	<1	4	6	2	<1	<1	3	5	9	3	16
Nonpoint	40	99	37	34	4	57	49	53	<1	82	3	49	39	<1	84
Southern Atlantic:															
Industrial	28	<1	10	14	73	58	27	12	100	7	89	25	8	98	<1
Municipal	35	<1	54	73	1	8	9	3	0	1	6	7	13	1	10
Nonpoint	37	99	36	13	26	34	64	85	<1	92	6	68	79	<1	90
Northern Atlantic:															
Industrial	6	<1	8	7	84	15	18	16	100	21	92	25	5	93	<1
Municipal	73	3	74	76	2	32	23	13	<1	10	5	19	46	7	12
Nonpoint	21	97	17	18	13	52	60	72	<1	69	3	56	49	<1	88
Total U.S. coastal:															
Industrial	11	<1	9	5	89	15	18	20	100	13	91	25	11	94	<1
Municipal	56	1	46	36	1	13	11	8	<1	5	5	9	41	6	16
Nonpoint	34	99	45	59	10	72	71	73	<1	82	3	66	47	<1	84

KEY: BOD—Biochemical oxygen demand   CD—Cadmium   AS—Arsenic   OIL—Oil and grease
TSS—Total suspended solids   CR—Chromium   FE—Iron   CHL HCS—Chlorinated hydrocarbons
TKN—Total Kjeldahl nitrogen   CU—Copper   HG—Mercury   FEC COL—Fecal coliform bacteria
TP—Total phosphorus   PB—Lead   ZN—Zinc

[a]Information regarding contribution of pollutants is aggregated for all maritime hydrologic units in each region. Hydrologic units are designated by the U.S. Geological Survey and represent natural and human-made drainage areas. Only pollutants that first enter surface waters in maritime hydrologic units (i.e., directly adjacent to marine waters) are included. Pollutants originating in upstream hydrologic units and flowing into the maritime units considered here are excluded, although in some instances the upstream units contribute a sizable portion or even a majority of the pollutants entering coastal waters. Regions are graphically illustrated in ch. 5 (see fig. 21). Here, the Northern Pacific excludes Alaska; the Southern Pacific includes California only and excludes Hawaii; the Southern Atlantic excludes Puerto Rico.
[b]The "industrial" category includes powerplants.

SOURCE: Office of Technology Assessment, 1987; based on Resources for the Future, *Pollutant Discharges to Surface Waters in Coastal Regions,* contract prepared for U.S. Congress, Office of Technology Assessment (Washington, DC: February 1986).

The largest quantities of waste material are introduced into marine waters by industrial and municipal pipeline discharges and from the dumping of dredged material. Pipeline discharges are generally expected to increase in association with increasing industrial development and the growth of coastal populations. Dumping of dredged material in coastal and open ocean waters has fluctuated widely (figure 2A), depending on the nature and timing of harbor development and maintenance activities. Only relatively small quantities of industrial wastes and sewage sludge are currently dumped in marine waters.[8] During the last 10 years, dumping of industrial wastes declined dramatically, while dumping of sludge increased (figure 2B).

Pollutants From Waste Disposal and Nonpoint Sources

Pollutants that enter marine environments from waste disposal activities and nonpoint sources are classified into three categories in the Clean Water Act.[9] *Conventional pollutants* include suspended solids, oxygen-demanding substances, pH, oil and grease, and fecal coliform bacteria. *Nonconventional pollutants* is a catch-all category that includes nutrients such as nitrogen and phosphorus. *Toxic or priority pollutants* include 126 metals and organic chemicals. Each of these categories of regulated pollutants has been linked with observed impacts on marine resources and humans.

Inputs of these pollutants from disposal activities are significantly greater in estuaries and coastal waters than in the open ocean because of the greater intensity of these activities in waters close to shore. This skewed distribution is even further accentuated because many of the same pollutants are introduced into estuaries and coastal waters by rivers and by nonpoint sources (i.e., agricultural and urban runoff).

On a national scale, available data allow a rough comparison of pollutant inputs from point source pipeline discharges and nonpoint runoff that directly enter marine waters. In this limited compar-

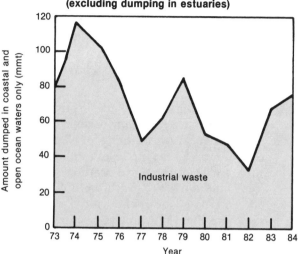

Figure 2A.—Amount of Dredged Material Dumped in Coastal and Open Ocean Waters Only, 1973-84 (excluding dumping in estuaries)

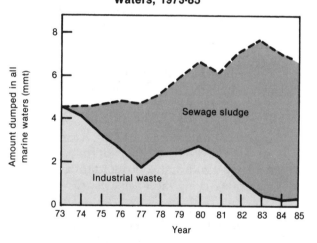

Figure 2B.—Amounts of Industrial Waste and Municipal Sewage Sludge Dumped in All Marine Waters, 1973-85

Amounts in million metric tons (mmt). All dumping of industrial wastes and municipal sewage sludge occurs in coastal and open ocean waters. Two-thirds of all dumping of dredged material occurs in estuaries, but data are not available on a yearly basis for such dumping; therefore, only the amounts of dredged material dumped in coastal and open ocean waters are shown in figure 2A. Note that the scale for dredged material is about 10 times greater than the scale for industrial wastes and sewage sludge.

SOURCES: U.S. Army Corps of Engineers, *1980 Report to Congress on Administration of Ocean Dumping Activities*, Pamphlet 82-P1 (Fort Belvoir, VA: Water Resources Support Center, May 1982); U.S. Army Corps of Engineers, *Ocean Dumping Report for Calendar Year 1981*, Summary Report 82-S02 (Fort Belvoir, VA: Water Resources Support Center, June 1982); U.S. Army Corps of Engineers, *Ocean Dumping Report for Calendar Year 1982*, Summary Report 83–SR1 (Fort Belvoir, VA: Water Resources Support Center, October 1983); U.S. Environmental Protection Agency, *Report to Congress, January 1981–December 1983, On Administration of the Marine Protection, Research, and Sanctuaries Act of 1972, As Amended (Public Law 92–532) and Implementing the International London Dumping Convention* (Washington, DC: Office of Water Regulations and Standards, June 1984): J. Wilson, U.S. Army Corps of Engineers, personal communication, 1986; R. DeCesare, Office of Water, U.S. Environmental Protection Agency, personal communication, January 1987.

[8]The total amount of dredged material dumped is about 10 times greater than the amount of sewage sludge and about 25 times greater than industrial wastes. In the New York Bight, however, the amounts of dredged material and sewage sludge are roughly comparable.

[9]MPRSA prohibits the disposal of substances that "unreasonably degrade" the marine environment. Unlike CWA, however, it does not explicitly classify substances, although it does include the lists of prohibited or regulated substances developed by the London Dumping Convention.

ison, the source contributing the majority of a particular pollutant varies with the pollutant (table 4).[10] It is also apparent that more than one source can be an important contributor of some pollutants (e.g., phosphorus). As can be seen from table 4, some generalizations at a national level are possible, however:

- Industrial discharges are, not surprisingly, the dominant sources of many organic chemicals and some metals, accounting for about 90 percent or more of the inputs of cadmium, mercury, and chlorinated hydrocarbons. Inputs of some other metals (e.g., chromium and lead) are dominated by nonindustrial sources in some areas of the country.

- Municipal point sources are major contributors of certain conventional pollutants, accounting for about half of biochemical oxygen demand, total nitrogen, and oil and grease. Surprisingly, however, municipal discharges contribute only one-sixth of the input of fecal coliform bacteria.[11] Municipal discharges are particularly dominant sources of biochemical oxygen demand and nitrogen in the northern Atlantic and in California.

- Nonpoint runoff dominates as a source of suspended solids, and also contributes half or more of total phosphorus, chromium, copper, lead, iron, and zinc. It is also the overwhelming contributor of fecal coliform bacteria in all areas of the country. Nonpoint runoff is a particularly significant contributor of a range of pollutants along the Pacific coast.

In addition, sufficient information is available to conclude that upstream sources of pollutants—whether originating from waste disposal or nonpoint pollution—are the largest sources in the Gulf of Mexico and appear to be important in the north-

ern Atlantic region. However, the absolute *quantity* of pollutants is only a partial measure of their subsequent *impact*; for example, many riverborne pollutants are considerably more diluted or degraded by the time they reach marine waters than they would be if they had been released directly into those waters. Thus, the magnitude of marine impacts due to upstream sources is not necessarily commensurate with the magnitude of their pollutant inputs.

It is difficult to compare pollutant inputs from pipeline discharges and runoff to those resulting from marine dumping of dredged material or sewage sludge because of the extreme variability in composition of dumped wastes and the intermittent and localized nature of dumping operations. Dumping—and resulting pollutant inputs—appears to be relatively minor in *most* estuaries and coastal waters; however, in those areas where dumping does occur, it can be a significant contributor of many pollutants. Table 5 compares inputs from various sources to the waters of the New York Bight based on estimates for the mid-1970s; more recent comprehensive data are not available. This example represents an extreme case, however, because the significance of dumping as a source of pollutants is probably greater in the New York Bight than in other estuarine or coastal regions of the United States.

On a local or regional scale, the relative importance of any source can vary from the above generalizations, depending on factors such as: the type of industrial development, the nature of industrial discharges to municipal sewage treatment systems, the relative amounts of urban and agricultural runoff, the extent of combined sewer overflow, the relative contamination of sediments by discharges and runoff, and the extent of port maintenance. The majority of total phosphorus, for example, is contributed by municipal pipelines along the east coast and by nonpoint sources along the west coast; in the Gulf of Mexico, roughly equal amounts are contributed by industrial discharges, municipal discharges, and nonpoint sources (table 4).

The amounts of specific pollutants in discharges or runoff can change over time. For example, regulations governing the production, use, or disposal of certain substances can affect the amounts of pol-

---

[10]The estimated amounts of pollutant inputs (478) are not included in table 4 because the purpose here is to examine the relative contributions by different sources and to illustrate the variability that is an important feature of pollutant inputs. The assumptions and uncertainties in the database are discussed in detail by RFF (477,478); information from other databases (particularly NOAA's) corroborate the general relationships portrayed in table 4 and support the importance of variability.

[11]This is *not* to imply that fecal coliforms are necessarily contributed primarily by natural sources. Sources such as combined sewer overflows, leakage from septic tanks, and other discharges of untreated sewage may well contribute to the high contribution of fecal coliforms by nonpoint sources.

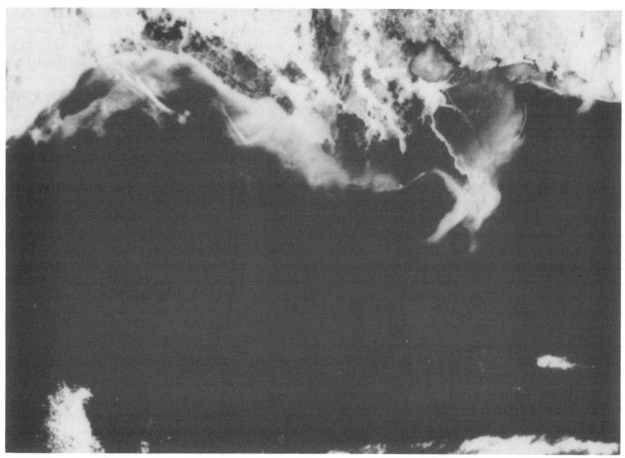

Photo credit: *National Oceanic and Atmospheric Administration*

Many pollutants are carried into the Gulf of Mexico by rivers, especially the Mississippi River, from areas far from the coast. This satellite photo shows river water laden with sediment and other matter appearing as whispy white plumes.

Table 5.—Relative Contribution of Various Pollutants by Major Sources in the New York Bight, Circa Mid-1970s

		Percent contribution by source			
	Total mass input	Dumping[a]	Atmospheric input	Pipeline discharges[b]	Runoff
Cadmium	880 mt/yr	82	2	6	10
Chromium	1,810 mt/yr	50	1	23	26
Copper	5,060 mt/yr	51	3	20	26
Lead	4,600 mt/yr	43	9	22	25
Mercury	110 mt/yr	9	—	73	18
Zinc	12,000 mt/yr	29	18	10	43
PCBs	7.4-8.6 mt/yr	55-64	—	3-13	39[c]
TSS	$8,800 \times 10^9$ mt/yr	63	5	4	28
TOC	950×10^9 mt/yr	25	12	30	33
Nitrogen	190×10^9 mt/yr	16	13	42	29
Phosphorus	50×10^9 mt/yr	50	0.7	36	13

ABBREVIATIONS: PCBs = polychlorinated biphenyls; TSS = total suspended solids; TOC = total organic carbon; mt/yr = metric tons per year.
[a]Includes dumping of both sewage sludge and dredged material.
[b]Includes both municipal and industrial discharges.
[c]Estimate is for upstream sources, which include both point and nonpoint sources.

SOURCES: Office of Technology Assessment, 1987; based on: J.A. Mueller, et al., "Contaminants in the New York Bight," *Journal of Water Pollution Control Federation* 48(10):2309-2326, 1976 (for metals, TSS, TOC, nitrogen, phosphorus); J.M. O'Connor, et al., "Sources, Sinks, and Distribution of Organic Contaminants in the New York Bight Ecosystem," in *Ecological Stress and the New York Bight: Science and Management*, G.F. Mayer (ed.) (Columbia, SC: Estuarine Research Federation, 1982) (for PCBs); A.J. Mearns, et al., "Effects of Nutrients and Carbon Loadings on Communities and Ecosystems," in *Ecological Stress and the New York Bight: Science and Management*, G.F. Mayer (ed.) (Columbia, SC: Estuarine Research Federation, 1982) (for TSS, TOC, nitrogen, phosphorus).

lutants, such as when restrictions on DDT and PCBs significantly reduced the levels of these substances in pipeline discharges. Regulations have resulted in some reductions in the amounts of oxygen-demanding pollutants and nutrients in industrial and municipal discharges.

Climatic factors also differentially affect the contribution of pollutants from various sources and their subsequent impacts. For example, municipal point sources may be more important contributors of nutrients during summer months, when rainfall and river flow (and thus nonpoint runoff) are generally lower; conditions conducive to eutrophication are also most prevalent in the summer.

MAJOR SOURCES OF POLLUTANTS TO U.S. MARINE WATERS

Two major source categories contribute pollutants to U.S. marine waters: waste disposal and nonpoint pollution. (In addition to this information, box H discusses the management of low-level radioactive waste, and box I summarizes information about the quantities of wastes dumped in the ocean by other countries.)

Waste Disposal

Waste disposal means the intentional release of wastes to marine waters, either through direct dumping or through pipeline discharges. Nonpoint pollution, in contrast, is more diffuse and includes, for example, runoff from rural and urban land surfaces.

Dumping Activities

Wastes dumped in marine environments include dredged material, municipal sewage sludge, and industrial wastes.

Dredged Material.—Very large amounts of dredged material—about 180 million wet metric tons (mt)—are disposed of each year in U.S. marine waters (table 3), accounting for some 80 to 90 percent of the volume of all material dumped in these waters. Approximately two-thirds of all dredged material is dumped in the Gulf of Mexico.

Almost two-thirds of marine dumping of dredged material occurs in estuaries (including intertidal areas). The remainder is divided more or less evenly between waters within the 3-mile territorial boundary and waters beyond this boundary.[12] The types of marine waters used most frequently for disposal vary considerably around the country. In the Gulf of Mexico and in California, for example, most material is dumped in estuaries; in the southern Atlantic region, in contrast, most material is dumped more than 3 miles from shore.

Over the last 10 to 15 years, the annual amount of dredged material disposed of in coastal and open ocean waters only has varied considerably (data are not readily available for the amounts of material dumped each year in estuaries). The total amount of material dumped in these waters showed a general decline from 120 million wet mt in 1974 to about 35 million wet mt in 1982 (figure 2A). It is difficult to predict how much material will be dredged in the future, but it could increase substantially if several harbor deepening projects that are now being considered by the Corps of Engineers and Congress are undertaken (see ch. 10).

The composition of dredged material also varies from one area to the next. In some areas, sediments have been contaminated by metals and organic chemicals originating from industrial and municipal discharges and nonpoint pollution. When these sediments are dredged and then dumped, the pollutants are carried along to the dumping site. Only a fraction of all dredged material is considered by the Corps of Engineers to be contaminated, although the absence of specific numerical criteria to define contaminated material is a source of controversy.

Municipal Sewage Sludge and Industrial Wastes.—Most waste other than dredged material that is dumped in marine waters consists of sewage sludge from municipal treatment plants and acid or alkaline liquid industrial wastes. These wastes can contain a variety of different pollutants.

---

[12]Most of the waters beyond this boundary can be classified as open but some—in particular, the New York Bight—are classified coastal waters.

For the last several decades, many marine dump-sites have been used for the disposal of sewage sludge and industrial wastes. However, most dumping of these materials has taken place in the coastal waters of the Northeastern United States. Currently, only a few sites are being used, all located either in the New York Bight or in open ocean waters about 100 miles east of the coast of Delaware.

The dumping of sewage sludge has steadily increased from 2.5 million wet mt in 1958 to 7.5 million wet mt in 1983; 6.6 million wet mt were dumped in 1985 (figure 2B). In 1980, EPA phased out dumping by over 100 municipalities (including one large city, Philadelphia); however, these municipalities together accounted for only 3 percent of all dumped sludge (292). The amount of sludge dumped continued to increase after 1980, partly because more secondary treatment plants, which produce more sludge, came into operation in the New York area. Most sewage sludge has been dumped either at the mid-Atlantic site off of Delaware Bay or at the 12-Mile Sewage Sludge Dump Site located in the New York Bight (figure 3). Sewage sludge currently dumped in marine waters originates from nine sewerage authorities in New York and New Jersey; most of it is currently dumped at the 12-Mile site, but over the next few years all remaining marine dumping will be moved to the Deepwater Municipal Sludge Site which lies just off the edge of the continental shelf (figure 3).[13]

Marine dumping of industrial wastes meanwhile has decreased dramatically over the last decade (figure 2B) from a peak of 4.6 million wet metric tons in 1973 originating from over 300 industrial firms (6,115, 292), to the current level of about 200,000 wet metric tons dumped annually by 3 firms (ch. 11; refs. 139,648). Most of this is dumped at the Deepwater Industrial Waste Site, located about 10 nautical miles west of the Deepwater Municipal Sludge Site.[14] The vast majority of these industrial

Figure 3.—Location of Current Municipal Sewage Sludge and Industrial Waste Dumpsites in the Northern Atlantic Ocean

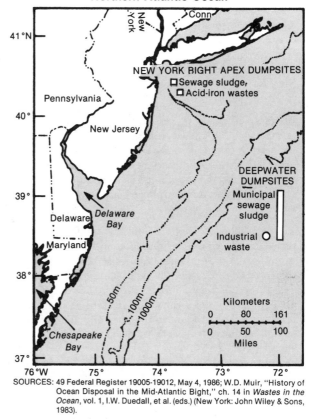

SOURCES: 49 Federal Register 19005-19012, May 4, 1986; W.D. Muir, "History of Ocean Disposal in the Mid-Atlantic Bight," ch. 14 in *Wastes in the Ocean*, vol. 1, I.W. Duedall, et al. (eds.) (New York: John Wiley & Sons, 1983).

wastes has been dumped in the northern Atlantic, although pharmaceutical wastes were dumped at a site north of Puerto Rico for almost a decade until 1981.

Pipeline Discharges

OTA obtained two different types of estimates for the number and flow of pipelines whose discharges may affect marine waters. *The first estimate includes all discharges located in coastal counties* of the United States; this clearly represents an overestimate because only a fraction (albeit unknown) of wastewater and associated pollutants discharged in inland areas of coastal counties will reach marine waters. *The second estimate includes only those discharges directly into marine (estuarine or coastal) waters*; this number probably underestimates the total number and flow of pipelines affecting marine waters because it excludes that fraction

---

[13]The Deepwater Municipal Sludge Site occupies an area of approximately 100 square nautical miles. It is located approximately 120 nautical miles southeast of Ambrose Light, New York, and 115 nautical miles from Atlantic City, New Jersey, in water depths ranging from 2,250 to 2,750 meters (49 FR 19005-19012, May 4, 1984).

[14]The Deepwater Industrial Waste Site occupies an area of approximately 30 square nautical miles. It is located approximately 125 nautical miles southeast of Ambrose Light, New York, and 105 nautical miles from Atlantic City, New Jersey, in water depths ranging from 2,250 to 2,750 meters (49 FR 19005-19012, May 4, 1984).

Box H.—Management of Low-Level Radioactive Waste

Low-level radioactive waste (LLW) is "defined" in the Low-Level Radioactive Waste Policy Amendments Act primarily by excluding certain materials from the category (the act excludes high-level radioactive waste, spent nuclear fuel, uranium and thorium tailings, and other wastes from ore mining and milling). As a result, LLW typically includes an assortment of discarded material that is contaminated with small amounts of radionuclides*—paper, glass, clothing, plastics, tools and equipment, wet sludges and resins, and organic liquids.

LLW can be generated by defense operations (from weapons production) and civilian commercial practices (e.g., nuclear powerplants, drug manufacturing, biomedical research, and hospital diagnostic tests and treatment). About two-thirds of the volume and about three-fourths of the radioactivity has been generated by defense operations. By 2020, however, the volume of commercial LLW shipped annually for disposal is expected to quadruple and its radioactivity is expected to triple, while the volume and radioactivity of defense wastes is expected to remain relatively stable (622). The difference probably will be greater because commercial decontamination of powerplants, which will become substantial around the turn of the century, is not reflected in these estimates.

Whether LLW poses significant risks to humans or the environment depends on several factors: 1) the concentration of the radionuclide in the waste and in the environment,** 2) the half-life and type of emitted radiation, and 3) potential exposure pathways. Regulations regarding disposal of some radionuclides reflect these factors. For example, vials commonly used in diagnostic testing generally contain a very small amount (less than 0.05 microcuries)*** of the radionuclide carbon-14.**** Carbon-14 from this source is not considered to pose a significant risk, despite its long half-life (5,730 years), because its concentration is low and it is naturally abundant in the environment; as a result, current regulations allow its disposal without regard to radioactivity. In contrast, special packaging is required for the disposal of cobalt-60, despite its relatively short half-life (5.26 years), because it does not occur naturally and exposure to it can cause adverse impacts on humans.

Land-Based Disposal

LLW is disposed of on land by burial, at sites owned by the Department of Energy (for defense waste) or by States (for commercial waste). In both cases, sites are operated under contract by private firms. The vast majority of LLW has been disposed of at six shallow sites, three of which are still operating (Barnwell, South Carolina; Richland, Washington; and Beatty, Nevada). By the end of 1985, the United States had buried about 3.34 million cubic meters (containing 16.5 million curies) of LLW (622).

Marine Disposal

From 1946 to 1970, the United States dumped approximately 94,000 curies of LLW (in a total of 89,472 drums) at 6 major sites, 4 in the Atlantic Ocean, 2 in the Pacific (378). Few records of these activities were kept, and only sporadic monitoring has been conducted at the known sites.

Marine dumping of LLW by the United States has been halted since 1970, largely in response to a recommendation by the Council on Environmental Quality (115). In 1981, the Navy considered disposing of obsolete, de-fueled nuclear submarines and concluded that the risks and costs of burying the radioactive components on land were slightly higher than those of dumping them at sea (629). In response, Congress enacted a 2-year moratorium on marine dumping of LLW.† Although this moratorium is not now in force, Congress

*A radionuclide is an atom that emits radioactive rays. Some radionuclides also are chemically toxic.
**Natural radioactivity in the environment is derived from the decay of radioactive elements in the Earth's crust and from radionuclides that form when cosmic rays collide with gas molecules in the upper atmosphere. Human-made radioactivity in marine waters is derived from accidental loss of nuclear submarines, intentional dumping of waste, discharged radioactive effluent, and fallout from atmospheric nuclear tests (378).
***A curie is a measure of the rate of radioactive decay, equivalent to 37 billion disintegrations per second. A microcurie is one-millionth of a curie.
****Biomedical Waste Disposal Rule, 10 CFR Part 20.
†The moratorium was included in an amendment to Section 104 of the Marine Protection, Research, and Sanctuaries Act, passed as part of the Surface Transportation Assistance Act (Public Law 97-424) in January 1983.

must still approve of any LLW dumping permit within 90 days of the initial application. As a result, the Navy has decided to bury the radioactive components on land, despite its conclusions about relative risks.

If dumping of LLW by the United States does occur again, current regulations would require greater containment than in the past.†† After dumping, for example, the waste would have to remain in a canister until it decayed to levels of radioactivity determined by EPA to be environmentally innocuous; in addition, a comprehensive monitoring plan would be required. EPA is updating and expanding these regulations (scheduled for completion in December 1987).

Some LLW has been discharged from pipelines into U.S. surface waters and carried downstream to marine waters. Prior to the mid-1960s, reactors at several defense sites had systems in which cooling water could be contaminated with radionuclides if a fuel leak occurred. Such incidents occurred at the Hanford site in Washington, for example, and the contaminated cooling water was discharged into nearby streams and carried into the Columbia River and the river estuary (456). Reactors with this design have been inoperative since the mid-1960s, yet estuaries downstream of them still contain residual radioactivity (e.g., of cobalt-60) (456). Reactors operating today are designed with a closed system, in which radioactive effluent is carried into cribs designed to trap radionuclides. Despite this, some radionuclides have migrated into nearby streams in concentrations higher than those acceptable under EPA drinking water standards (C. Welty, U.S. Department of Energy, pers. comm., September 1986); some of these radionuclides subsequently have migrated into estuaries.

Internationally, no LLW has been dumped since 1983, and an "indefinite" moratorium on such disposal was adopted in 1985 (see box I).

Issues Affecting Future Marine Disposal

Several issues affect the potential for future marine disposal of LLW: 1) environmental feasibility; 2) economical, political, and social acceptability; and 3) potential problems in siting land-based disposal facilities. Most analyses to date have focused on the environmental issue.

The environmental consequences of dumping at the most heavily used international marine site (in the northeastern Atlantic, see box I) have been investigated by a panel of experts appointed by the Organization for Economic Cooperation and Development's Nuclear Energy Agency. The panel concluded that: 1) maximum exposure of humans to radioactivity from marine dumping of LLW has been extremely low, substantially lower than exposure from land-based disposal; 2) no significant damage has occurred to marine organisms at or near the site; and 3) accelerated use of the site would not change these conclusions (312). However, the panel also noted that monitoring the site was difficult, that impacts could not be detected outside of a certain stated range, and that impacts in the deep sea could not be adequately assessed without improved biological and radioecological information (312). In addition, poor recordkeeping at many LLW dumping sites, including the early years of dumping at the northeastern Atlantic site, has made assessments of environmental impacts more difficult.

In September 1985, a majority of members of the London Dumping Convention (see box Q in ch. 7) passed a resolution that required additional studies of: 1) the political, legal, economic, and social aspects of dumping at sea; 2) costs and risks of land-based and marine disposal; 3) whether it can be proven that dumping will not harm human life or cause significant environmental damage; and 4) assessment by the International Atomic Energy Agency of several technical questions. A decision to terminate the moratorium on dumping was tabled (at the Tenth Consultative Meeting of the Contracting Parties to the LDC, held in October 1986) pending the outcome of these studies.

Problems in siting, developing, and using land-based disposal facilities could also influence decisions about marine disposal of LLW. In the United States, for example, the Department of Energy has been unable to find a disposal site on land that is acceptable to the public for Manhattan Project waste (mostly contaminated

---

††40 CFR 227.11, adopted after dumping practices ceased.

soil and building rubble containing very low concentrations of radionuclides) and is therefore considering the feasibility of dumping the waste at sea (E. Delaney, U.S. Department of Energy, pers. comm., September 1984). In addition, the governors of the three States (South Carolina, Nevada, and Washington) with land-based sites for LLW disposal have threatened to stop accepting commercial LLW from other States. To address this problem, Congress passed the Low-Level Radioactive Waste Policy Act of 1980, under which States are encouraged to form multi-State "compacts," with each compact having its own disposal site. No new sites had been selected by 1986, however, and Congress amended the act to further encourage States to take responsibility for their LLW. If new land-based disposal sites are not developed in the future, interest in marine disposal possibly could increase.

of upstream discharges that does reach marine waters. Table 6 presents a comparison of these two estimates for the number and flow of municipal and industrial pipelines.

Using the conservative data, almost 2,000 municipal and major industrial pipelines discharge effluent directly to estuaries and coastal waters. Almost all of these pipelines (about 96 percent) are located in estuaries, and over two-thirds are industrial (table 2).[15] The largest share (43 percent) of these discharges are concentrated in the northern

[15]In addition to these discharges, a larger number of minor industrial and commercial facilities also discharge into these waters, but they account for only a small fraction of total pollutant inputs.

Atlantic region. The Gulf of Mexico, in particular the western Gulf, also has a high concentration of pipelines.

There are, of course, substantial variations in the amounts of municipal and industrial discharges into individual waterbodies. In one analysis of four estuaries and coastal waterbodies, the number of major industrial dischargers was estimated to be three to five times higher than the number of municipal dischargers in three waterbodies (Puget Sound, San Francisco Bay, and Narragansett Bay). In contrast, in the Chesapeake Bay municipal dischargers were three times as numerous as major industrial dischargers (139).

Table 6.—Comparison of All Discharges in Coastal Counties and Those Discharges Directly to Marine Waters

	Number of dischargers	Flow (bgy)	Database, source	Reference
Municipal dischargers:				
Coastal county[a]	2,207	3,620	NCPDI, from NOAA	1
Direct marine[b]	578	2,306	IFD and Needs Survey, from EPA	2
Industrial dischargers:				
Coastal county (major and minor)[c]	4,592	4,914	NCPDI, from NOAA	1
Direct marine (major only)[d]	1,332	4,136	IFD, from EPA	3

ABBREVIATIONS: bgy = billion gallons per year
NCPDI = National Coastal Pollutant Discharge Inventory
IFD = Industrial Facilities Database
NOAA = National Oceanic and Atmospheric Administration
EPA = Environmental Protection Agency.

[a]Dischargers located in coastal counties of the United States.
[b]Dischargers actually discharging wastewater directly into marine (estuarine or coastal) waters of the United States.
[c]Estimates include both major and minor dischargers.
[d]Estimates include only major dischargers (defined by EG&G, 1986 (ref. 3 below), as those with wastewater flows greater than 0.01 million gallons per day).

REFERENCES:
1. Data from National Coastal Pollutant Discharge Inventory (NCPDI) received through personal communication, D.J. Basta, Chief, Strategic Assessments Branch, NOAA, Washington, DC, Nov. 14, 1986.
2. Adapted from Science Applications International Corp., *Overview of Sewage Sludge and Effluent Management*, contract prepared for U.S. Congress, Office of Technology Assessment (McLean, VA: 1986); based on analysis of data from EPA's Industrial Facilities Database (IFD) and a 1982 EPA Needs Survey of municipal sewage treatment facilities (U.S. Environmental Protection Agency, Office of Municipal Pollution Control, *Assessment of Needed Publicly Owned Wastewater Treatment Facilities in the United States*, EPA 430/9-84-011 (Washington, DC: February 1985)).
3. Adapted from EG&G Washington Analytical Services Center, Inc., Oceanographic Services, *Industrial Waste Disposal in Marine Environments*, contract prepared for U.S. Congress, Office of Technology Assessment (Waltham, MA: 1986); based on analysis of data from EPA's Industrial Facilities Database (IFD).

SOURCE: Office of Technology Assessment, 1987.

Municipal Discharges.—Of the approximately 15,500 publicly owned treatment works (POTWs) in the United States, only about 3.5 percent (i.e., a total of 578) discharge directly into estuaries and coastal waters (ref. 503). The POTWs that discharge into marine waters, however, account for one-fourth of the Nation's municipal wastewater; moreover, almost 90 percent of them (509) discharge into estuaries (table 2). POTWs discharging into marine waters account for such a large portion of total wastewater because many of them are large and serve densely populated coastal areas. On an annual basis, they discharge a total of about 2.3 trillion gallons of effluent into marine waters—2 trillion gallons into estuaries and 0.3 trillion gallons into coastal waters (503).

The amount of municipal effluent discharged to marine waters varies considerably among different regions of the country (figure 4). More than 60 per-

Figure 4.—Amount of Effluent Discharged From Major Municipal Sewage Treatment Plants Directly Into Marine Waters, By State, Circa 1982 (amounts in million gallons per day, MGD)

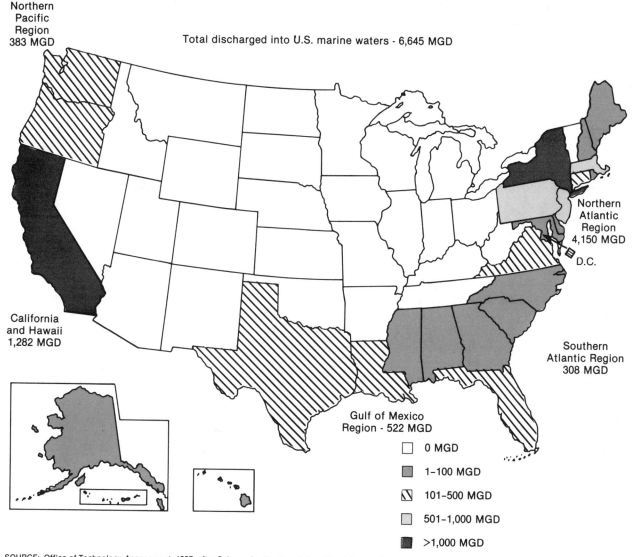

SOURCE: Office of Technology Assessment, 1987; after Science Applications International Corp., "Overview of Sewage Sludge and Effluent Management," contract prepared for U.S. Congress, Office of Technology Assessment (McClean, VA: 1986).

cent of all municipal discharges to marine waters occurs in the waters of the northern Atlantic region, especially from New York. Almost 20 percent is discharged from California. The magnitude of municipal discharges has increased roughly in parallel with population growth and as previously unsewered sources have been connected to municipal systems.

A few sewerage authorities in Los Angeles and Boston discharge sludge through POTW outfalls into marine waters (see ch. 9). Such discharges are scheduled to be terminated by 1987 for Los Angeles and the mid-1990s for Boston.[16] In 1980, some 107,000 dry metric tons of sludge were discharged by POTWs in southern California (U.S. General Accounting Office, 1983, cited in ref. 503).

The quantities of different pollutants in municipal effluent and sludge depends primarily on the nature of any industrial discharges to POTWs and the degree of treatment used by POTWs. A significant portion of the wastewater entering POTWs consists of indirect industrial discharges. Nationally, some 160,000 indirect discharges account for about one-eighth of the wastewater flow through all POTWs (ref. 666). For those POTWs that discharge into marine waters, indirect industrial discharges account for a slightly larger portion, about one-seventh (0.33 trillion gallons per year, or tgy) of wastewater flow; most of this (about 0.31 tgy) enters estuaries rather than coastal waters (ref. 503). In addition, the concentration of pollutants in municipal discharges depends on the degree of treatment because higher levels of treatment remove (either intentionally or incidentally) greater amounts of pollutants. About two-fifths of the effluent discharged into marine waters receives less than secondary treatment (see ch. 9).

Industrial Discharges.—Over 1,300 major industrial facilities (excluding powerplants) discharge effluents directly into marine waters; about 98 percent of these discharge into estuaries (table 2). This

estimate excludes minor dischargers, facilities located upstream whose discharges reach marine waters, and indirect industrial discharges into municipal sewers; lack of data on these additional industrial sources introduces considerable uncertainty into the estimation of these contributions.

As seen in table 2, the number of industrial dischargers varies significantly among different geographic regions, not surprisingly showing a strong correlation with the density of industrial development. The quantity and composition of industrial discharges also varies from one area to the next, depending on the degree and nature of industrial development. Because of the wide variations resulting from these factors, it is very difficult to assess the relative importance of pollutant inputs from industrial discharges in different regions of the country. Information about the amounts of metals and organic chemicals in industrial discharges is discussed in detail in chapter 8.

Nonpoint Pollution

Nonpoint pollution is an important contributor of pollutants to marine waters in all parts of the country. Sources of nonpoint pollution include:

- runoff from cities, industrial sites, and farmland, caused mostly by precipitation and subsequent drainage;
- precipitation itself;
- atmospheric deposition;
- underground transport through aquifers; and
- other releases of pollutants (e.g., leaching of pollutants such as tributyltin from ship hulls; see box J).

Nonpoint pollution also can originate from septic tank systems and from combined sewer overflows (CSOs) (figure 6). Sewage from septic tanks, for example, can drain either directly or through aquifers into marine waters or into rivers flowing into marine waters.

Generally, the only data available on the contribution of pollutants by different nonpoint sources are for runoff. Runoff tends to be an especially large source of fecal coliform bacteria, suspended solids, and, to a lesser extent, oxygen-demanding pollutants and nutrients (table 4). *Urban runoff* contributes large quantities of oil and grease, lead, and

---

[16]In the Ocean Dumping Amendments Act of 1985, passed by the House of Representatives but not considered by the Senate, a provision was included which would have allowed Boston to dump its sewage sludge on an interim basis in the open ocean beyond the edge of the continental shelf (146,581). However, Boston has since announced its intention to develop land-based options and not pursue ocean dumping, either at the Deepwater Municipal Sludge Site or a new site (153).

Box I.—Quantities of Wastes Dumped in the Ocean by Other Countries

The United States is not the only country that dumps wastes into marine waters. According to the most recent data available from records maintained by the London Dumping Convention (LDC), an annual average of 300 to 400 million tons of waste was dumped into marine waters between 1976 and 1982 by Nations that are members of the LDC (including the United States) (figure 5).* About 90 percent of this is dredged material generated by the deepening or maintenance of ports, harbors, and shipping channels. Of the remaining 10 percent, about half is industrial waste and half is sewage sludge. This material was disposed of under some 400 to 600 individual annual permits.

No data are available on numbers and amounts of pipeline discharges of industrial and municipal effluents worldwide, and virtually no data exists on the practices of Nations that are not parties to the LDC. Information on the incineration of hazardous wastes at sea is reviewed in reference 586.

Dredged Material

About 1.3 billion metric tons of sediment are dredged each year worldwide. The United States accounts for about 35 percent of this material. Of the total amount of material dredged worldwide, a large portion—about 1.1 billion tons—is disposed of in or near marine waters.** Some 23 percent is disposed of in open ocean waters, 36 percent in nearshore and intertidal sites (and behind bulkheads), 27 percent in wetlands or in open-water areas in estuaries, and the remainder in upland areas and other environments. About three-fourths of this total was from new projects and one-fourth from maintenance dredging (442).

Industrial Wastes

The types of industrial wastes disposed of in the ocean vary greatly among different countries. The most toxic industrial wastes are banned from ocean disposal by all of the international conventions (see box Q in ch. 7), and some countries are phasing out all ocean dumping of industrial wastes. Little if any hazardous waste (as they would be defined by the

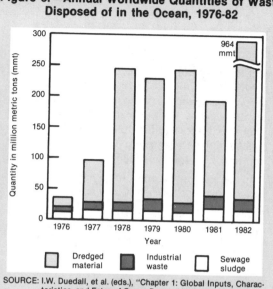

Figure 5.—Annual Worldwide Quantities of Waste Disposed of in the Ocean, 1976-82

SOURCE: I.W. Duedall, et al. (eds.), "Chapter 1: Global Inputs, Characteristics, and Fates of Ocean-Dumped Industrial and Sewage Wastes," *Wastes in the Ocean*, vol. 1 (New York, NY: John Wiley & Sons, 1983).

U.S. Resource Conservation and Recovery Act) is disposed of in the ocean, other than certain corrosive wastes (acid or alkaline liquids) which are neutralized by the natural buffering capacity of seawater.

Many nations, however, still dump some "nonhazardous" industrial wastes in the ocean. Between 1977 and 1982, an annual average of about 17 million metric tons of industrial waste was dumped in the ocean. The largest amount was dumped by the United States, followed by France, the United Kingdom, Hong Kong, Germany (FRG), Ireland, the Netherlands, Belgium, Italy, Spain, New Zealand, Canada, Australia, and Denmark (132).

Sewage Sludge

Between 1977 and 1982, an average of 17 million wet metric tons of sewage sludge was dumped in the ocean each year. The United States and the United Kingdom contribute roughly equal shares, and together account for more than 95 percent of this total. In many countries, both treated and untreated

*The LDC is an international agreement that governs the deliberate dumping of wastes into the world's oceans (see box Q in ch. 7). Member Nations are required to report annually to the LDC the number of permits granted for ocean dumping and the types and tonnages of wastes disposed of in this manner. It is not possible to discern whether dumping of waste into *all* marine waters (estuaries, coastal waters, and the open ocean) is included in the LDC estimates. This accounts for any discrepancies between these figures and others cited here.
**These figures are based on a survey of 108 ports in 38 countries conducted in 1980 by the International Association of Ports and Harbors.

sewage is discharged into marine waters through pipelines or into rivers that directly enter marine waters.

Radioactive Wastes

No nation has yet used marine waters for the intentional disposal of high-level radioactive waste. The concept of intentionally disposing of such waste within deep-sea sediments, however, has been cooperatively investigated by the Subseabed Working Group, a group of countries within the Nuclear Energy Agency (NEA) of the Organization for Economic Cooperation and Development (585).

In contrast, intentional dumping of low-level radioactive waste has occurred in marine waters. Since 1950, European countries have dumped almost 1 million curies of low-level radioactive waste at a site approved by NEA, northwest of Spain in the northeastern Atlantic Ocean (378). The dumping of low-level radioactive waste in marine waters has been curtailed since 1983 by all European countries (as well as by the United States), pending the completion of several studies identified by the London Dumping Convention (see box H). Several European countries (e.g., France, Switzerland, and the United Kingdom) and Japan have expressed interest in resuming such disposal should it be allowed (refs. 378,559; J.P. Olivier, Division of Radiation Protection and Waste Management, NEA, pers. comm., May 1986).***

***Effluent containing low-level radioactive waste from two fuel reprocessing plants in Europe (Sellafield in the United Kingdom and La Hague in France) continues to be discharged into marine waters. The effluent from Sellafield has been discharged into the Irish Sea since 1957; between 1957 and 1980, it contained 2.3 million curies of radioactivity, considerably more than the total amount of curies dumped at the northeastern Atlantic site.

Figure 6.—Typical Combined Sewer Collection Network During a Storm

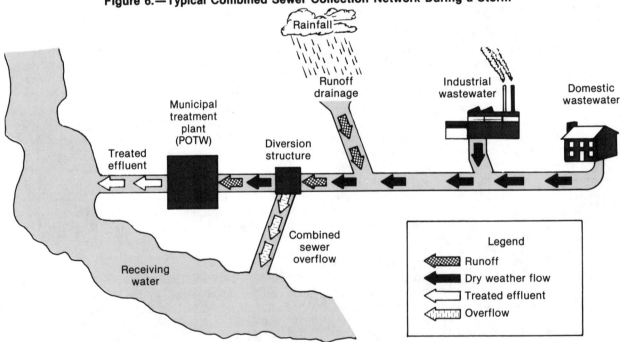

The capacity of municipal sewage treatment plants is usually not adequate to handle the large volumes of combined wastewaters (domestic wastewater, industrial wastewater, and storm water runoff) that may result during storms. In such situations, the wastewater that cannot be handled by the plant is not treated and is diverted to the receiving waters. This diversion is known as a combined sewer overflow.

SOURCE: After U.S. Environmental Protection Agency, Office of Water, *Combined Sewer Overflow Toxic Pollutant Study*, EPA 440/1-84/304 (Washington, DC: April 1984).

chromium; *agricultural runoff* contributes large quantities of pesticides and herbicides, including various chlorinated hydrocarbons (478,608). Runoff is highly variable in different areas and at different times, although this fact can be obscured in average annual statistics.

Some information has been collected to address the importance of nonpoint sources in general. According to an analysis of State and EPA data for 10 States, nonpoint sources were considered the most important contributor of damaging pollutants in 48 percent of the cases where estuaries failed to support key uses (e.g., fishing, swimming, and the propagation of marine life) (658). Furthermore, in all regions but the Northeast, nonpoint sources were considered more important than point sources; 78 percent of the States considered the magnitude of water problems associated with nonpoint sources to be greater than that relating to point sources (658). Even in the Northeast, there are numerous instances where nonpoint sources are the most important sources of specific pollutants and major contributors to serious problems. In the Chesapeake Bay, for example, roughly 60 to 80 percent of the nitrogen (which contributes to eutrophication) in the Bay originates from nonpoint sources (624).

Additional evidence from State reports issued in 1986 provides ample support for the conclusion that nonpoint sources are very significant. In Florida, for example, nonpoint sources were the primary factor in 43 percent of the estuaries which failed to support their designated uses (220). These State reports also indicate that septic systems can be important contributors of certain pollutants (in particular, fecal coliform bacteria) in coastal areas with a high portion of unsewered households—e.g., the Gulf of Mexico and the southern Atlantic coast. In addition, CSOs tend to be more frequent in the older cities of the Northeast that rely to a greater extent on combined sewer systems (631), but are also major problems in areas such as Puget Sound (463) and coastal Florida (220).

Pollutants such as metals and organic chemicals can also be carried from some hazardous waste sites into marine waters through contaminated runoff or transport through aquifers (610). At least 75 hazardous waste sites in coastal counties are considered to present some threat to marine resources and human health.

Photo credit: S. Dollar, University of Hawaii

After treatment, sewage effluent is often discharged from underwater pipes. No major effects to the surrounding coral reef communities have been observed at this discharge site in Hawaii.

Box J.—Other Pollutants in Marine Environments: Plastics and Tributyltin

Several sources or types of pollutants that affect marine waters and organisms are not covered in detail in this report. These include: deposition from the atmosphere, underground transport through aquifers, disposal of plastic objects, and leaching from ship hulls. Some information is briefly summarized here about two of these types of pollution, the disposal of plastics and leaching of tributyltin from ship hulls, which have recently gained considerable attention as potentially serious problems in marine waters.

Disposal of Plastics

The accumulation of nonbiodegradable plastic objects (manufactured pieces such as fishing nets and six-pack beverage can yokes, or raw particles such as pellets and beads) in marine waters has drawn the attention of scientists, environmental groups, the media, several State legislatures, and Congress. The obvious problem is that these objects do not readily break down but instead persist in the environment; for example, some can float on the ocean's surface for an estimated 50 years. In addition, plastics disposed of in one country's territorial waters can be transported to another country's waters.

The main effects of plastics are on marine animals. Tens of thousands of sea birds and an estimated 100,000 marine mammals die each year by ingesting or becoming entangled with plastic debris. A massive kill of green sea turtles off of Costa Rica was caused by the turtles' ingestion of plastic banana bags which had been thrown off of a dock (328). Although the actual plastic may not be toxic, it is indigestable and interferes with feeding and digestion.

Plastic objects come from both sea-going vessels and land-based activities, but the actual amount disposed of in marine waters is not known; a survey by the Center for Environmental Education, under contract from the Marine Mammal Commission, of different involved interests around the world may help clarify this issue (A. Blume, Center for Environmental Education, pers. comm., January 1987). The world maritime industry and the U.S. Navy fleet, for example, annually dispose of millions of tons of trash, including plastics, and an estimated 690,000 plastic containers are disposed of daily from ships (22). In the 1970s, commercial fishing fleets annually dumped an estimated 26,000 tons of plastic packing material and lost about 150,000 tons of plastic fishing gear (including nets, lines, and buoys) (104,

380). These amounts may have increased, since the fishing industry is using more plastic in its handling and packaging processes. Recreational boaters and beachgoers often improperly dispose of plastics such as picnic utensils and beverage can yokes. Antiquated sewer systems sometimes discharge plastic debris that subsequently washes up on beaches or is carried further into marine waters.

The most important international law concerning the release of plastics from *sea-going vessels* is the International Convention for the Prevention of Pollution from Ships (known as the MARPOL Convention). Specifically, optional Annex 5 prohibits the disposal of "all plastics, including but not limited to synthetic ropes, synthetic fishing nets and plastic garbage bags." Annex 5, however, has not been ratified by enough nations to be binding (328). Ratification by the United States and U.S.S.R. would be sufficient to make it binding on all other ratifying countries; U.S. ratification is expected to be submitted in the near future (328) (see ref. 303 for further discussion of international aspects).

In contrast, the London Dumping Convention controls the deliberate transport to and dumping at sea of plastics and other wastes generated by *land-based sources*. In order for any control action to take place, such dumping must be shown to have a significant adverse impact on marine life. The Law of the Sea Convention (not yet ratified by the United States) and the United Nations Regional Seas Programme (see box Q in ch. 7) could also be used to control land-based sources of plastics.

In the United States, the disposal of plastic objects and the raw materials used to manufacture them fall under the authority of the Clean Water Act and the Marine Protection, Research, and Sanctuaries Act. Several alternatives for controlling plastics pollution have been suggested (104,329), including: 1) requiring that plastic products such as beverage can yokes be biodegradable or photodegradable; 2) allowing only biodegradable plastics on ships; 3) banning some plastic products until biodegradable alternatives are developed (11 States currently prohibit the use of six-pack beverage can yokes); and 4) imposing a tax on manufacturers, wholesalers, and retailers of designated products (in New Jersey, for example, revenues from taxes on "litter generating products" are used for recycling projects by municipalities and industries).

While most observers agree that recycling and the development of biodegradable plastics would help reduce the problem, opinions differ about the availability of technologies to carry out these processes, the degree to which they would help, whether biodegradable plastics will break down into nonharmful substances, and which institutional methods (e.g., taxation) would provide effective incentives. Environmental groups (e.g., Environmental Defense Fund, The Oceanic Society) and the plastics industry (e.g., Society of the Plastics Industry), along with the U.S. Coast Guard and several other governmental agencies, support the adoption of Annex 5 of MARPOL (580). Legislation was introduced in Congress in 1986 to begin addressing the issue.

Leaching of Tributyltin From Ship Hulls

Another source of pollution to marine waters is the leaching of constituents of antifouling paint from ship hulls. One of these constituents, tributyltin (TBT), is a highly effective antifouling agent used on thousands of recreational and commercial vehicles. By preventing the attachment of barnacles and tube worms to hulls, for example, TBT can save vessel operators hundreds of millions of dollars annually; in addition, it is used in many consumer products, such as household cleaners.

TBT also is one of the most toxic antifouling agents used. Although it has been used for decades, its potentially deleterious impacts have only recently received attention (190,332). It can damage nontarget marine organisms, some of which have commercial value or are important in estuarine food webs. In Europe, for example, TBT has been considered responsible for malformations and deaths of oysters in the coastal waters of Britain and France. There is also concern over worker exposure to TBT in shipyards; some studies indicate that it may cause health problems ranging from skin disorders to some forms of cancer.

As a result of these and other findings, some countries now regulate the use of TBT. For example, the United Kingdom regulates the composition of antifouling paint, while France prohibits its use on ships below a certain size (190,474).

In the United States, most attention to date regarding the control of TBT has occurred at the State level (D. Bailey, Environmental Defense Fund (EDF), Richmond, VA, pers. comm., January 1987). North Carolina has adopted a strict water quality standard (2 parts per trillion) for the presence of TBT in saltwater. Concern that TBT may harm shellfish in Chesapeake Bay has led to calls for controls on the use of TBT (D. Bailey, EDF, pers. comm., January 1987; Reid, 1986; EDF, 1986). In Virginia, for example, EDF petitioned the State Water Control Board to adopt a water quality standard for TBT; the Board may consider such an action after EPA's water quality criteria document for TBT is issued (tentatively scheduled for draft release in spring 1987). The State has considered banning the use of TBT on recreational vessels less than 25 meters in length and restricting its use on larger vessels; Maryland may consider similar legislation. As a result of these types of concerns, Congress held hearings in 1986 on the use and effects of TBT and EPA has conducted risk-benefit analyses of its use.

Runoff from agricultural lands can carry soil particles, pesticides, bacteria, and other pollutants directly into estuaries and coastal waters or into rivers that later flow into these waters.

Chapter 4
Marine Environments and Processes, and Fate of Pollutants

CONTENTS

Marine Environments and Processes, and Fate of Pollutants

Marine waters are classified in this report into three categories: estuaries, coastal waters, and the open ocean.[1] The multitudes of different organisms that inhabit these waters require certain environmental conditions to survive and reproduce (box K). Changes in these conditions can occur as a result of waste disposal activities, depending on the fate of the wastes and associated pollutants and on the susceptibility of the waters and organisms to impacts.

The fate of wastes and associated pollutants, and the susceptibility of organisms to impacts from these substances, is influenced by features of marine environments and the organisms. For example, wastes and associated pollutants generally are dispersed and diluted by physical processes that transport and

---

[1]Some waterbodies may exhibit a combination of characteristics.

mix water; these physical processes, however, vary greatly in relation to a waterbody's degree of enclosure, currents, volume, and depth. In contrast, the concentrations of some pollutants can increase after disposal because of chemical processes (e.g., adsorption of pollutants to solid material), physical features (e.g., salinity gradients), or biological processes (e.g., the ability of organisms to accumulate some pollutants).

The three marine environments differ in their general susceptibility to adverse impacts. To understand how the various marine environments are affected by wastes, it is important to consider three issues: 1) the fate of wastes and associated pollutants after disposal, 2) the relationships between pollutants and specific impacts, and 3) whether marine environments can assimilate wastes or recover from adverse impacts.

TYPES OF MARINE ENVIRONMENTS AND THEIR SUSCEPTIBILITY TO ADVERSE IMPACTS

Estuaries

An *estuary* is a semi-enclosed waterbody with a free, but often small, connection to coastal waters or the open ocean, within which seawater is measurably diluted with freshwater from land drainage (fig. 8; ref. 612). Estuaries can vary greatly in size, shape, and degree of freshwater influence. For example, the category includes tidal marshes and lagoons, as well as the mouths of large rivers such as the Connecticut and Mississippi rivers. The Chesapeake Bay is considered an estuary because it is semi-enclosed and influenced by several rivers (including the Susquehanna and Potomac rivers).

Estuaries are among the most ecologically and economically important of all aquatic environments. Their productivity is generally very high because nutrients are plentiful (figure 9); large quantities of nutrients stimulate growth, reproduction,

and photosynthesis in phytoplankton, and large populations of phytoplankton support large populations of other organisms. The nutrients are generally the result of discharges and runoff from the land, which tend to get trapped when they enter estuaries. Many marine organisms, including some that live primarily in coastal or open ocean waters and some that are very important commercially, migrate into and use estuaries during critical parts of their life cycles. For example, estuaries provide critical breeding, spawning, and nursery habitat for many fish and shellfish, and they also provide important habitat for many birds and mammals; some of these organisms (e.g., bald eagles and several types of whales) are endangered species. The submerged aquatic vegetation (SAV) in estuaries, such as seagrass beds or kelp forests, is important ecologically as a source of food and shelter for many organisms and in stabilizing sediments.

Box K.—Ecological Conditions and Concepts

Marine organisms require adequate habitat and water and sediment quality for survival, proper growth, and reproduction. When ecological conditions change, organisms not able to tolerate or adapt to the new conditions may reproduce poorly, die, or move from the area. If the impacts on individual organisms are sufficient, entire populations and communities can be affected. Some important concepts to understand include:

- **Adequate Water Quality:**
  —Depends on appropriate levels of dissolved oxygen, light, and nutrients; and on proper chemical composition (including pH and salinity) and temperature.
  —Without adequate water quality, declines in levels of dissolved oxygen can result in fish kills; in addition, waters that lose their clarity restrict the light penetrating to submerged aquatic vegetation.
- **Adequate Sediment Quality:**
  —Depends on proper chemical and physical composition.
  —Sediment composition can be changed, for example, by accumulation of particles from dumping of dredged material or by accumulation of toxic pollutants.
  —Without adequate sediment quality, the types of bottom-dwelling organisms such as shellfish can change dramatically, or submerged aquatic vegetation may not be able to use the sediments as a substrate.
- **Portions of Marine Waters That Concentrate Pollutants:**
  —*Sediments* provide habitat for numerous marine organisms, including some commercially important fish and shellfish, and act as a substrate for submerged aquatic vegetation. Sediment concentrations of some pollutants (such as metals and persistent organic chemicals) can be very high because: 1) the pollutants often are attached to particulate material that becomes incorporated into sediments, and 2) the chemical conditions in many sediments inhibit the release of such pollutants back into the water column.
  —*Pycnoclines* refer to the boundaries between layers of water that have different densities. For example, freshwater is less dense than and rises above saltwater. Pycnoclines also can occur in other situations not involving freshwater. For example, they can form in saltwater during the summer, when water near the surface becomes warmer and less dense and forms a distinct layer above colder and denser water from greater depths. These types of pycnoclines typically vary seasonally in response to temperature differences in the water; the boundary between water layers with different temperatures is commonly known as a thermocline.
  —The *surface microlayer* is a thin layer (less than one-tenth of a millimeter to several centimeters thick) at the surface of all marine waters that, while poorly understood, seems to play an important role in the lives of many marine organisms. It appears to serve as essential, though temporary, habitat for the embryonic life stages of many fish and crustaceans, including commercially important species, and therefore may play an important role in marine food chains. At the same time, high concentrations of many materials (e.g., organic matter, metals, organic chemicals) and organisms typically can be found in the surface microlayer (357). Some of these increases are beneficial; for example, increased amounts of organic matter and nutrients result in large populations of many small organisms. However, toxic metals and organic chemicals also exhibit increased concentrations in the microlayer and these pollutants can create problems. (For further information, see ref. 586.)
- **Ecological Concepts About Populations and Communities:**
  —*Primary productivity* is the amount of organic material produced by plants (e.g., phytoplankton, algae, seagrasses). Plants produce this material in the form of plant tissue during photosynthesis, when the energy in sunlight is used to convert water and carbon dioxide into organic material.
  —*Community structure* refers to relationships between populations of different organisms in an area.
  —*Species diversity* refers to the number and abundance of different species.
  —*Food chains* (or webs) consist of organisms that produce, consume, and decompose organic material (figure 7). They include the smallest bacteria and microscopic plantlife (e.g., phytoplankton) and the largest consumers (e.g., fish such as salmon and flounder, marine mammals and birds, and humans).

Figure 7.—Generalized Marine Food Web

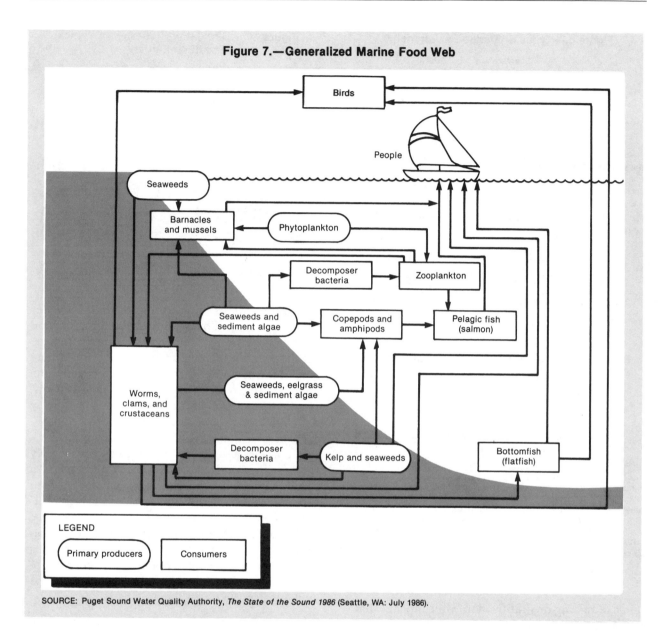

SOURCE: Puget Sound Water Quality Authority, *The State of the Sound 1986* (Seattle, WA: July 1986).

Estuarine circulation systems can be extremely complex and variable, depending on freshwater flow, tidal action, wind, depth, and shape (100). Tidal action and seasonal circulation patterns, for example, transport and mix freshwater and saltwater to varying degrees. Freshwater is lighter than seawater, so it tends to flow into estuaries along the surface, while seawater tends to enter below it.

It is useful to distinguish three types of estuaries based on the degree of mixing or stratification of freshwater and saltwater (165): 1) highly stratified (or salt-wedge) estuaries (figure 10), exemplified by the mouth of the Mississippi River; 2) partially mixed estuaries such as the James River estuary in Virginia and the Chesapeake Bay; and 3) well-mixed or homogeneous estuaries, which are rare but do illustrate an extreme in mixing and stratification patterns. The stratification of the water column creates water layers of different densities, with pycnoclines marking the layers. During the course of a year, a single estuary may exhibit a

Figure 8.—Types of Marine Environments

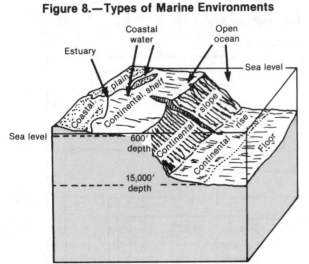

SOURCE: A.W. Reed, *Ocean Waste Disposal Practices* (Park Ridge, NJ: Noyes Data Corp., 1975).

Figure 9.—Primary Productivity In and Near a Typical Estuary

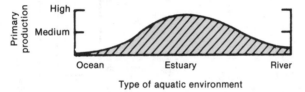

Primary productivity represents the amount of organic material produced by photosynthesis. The peak in primary productivity typically occurs in estuaries; as a result, estuaries contain abundant food resources and serve as nursery areas for many species (including some saltwater and freshwater organisms that move into estuaries during the breeding portion of their life cycle).

SOURCE: J.R. Clark, *Coastal Ecosystem Management: A Technical Manual for the Conservation of Coastal Zone Resources* (New York, NY: John Wiley & Sons, 1977).

range of mixing or stratification, which means that organisms inhabiting it must be capable of tolerating large changes in environmental conditions.

Estuaries and the organisms using them are highly susceptible to adverse impacts. One reason for this is that many organisms use these waters during particularly critical periods of their lives. Second, estuaries often trap particulate matter and nutrients that enter them via rivers, runoff, and waste disposal (effluent discharge or dumping). These waters bear the brunt of marine disposal activities.

Trapping varies among estuaries. In stratified or partially stratified estuaries, material begins to

Figure 10.—Circulation Pattern in a Typical Stratified Estuary

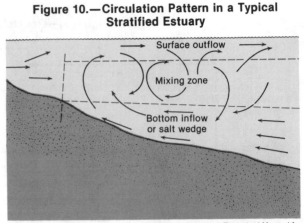

SOURCE: J.R. Clark, *Coastal Ecosystem Management: A Technical Manual for the Conservation of Coastal Zone Resources* (New York, NY: John Wiley & Sons, 1977).

settle out from incoming freshwater into saltwater and then is carried back toward the upper portion of an estuary by the landward-flowing saltwater. This happens in the upper Chesapeake Bay, where most of the sediment carried by the Susquehanna River is trapped (139). In shallow or relatively unstratified estuaries, particles can descend quickly to the seafloor, where they are less subject to physical processes that might flush them from the estuary.[2] Particles also descend more quickly when they stick together and increase in size and density, a process called *flocculation* which is enhanced when freshwater mixes with saltwater. Well-mixed estuaries generally trap less particulate matter because they are more subject to physical processes that can flush water from an estuary. In addition, flooding in the watershed of a small estuary may reduce trapping because the high influx of freshwater transports most particles to the sea (349).

In most estuaries, trapping plays an important role in determining the fate of pollutants and the likelihood of impacts. For example, although nutrients can stimulate productivity, trapping of nutrients can lead to nutrient enrichment (eutrophication), which can result in low levels of dissolved oxygen. In addition, many metals, organic chemicals, and pathogens bind to the surfaces of particulate material (e.g., suspended solids) and thus are frequently trapped in estuaries. These pollutants can cause problems when they are ingested by

---

[2]The rate at which water in an estuary is replaced is called the flushing rate.

organisms, and the problems can be magnified when they are passed up the food chain to other organisms, including humans. If inputs of toxic pollutants are frequent or large enough, "hot spots" of contaminated sediments such as those in San Francisco Bay can occur (395).

Estuaries and estuarine organisms can recover from certain impacts if the inputs of pollutants are reduced or terminated. For example, some water quality impacts such as low dissolved oxygen levels or eutrophication can be reversed; similarly, areas where populations or communities have been destroyed by physical burial can be recolonized if individuals of the same species are capable of migrating back into the areas. Other impacts, however, may require more time to be reversed or may in some cases be irreversible. For example, contamination of sediments with metals or persistent or-

ganic chemicals or major changes in community structure can be impossible to correct.

Coastal Waters

Coastal waters include those waterbodies lying over the inner portion of the continental shelf that are less enclosed and more saline than estuaries; these waters generally, but not always, lie within 3 miles of shore (i.e., within the boundary of the territorial sea) (figure 8). The movement of water and material in coastal waters is heavily influenced by tidal action and wind, as well as oceanic forces such as longshore currents, coastal upwelling of bottom waters, eddies, and riptides (99). Rivers may influence these waters to some degree, but this effect is generally less than in estuaries.

Coastal waters also vary greatly in shape, size, and configuration. They include bays (e.g., Mon-

Photo credit: U.S. Fish and Wildlife Service

Submerged aquatic vegetation (SAV) provides vital shelter and food for many marine organisms and also performs other important ecological functions such as helping to stabilize sediments. Substantial impacts on SAV in many areas of the country have been linked, with varying degrees of certainty, to waste disposal activities. Perhaps the best known example is the Chesapeake Bay, where SAV has declined precipitously over the last two decades.

terey Bay), sounds (e.g., Puget Sound), and open waters along the shoreline (e.g., the Southern California Bight). The New York Bight is considered a coastal water, even though it extends beyond the 3-mile boundary, because it is located on the inner edge of the continental shelf and is heavily influenced by river drainage.

Coastal waters tend to be moderately productive, although less so than estuaries because they receive fewer nutrients; some coastal waters such as Monterey Canyon off of central California are extremely productive. Again, submerged aquatic vegetation provides food and shelter for many organisms. As a result, coastal waters also tend to be ecologically and economically important, and often contain important fishing grounds that can be affected by pollutants from waste disposal activities. Many organisms that normally inhabit coastal waters migrate into estuaries and rivers during portions of their lives (e.g., for spawning). In addition, some endangered birds and mammals migrate through and use coastal waters, primarily to obtain food.

In comparison with estuaries, coastal waters are more directly linked to the open ocean and tend to disperse and dilute pollutants somewhat more readily. Trapping is less important than in estuaries, but it can be significant in small inlets and embayments (e.g., in Puget Sound where several toxic ''hot spots'' have developed). In addition, coastal currents can transport pollutants toward or along the shoreline instead of out to sea. These pollutants can accumulate in organisms and sometimes be transferred to other organisms in the food chain.

Coastal waters can recover from impacts such as eutrophication and hypoxia if the problem inputs are reduced or terminated. In many cases, the more open and dispersive coastal waters (relative to estuarine waters) restrict buildups of excess nutrients or bring in waters with higher oxygen levels. On the other hand, seasonal pycnoclines and thermoclines are typical of many coastal waters, and these features can contribute to increased eutrophication and hypoxia, at least temporarily. Some impacts, such as contamination of sediments with toxic pollutants, can be difficult or impossible to reverse.

The Open Ocean

The open ocean refers to waters overlying the outer portion of the continental shelf, the continental slope, and beyond (figure 8); together, these waters comprise about 92 percent of the Earth's surface water. Open ocean waters are deeper, more open, and more saline than coastal waters and estuaries. In comparison with estuaries and coastal waters, the open ocean is influenced less by tidal flow and more by permanent ocean currents. Open ocean processes are more capable of dispersing and diluting wastes and associated pollutants.

The open ocean typically is not as biologically productive as are estuaries and coastal waters, primarily because of a general lack of nutrients. An exception to this occurs in localized areas of upwelling, where nutrients and oxygen are transported up to surface waters, for example in the Georges Bank off the New England coast. This increase in nutrients can result in increased productivity of phytoplankton and corresponding increases in local fish populations. As a result, resources in the open ocean can be commercially important even though they are distributed unevenly (i.e., concentrated in certain areas and relatively absent in others). Many open ocean organisms, particularly some fish and mammals, also migrate to and spend portions of their life cycles in estuaries and coastal waters.

The open ocean is generally less vulnerable than other marine waters to many impacts because of its large volume and free exchange of water; open ocean currents generally have a considerable capacity to transport, disperse, and dilute wastes or pollutants. As a result, problems such as hypoxia and eutrophication (which generally occur only when certain conventional pollutants and nutrients are present in high concentrations) and physical burial of organisms are less likely to occur.

However, some metals, organic chemicals, and pathogens are of concern, even though they also are dispersed, because they can: 1) cause impacts at low concentrations, 2) persist in the environment, 3) accumulate in organisms, and 4) increase in concentration in successive levels of marine food chains. Some of these pollutants have been detected in significant concentrations in the water and in the tissues of fish, seabirds, and marine mammals. The significance of such contamination, and whether organisms and populations can recover from associated effects, is not always clear because of insufficient knowledge in several areas. For instance, questions remain about: how open ocean food chains operate, what concentrations of chemicals cause reproductive failure in marine organisms, how tolerant open ocean organisms are to change, and the likelihood of pollutants being transferred to humans. In addition, detection of such impacts is difficult and the impacts may not be observed until long after the polluting incident is over.

FATE AND DISTRIBUTION OF WASTES AND ASSOCIATED POLLUTANTS

The ultimate fate of wastes and associated pollutants depends to a large extent on the many different processes that affect dispersion and deposition. Initially, wastes are diluted immediately after entering the water. Simultaneously, other processes begin to transport wastes and pollutants over longer distances and over time to modify their chemical and biological nature. Besides initial dilution, waste particles and pollutants are affected by: 1) physical transport (e.g., transport as part of a water mass, whether they are suspended or dissolved in the water); 2) biological transport[3] (e.g., extraction by plants and animals and subsequent movement with the organisms or their remains or excretions); and 3) sedimentation (e.g., attachment to clays and organic materials as they settle through the water column and eventual incorporation into bottom sediments) (262). Figures 11 and 12 depict the general fate of dumped sludge and discharged effluent.

Initial Dilution

When a waste enters marine waters, either via discharge or dumping, it mixes with and entrains seawater; in addition, the wastestream begins to broaden and particles begin to disperse. As a result, the waste is diluted substantially, often by a factor of 5,000 or more. This *initial dilution* takes place within the first few hours after disposal and lasts until waste particles either cease moving vertically in the water column or reach the bottom (280,385).

Figure 11.—General Fate of Sewage Sludge Dumped in Marine Waters

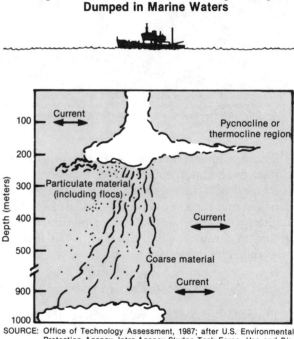

SOURCE: Office of Technology Assessment, 1987; after U.S. Environmental Protection Agency, Intra-Agency Sludge Task Force, *Use and Disposal of Municipal Sludge*, EPA 625/10-84-003 (Washington, DC: September 1984).

Vertical movement of waste particles in the water column depends on the waste's bulk density[4] and the presence of pycnoclines (286). Waste particles that are more dense than the surrounding water will sink, while less dense particles will rise. This vertical movement continues until the waste particles reach a water layer or pycnocline of the same density, or the bottom. Particles that accumulate along pycnoclines eventually are transported by other processes or settle to the bottom.

---

[3]These biological transport processes are distinct from the biological processes of bioaccumulation and biomagnification discussed later.

[4]Bulk density is the weight of a waste per unit of volume.

Figure 12.—General Fate of Effluent Discharged Into Marine Waters

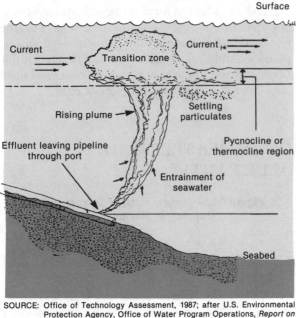

SOURCE: Office of Technology Assessment, 1987; after U.S. Environmental Protection Agency, Office of Water Program Operations, *Report on the Implementation of Section 301(h)*, EPA 430/9-84-007 (Washington, DC: August 1984).

This description is simplistic because wastes usually are not homogeneous; instead, they usually are composed of a variety of particles with different densities and are not uniformly diluted (132,286). Lighter, floatable materials (e.g., plastics, oil, or grease) generally rise to the surface, while heavier materials remain at lower depths.

Physical Transport

Physical transport generally results in the dispersion of particles and pollutants away from the disposal site. Two major physical processes are largely responsible for dispersion: *currents* that move water or other matter from one place to another and *mixing* of waters with different characteristics.

There are two general categories of currents: permanent and transient (473). Permanent currents are found in coastal and open ocean waters (e.g., the strong Gulf Stream off the east coast and the weak California Current off the west coast). Portions of these currents (called eddy rings or jets) can meander from the main current and modify the net direction of waste material transport (120). The effects of long-term currents are particularly important when evaluating potential dumping sites.

Transient currents occur over smaller distances and time periods than permanent currents. They are found in all marine environments, including estuaries, and are caused by factors such as winds, tides, and waves. Their effect on the transport of waste material tends to be most important when wastes are discharged or dumped near shore. For example, currents close to shore and at or near the surface can move sediment such as dredged material along the shoreline (210,473).

Mixing occurs when two masses of water with different densities (e.g., two currents, or a current and a relatively stable water mass) intermingle along their common boundary (262). This is generally caused by random motion along density gradients, but it can be enhanced by wave action and tides (48). Its effects vary in different marine environments. For example, mixing of bottom and surface waters in the open ocean can take hundreds of years (48,120). In estuaries and some coastal waters, mixing can occur immediately after disposal and increase dilution of the waste material.

Currents and mixing can transport and disperse wastes over hundreds of kilometers (120,286). The depth at which such transport occurs depends on the presence of pycnoclines, because particles tend to fall until (and if) they reach a layer of water denser than the particles themselves. For example, particulate matter from dumping activities in the New York Bight generally descends to the depth of a seasonal pycnocline and then is transported laterally (426,709).

Biological Transport

Waste particles and pollutants can be extracted from the water column by plants and animals in various ways, including direct ingestion, passing through gills, or other mechanisms. Once extracted, the wastes can be transported from one location in the water column to another by several processes that are associated with individual organisms.

Migration by Organisms.—Many organisms move relatively long distances as part of their normal living pattern and, as a result, they can carry waste particles and pollutants to new locations.[5] The organisms can move either vertically (i.e., up and

_____
[5]Pollutants can also be transported out of the water, for example when predatory birds consume contaminated marine organisms.

down in the water column) or horizontally (e.g., from coastal waters into estuaries). For example, some plankton move vertically in the water column on a daily basis and certain fish exhibit similar movements on a seasonal basis. Tuna and certain marine mammals are well-known for their lengthy horizontal migrations.

Many organisms must move from coastal or open ocean waters into estuarine waters to breed and/or to provide proper nursery habitat for their offspring. Some organisms such as salmon migrate from marine waters to rivers to breed; fish that undertake such migrations are known as anadromous fish. These migratory organisms can be exposed to a variety of pollutants while they migrate.

Movement of Eggs and Body Remains.—Many organisms lay eggs that are carried by currents to other areas or that move vertically depending on their relative buoyancy; for example, some eggs sink to deeper waters while others rise to the microlayer. In some cases, pollutants may have been incorporated into the eggs as they were formed, which occurs frequently in the surface microlayer. In addition, once an organism dies, its remains can be moved to new areas by similar means and these remains also can contain pollutants.

Excretion.—When a marine organism ingests waste materials, they often can eventually be excreted. Since many organisms move substantial distances in the course of their daily activities, excretion can occur in locations far from the point of ingestion. If excreted by organisms inhabiting the water column, excretory products often settle to the seafloor; if excreted by benthic organisms, the products generally remain on the seafloor.

Sediment Deposition

The processes discussed so far transport substances throughout the water column. Other processes, particularly sedimentation and flocculation, influence the manner in which particulate material and associated pollutants are deposited on the bottom. Two additional processes, resuspension and bioturbation, can counter this, but generally only to a small degree. The combined effects of these processes determine the overall rate of accumulation of particulate matter on the bottom.

Sedimentation is the settling of particulate matter through the water column and onto the bottom. The rate of sedimentation depends on particle size and density, water density, mixing, and flocculation. For example, larger particles tend to be heavier and they settle faster than smaller particles. The presence of pycnoclines can alter the rate of movement of particles; particles that reach a layer of similar water density generally cease moving and remain at the pycnocline, at least until some other transport processes move them elsewhere.

Flocculation refers to chemical reactions that result in the aggregation of small particles into larger particles, thus increasing particle size, density, and settling rate. Flocculation often occurs at the boundary of freshwater and saltwater, such as when freshwater effluent containing metals encounters more saline estuarine water.

The method of disposal and the concentration of solids in the waste also can affect both sedimentation and flocculation (385,596). For example, wastes dumped from a ship often have a higher concentration of particles (even after initial dilution) than do pipeline effluents and may exhibit higher rates of flocculation and sedimentation. In addition, the settling rate may be increased because dumped materials have a greater downward momentum.

In well-mixed waters that do not exhibit pycnoclines, particles can settle directly to the bottom. If physical transport processes are sufficiently strong to widely disperse the particles, however, the rate of particle accumulation on the bottom in a given area will be relatively low. In contrast, if wastes are disposed of in shallow or quiescent estuarine and coastal waters, or even in some deeper coastal waters at a high and continuous rate, sedimentation and flocculation can be significant and cause an accumulation of particulate material on the bottom. Such accumulations can alter the particle size or chemical composition of bottom sediments, which can in turn affect benthic organisms and the structure of benthic communities.

Even after it settles to the seafloor, particulate matter can be *resuspended* back into the water column. For example, bottom currents and storms are capable of stirring up sediments and moving settled particles back into the water (139,385). The physical characteristics of the particles influence resuspension; for example, silt and clay particles are more cohesive than sand particles and are less likely to be resuspended (596). Resuspension can also be caused by chemical and biological processes. For example, when bacteria in the sediments decompose particulate matter with a high organic content, some byproducts of decomposition can be released into the water column. In some cases, this can include pollutants such as metals. Predicting resuspension rates generally is difficult and site-specific (262).

In another biological process that affects sediment deposition, burrowing animals can move particles and associated pollutants toward or away from the boundary between the sediment and the water column, a process known as *bioturbation* (385). As a result, pollutants such as polycyclic aromatic hydrocarbons and polychlorinated biphenyls (PCBs) can either be brought into contact with the water column, where non-burrowing organisms can be exposed to them (295,385), or buried more deeply in the sediments.

POLLUTANTS AND THEIR RELATIONSHIP TO IMPACTS

The wastes and pollutants disposed of in marine waters can have varying impacts on marine environments and organisms. The relationship between different pollutants (i.e., oxygen-demanding substances, nutrients, suspended solids, pathogens, metals, and organic chemicals) and impacts can be complex, depending on what biological, chemical, and physical processes occur to catalyze the impacts.

Oxygen-Demanding Substances

Minimum levels of oxygen are critical for the maintenance of most forms of life. Oxygen levels in the water and sediments can decline, however, as a result of several common processes. First, microorganisms use oxygen to decompose or transform organic material, which is contained in most wastes, into compounds such as carbon dioxide, water, and nitrates. The amount of oxygen used during this process is termed the biochemical oxygen demand (BOD) (132,418). Second, waste material is also broken down by chemical processes (independent of organisms) that use oxygen; the total amount of oxygen that can be used in biological and chemical processes is termed chemical oxygen demand (498). Oxygen levels also can decline as a consequence of nutrient enrichment, which is discussed below.

When the amount of dissolved oxygen in water falls below a critical level, often around 2 parts per million, the water is said to be hypoxic. Water that is completely depleted of dissolved oxygen is called anoxic. Hypoxic and anoxic conditions can cause massive fish kills; if oxygen levels are reduced only slightly (but not enough for conditions to be considered hypoxic), organisms can still be stressed and chemical reactions in the water column can be modified. These problems are more common in estuaries and coastal waters than in the open ocean, where waters are generally more dispersed and mixed.

Nutrients

Nutrients are essential for the proper growth and reproduction of individual organisms and, consequently, for the general productivity of marine environments. *Eutrophication* refers to an increase in nutrient levels in a body of water. This can occur naturally (e.g., in the open ocean through the upwelling of nutrients from deep waters) or as a result of human activities (e.g., runoff from fertilized farmlands or discharges from sewage outfalls). In either case, the addition of nutrients such as nitrogen and phosphorus can lead to increases in the productivity of marine organisms, particularly some algae. Up to a certain limit, such increases in productivity can be beneficial, for example, by leading to corresponding increases in fish populations in areas where nutrients are relatively lacking. However, if nutrient levels are too high, several adverse impacts can occur:

- increased turbidity or cloudiness of the water, which can keep light from reaching submerged vegetation;
- changes in the distribution, abundance, and diversity of species (e.g., the replacement of typically abundant species with less common species);

- subsequent changes in food chain relationships (165); and
- depletion of oxygen levels when large numbers of algae die and are decomposed by microorganisms; oxygen depletion in turn can cause other changes including fish kills.

Increased nutrient levels are more common in estuaries and coastal waters than in the open ocean. Large quantities of nutrients enter estuaries and coastal waters from rivers, runoff, and disposal, as well as from coastal upwelling. In addition, processes that dilute and disperse nutrients are weaker in these waters.

Suspended Solids

Marine plants generally grow best in relatively clear water where sunlight, which is critical for photosynthesis, can penetrate to and be used by the plants. Since the introduction of suspended solids into shallow waters can increase turbidity and block out sunlight, it can reduce the rate of photosynthesis significantly and harm submerged vegetation. In addition, solid particles can settle out of the water column and accumulate on the bottom, sometimes changing the nature of the sediments and associated benthic community. In addition to solids suspended in the water column, the solid material in a waste can bury and kill benthic organisms (e.g., if large amounts of dredged material are disposed of in a small area).

Pathogens

Many pathogens—including viruses, bacteria, fungi, and parasites—that can cause human disease are found in marine environments. The viability of pathogens depends primarily on their survival after they enter a particular marine environment. While many pathogens die quickly after exposure to various environmental factors (e.g., light), some can persist, especially if adsorbed onto particulate matter that provides protection from the surrounding environment (183,205). In addition, some pathogens can be ingested by and survive in marine organisms without harming the organisms, but they can cause serious human health effects if a person consumes an organism that has ingested but not yet excreted the pathogens. For example, shellfish "filter" water to obtain food and, in the process,

can ingest pathogens; the concentration of pathogens in the gut of a shellfish can be quite high. Detecting pathogens that survive in marine waters is often difficult, as is predicting the impacts they can cause.

Organic Chemicals and Metals

Whether toxic pollutants such as metals and organic chemicals cause adverse effects depends on the interaction of many chemical and biological factors. Only some toxic pollutants are *bioavailable* to organisms, that is, present in a form to which organisms can actually be exposed. Many metals, for example, are attached to particulate material and buried deep in sediments where they are exposed to few organisms. If a pollutant is bioavailable, the effects of exposure will vary depending on its concentration and the length of exposure (270), as well as the stage in the organism's life cycle (figure 13).

Not all metals and organic chemicals cause adverse effects, even if an organism is exposed to and takes up a pollutant. In fact, organisms require small amounts of some metals that occur naturally in marine waters for important physiological functions. Other pollutants may be present in forms that

Figure 13.—Life Cycle of the Winter Flounder and Potential Contact With Chemical Pollutants

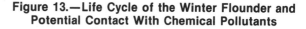
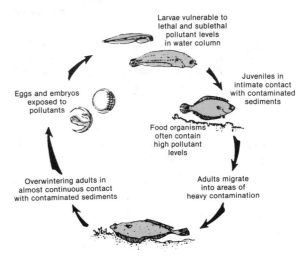

NOTE: An individual flounder could come into contact with pollutants at one or more stages of its life cycle.

SOURCE: C.J. Sindermann, "Fish and Environmental Impacts," *Archiv für Fischereiwissenschaft,* 35:125-160, 1984.

Figure 14.—The Transfer of DDT Through a Portion of a Food Web in the Long Island Estuary

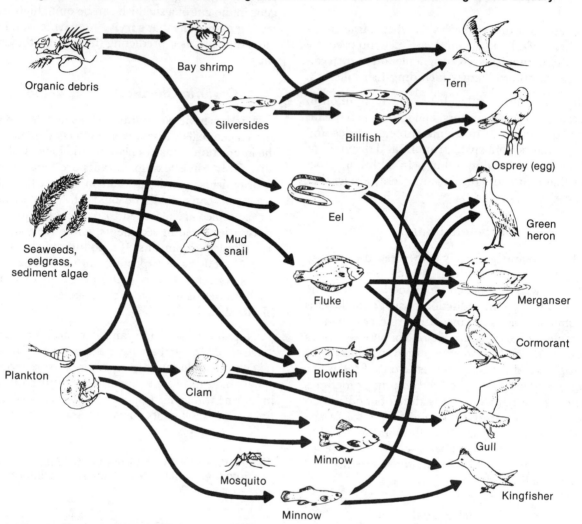

Persistent, lipid-soluble pollutants such as DDT can be taken up from the water column or sediments by many organisms. This diagram depicts the many pathways by which a pollutant such as DDT can be transferred up the food chain and eventually accumulate in high concentrations in the tissues of fish-eating vertebrates.

SOURCE: Office of Technology Assessment, 1987; after B.W. Pipkin, et al., *Laboratory Exercises in Oceanography* (San Francisco, CA: W.H. Freeman & Co., 1977).

are not toxic to organisms; they may stay in the gut for varying periods and then be excreted without causing harm (262,385).

Two important processes determine the ability of many metals and organic chemicals to cause adverse impacts on marine organisms or humans: bioaccumulation and biomagnification. *Bioaccumulation* is the process whereby a substance enters an aquatic organism, either from the water or from consumed food, and is stored within the organism's tissues. *Biomagnification* refers to increases in the concentrations of bioaccumulated substances in the tissues of consumers and predators occupying successive levels of a food chain (58,260). The likelihood of adverse impacts on organisms such as invertebrates, fish, or humans is increased if toxic pollutants biomagnify in a food chain (figure 14). The effects might show immediately after exposure or take long periods to appear, and correlating them with degree and duration of exposure is difficult.

Bioaccumulation can occur at any level of the food chain. For example, phytoplankton can be contaminated by pollutants present in the water column, while benthic zooplankton and shellfish

Healthy striped bass have long supported a valuable commercial and recreational fishery along the Atlantic Coast. The fishery has declined significantly during the last 15 years, and many individual fish are severely contaminated with PCBs.

that filter bottom sediments in search of food can ingest pollutants that are adsorbed onto sediment particles. These organisms can then contaminate organisms in higher trophic levels that ingest them as food (410). However, bioaccumulation does not always cause adverse impacts. Some organisms can internally degrade or regulate the levels of certain toxic pollutants in their systems.

It is also possible for some toxic pollutants to cause problems even if they do not bioaccumulate. For example, some pollutants can be concentrated in the gut of an organism (but never bioaccumulate in the organism's tissues); these pollutants, however, can often cause adverse effects on predatory organisms higher in the food chain. In addi-

tion, some behavioral effects can occur even if a pollutant is not ingested. For example, herring have avoided waters they typically use for spawning because of pulp mill effluents discharged into a river (398).

Organic Chemicals

Organic chemicals, whether dissolved in the water column or adsorbed onto sediment particles, can be taken up by an organism in several ways, such as through filtering water or sediment or ingesting other organisms. Whether bioaccumulation occurs depends primarily on a chemical's ratio of lipid solubility to water solubility (165,195,262, 385). Some organic chemicals are more soluble in water and can pass through an organism (or be metabolized by the organism), while others are more soluble in lipids and will bioaccumulate in fatty tissues. For example, PCBs are soluble in lipids and are only slowly metabolized by marine animals, so they tend to bioaccumulate in an animal's fatty tissues (139). Birds (such as bald eagles and pelicans) and mammals (such as seals and sea lions) that occupy higher levels of marine food chains usually have large amounts of fatty tissues and are particularly likely to bioaccumulate organic chemicals.

Bioaccumulated organic chemicals can affect organisms at all levels of food chains. Halogenated organic chemicals (including pesticides such as DDT) are the most likely organic chemicals to be biomagnified because they are persistent and highly soluble in fatty tissues.[6] PCBs cause decreases in plantlife and changes in community structure, and they have been implicated in fish mortality and physiological abnormalities (132,133,139).

Some organisms have the ability to degrade some organic chemicals into other forms. For example, microorganisms can degrade some highly persistent and toxic chlorinated compounds under certain conditions (40,41,540). Some invertebrates, fishes, and mammals can partially degrade polycyclic aromatic hydrocarbons and PCBs, but the byproducts can sometimes be more toxic than the parent compound. The bioavailability and toxicity of organic chemicals can also be modified prior to ingestion by processes such as bioturbation or degradation by light (385,710).

Metals

The ability of a metal to affect marine organisms depends primarily on its *form* (e.g., dissolved or particulate, bound to another substance or free), and this is greatly affected by site-specific conditions. In their *particulate* form, most metals tend to adsorb onto other particles that eventually settle from the water column and are deposited as sediment (139,165). Once deposited in typical oxygen-poor sediments, the chemical form of these metals is generally stable.[7]

---

[6]The term halogenated refers to compounds that include certain elements such as fluorine, chloride, and bromine.

[7]Microorganisms in sediments, however, can modify the form of some metals; for example, bacteria can convert slightly toxic inorganic mercury to highly toxic and volatile methyl mercury (26).

If the sediments are subsequently oxygenated, however, some metals (such as cadmium, copper, nickel, and zinc) may dissolve and be slowly released back into the water column, where they may be taken up by non-benthic organisms (124). For example, zinc is very insoluble when combined with sulfide in oxygen-poor environments but is soluble in oxygen-rich environments (385). Sediments can be oxygenated (and also resuspended) by bioturbation, storms, and other disturbances (139). Metals also can be released as a result of other changes such as salinity fluctuations in estuaries.

Metals present in the water column as *dissolved* ions often bind with other molecules to form soluble complexes that can affect bioavailability and toxicity to organisms (385). For example, the toxicities of lead and zinc are reduced when they are in complexes, while the bioavailability and toxicity of methyl mercury is greatly enhanced (75,247).

Marine organisms can ingest metals that are dissolved in the water or they can ingest particulate matter onto which metals are adsorbed (46,232, 260). Once ingested, some metals can pass through the gut and be excreted, while others cross the gut membrane and bioaccumulate in organismal tissue (410). Of the four metals of primary concern to humans (ch. 6), cadmium and mercury tend to bioaccumulate in marine organisms, while neither arsenic nor lead have been shown to bioaccumulate significantly in seafood (lead, however, can be present in high concentrations in the guts of some shellfish). Mercury in its methylated form, however, is the only metal known to biomagnify in successive levels of aquatic food chains.

Some organisms can regulate the internal availability or level of certain metals (133,139). In some fish, for example, metals that are toxic as free-floating ions (e.g., cadmium, copper, mercury, and zinc) can be bound to the protein metallothionein and the bound metals are generally not toxic (250, 410). Proteins with similar capabilities occur in crustaceans and mollusks (74). Arsenic is largely converted to nontoxic forms by marine organisms. Some metals also may be deposited in skeletal material and intracellular spaces, or removed from an organism via excretion, diffusion, molting, or egg production.

CAN MARINE ENVIRONMENTS ASSIMILATE WASTES OR RECOVER FROM IMPACTS?

Two concepts are often used in discussions about the role of marine environments in waste disposal: assimilative capacity and recovery. *Assimilative capacity* has been defined as the amount of material that could be contained within a body of water without producing an unacceptable biological effect (188). *Recovery* refers to the degree to which a condition that existed prior to an impact is restored. In essence, the assimilative capacity concept asks how much waste can be added to a marine environment before an impact occurs, while the recovery concept asks what happens after inputs of wastes and pollutants cease.

Both concepts have an intuitive appeal, and even some utility as concepts that can illuminate general discussions, but applying them in actual practice is difficult. This difficulty stems primarily from the inability to develop quantitative criteria for assessing questions such as:

* What is an impact and what is an unacceptable level of such an impact?
* At what ecological level should impacts be measured (e.g., on cellular systems, organisms, populations, general habitat qualities)?
* How many impacts must be reversed (and to what degree) before an area is considered to have recovered?
* Over what time scales and sizes of areas should assimilative capacity or recovery be measured (20,165,188)?

Both assimilative capacity and recovery depend on many site-specific factors including: the physical and chemical characteristics of the receiving waters (e.g., strength of currents and mixing, sedimentation, adsorption of pollutants); the resiliency of organisms that inhabit the area (e.g., their ability to survive, grow, and reproduce); and the ability of other organisms to migrate into the area. These questions might be answered at a specific site *if acceptable quantitative criteria could be developed*, but both assimilative capacity and recovery would still vary from site to site and would have to be determined on a case-by-case basis.[8]

Clearly, some wastes can be disposed of under certain conditions without causing severe adverse impacts. Furthermore, some marine environments are not as likely as others to suffer certain impacts and therefore could be considered to have a greater assimilative capacity with respect to those impacts. For example, the open ocean is less likely to exhibit hypoxia and eutrophication and therefore might be able to assimilate oxygen-demanding substances and nutrients to a greater extent than can estuaries and coastal waters.

Marine environments also can recover from some impacts, particularly when inputs of oxygen-demanding substances, suspended solids, and nutrients are reduced. For example, dissolved oxygen levels have increased markedly and anoxic conditions have declined following improvements in municipal sewage treatment plants in the upper Delaware Bay estuary and in Newark Bay (346, 683). In addition, some estuaries and coastal waters are subject to large natural fluctuations in physical conditions (e.g., tides, waves, storms). Organisms in these environments show large natural population fluctuations, as well as rapid growth and migration rates. These organisms may be able to survive catastrophic events or repopulate an area quickly, often in as little as one year (165).

In contrast, marine waters cannot readily assimilate many organic chemicals or recover from their impacts, primarily because of the persistence of many of these pollutants. For example, "hot spots" of DDT still exist in some coastal sediments and some fish species contaminated with DDT are still being caught (e.g., the white croaker off the southern California coast), even though most uses of DDT were prohibited in the United States in 1972. For these persistent organic chemicals, the length of time under consideration greatly influences any assessment of assimilative capacity and recovery potential, and in some cases, even if all pollutant inputs were halted, the environment still

---

[8]Another concept, *accommodative capacity*, has been proposed as a replacement for assimilative capacity (284). Accommodative capacity has a slightly different focus in that it emphasizes how an environment "adjusts" to overall inputs of pollutants. However, it suffers from similar shortcomings, for example, its dependence on defining the term "adjustment."

may not ever return to its unpolluted state. As an illustration, consider the sediments in portions of Buzzards Bay, Massachusetts, which are contaminated with highly persistent PCBs. As the contaminated sediments are covered (naturally or artificially) with cleaner sediment, the potential for organisms to be exposed to the PCBs could decrease, and this could be considered a reversal of contamination. However, subsequent dredging of the harbor or the effects of severe storms could resuspend the sediment and PCBs and re-expose organisms to these pollutants.[9]

Moreover, most situations are complicated because multiple impacts and pollutants are typically involved. Assimilative capacity or recovery potential might be assessed for one impact, but they are much more difficult to determine for interrelated impacts. For example, it may be possible to determine how much waste can be disposed of before oxygen is depleted (i.e., assimilative capacity) or how quickly adequate oxygen levels will return following disposal (i.e., recovery rate). However, oxygen declines can cause other impacts such as fish kills and it can be difficult to ascertain when populations would recover to "acceptable" levels. When multiple pollutants are involved it can be difficult to determine the effect of removing or reducing one pollutant from the total waste stream.[10]

[9]In some situations, however, impacts from organic chemicals clearly have lessened over time. For example, the population of brown pelicans in southern California, which declined because DDT caused eggshell thinning and subsequent reproductive failure, has steadily increased since the discharge of DDT was sharply curtailed (54).

[10]Tests to assess the toxicity of "whole" effluents (i.e., the cumulative toxicity of all pollutants in an effluent) are being developed by EPA.

Photo credit: Stephen C. Delaney

Chapter 5
Impacts of Waste Disposal on Marine Resources

CONTENTS

Table

Figures

Impacts of Waste Disposal on Marine Resources

OVERVIEW

Marine resources are affected by a wide range of natural and human perturbations, including pollutants from waste disposal. Waste disposal occurs directly in marine waters, but also indirectly as wastes are carried to the sea by rivers. It can be difficult, however, to establish a clear understanding of the precise connections between pollutants from these activities and impacts on marine resources. Nevertheless, sufficient evidence is available to conclude that pollutants from disposal activities have resulted in a wide variety of impacts on water quality, sediment quality, and marine organisms.[1]

[1]Space limitations preclude a detailed presentation of all evidence documenting these impacts. The reader interested in an extensively documented, detailed discussion of this evidence is referred to ref. 588, prepared by OTA and available from the National Technical Information Service.

Most of the impacts that are attributable to pollutants from waste disposal have been observed in estuaries and coastal waters, often the most productive marine waters. The degree and distribution of these impacts vary widely among different waterbodies and organisms, but no region of the country is immune to serious adverse impacts from pollutants. Even small quantities of certain pollutants can result in chronic, persistent, and serious effects on organisms.

Where trends in impacts over the past 10 to 15 years are discernible, they have been mixed. They have varied among specific pollutants, species, and locations. Some improvements have been observed, while in other cases deterioration is evident. Sometimes no clear trend appears.

ESTABLISHING LINKS BETWEEN POLLUTANTS AND IMPACTS

Determining the causes of impacts on marine resources can be difficult.[2] Changes can result not only from waste disposal activities and runoff, but also from natural perturbations, fishing, or other human-induced changes such as habitat destruction or freshwater diversions. Even when pollutants are correlated with impacts, the ultimate source of the pollutants may be unclear—they may emanate from any combination of surface runoff, various industrial discharges, municipal discharges, dumping activities, and atmospheric deposition, and they may come from sources in or adjacent to marine waters or from far upstream.

Another complicating factor is that impacts caused by pollutants may not be observed for years or decades after the pollutants are released, or they may occur far from the release area. For example,

[2]This is assuming that the impacts are detected in the first place; while many effects can be detected rather easily, others can be detected only with great difficulty and expense or may escape detection altogether.

when a pollutant is extremely persistent in the environment or when water flow and circulation are great, pollutants can be transported great distances. In addition, some impacts on organisms may not occur until the affected organism is far from the original point of contamination.

Consequently, establishing the causes of past and present impacts and predicting future long-term impacts on marine communities is a formidable task. These difficulties are frequently aggravated by a lack of information. The picture that emerges from an analysis of the available information looks like a jigsaw puzzle with many pieces missing.

Thus, although waste disposal activities may be fully or partly responsible for many marine impacts, it is often difficult to assess their precise involvement. **Despite these problems of documentation, a strong overall case can be established that waste disposal activities are contributing significantly to substantial declines in the quality of marine waters and harming marine organisms, and in some cases having effects on humans.**

IMPACTS ON WATER AND SEDIMENT QUALITY

Enrichment With Organic Matter and Nutrients

Perhaps the most conspicuous and widespread impact that pollutants have on marine environment is eutrophication, a process associated with the introduction of nutrients. Eutrophication is evident in every region of the country. The impacts of eutrophication range from stress on individual organisms (which in turn may increase the incidence of disease or abnormalities) to major ecological changes. Nutrient enrichment sometimes contributes to massive blooms of tiny photosynthetic organisms,[3] sometimes dubbed "green tides," "brown tides," or "red tides."[4] These organisms can harm—and even kill—other marine organisms and humans (343,536,545). Under extreme conditions, eutrophic conditions can lead to a severe depletion of dissolved oxygen called hypoxia. The most dramatic consequences of extreme hypoxia are mass kills of organisms.

Eutrophication and hypoxia often have been linked to human activities, including waste disposal. Waste disposal activities (particularly municipal discharges) contribute large quantities of nutrients to marine environments, and hypoxia can be caused or aggravated by the introduction of oxygen-demanding pollutants (e.g., organic matter) from these same sources. Other pollutant sources, such as runoff, also contribute to eutrophic and hypoxic conditions, and natural factors such as seasonal stratification of the water column can also cause hypoxia.

Eutrophication and hypoxia are serious and regularly recurring problems in many major waterbodies (695). Hypoxic areas vary widely in magnitude, from a fraction of a square mile to thousands of square miles. Examples of large and regularly occurring hypoxic waters are an area (up to 8,000 square kilometers (km^2)) off the Louisiana coast (Rabalais (465) and a portion (up to 3,000 km^2) of the Chesapeake Bay (419).

Trends in the occurrence of hypoxia around the country are mixed. In some areas, the problems have been alleviated because discharges of organic matter and nutrients have been reduced (395,463, 554,703). In other instances, the problems have grown in severity, either because quantities of nutrients and organic matter have increased or because other changes have reduced the natural system's capacity to accommodate the discharges without major ecological impacts (315,419,486).

Hypoxia is least extensive along the Pacific coast; conversely, the Atlantic coast and particularly the Gulf of Mexico are greatly affected by hypoxia. Extensive hypoxia has been found along the southern coast of Louisiana (figure 15), and it is common in the Chesapeake Bay (figure 16) and the New York Bight. Its causes are multiple and include natural factors as well as pollutant inputs from raw sewage, sewage sludge, and other waste materials.

Elevated Concentrations of Other Pollutants in the Water Column

In addition to organic matter and nutrients, many other pollutants are discharged into marine waters in large quantities. Among these are pathogens, metals, and organic chemicals such as chlorinated and aromatic hydrocarbons. Contamination can vary from levels scarcely above the threshold of detectability to extremely high levels. Contamination tends to be greatest in the vicinity of estuaries flanked by heavy urban or industrial development, or near estuaries that receive pollutants from developed areas upstream. Contamination also tends to be most serious near municipal and industrial outfalls, and in the vicinity of major dumpsites for sewage sludge or other contaminated materials (figure 17).

Even at a given location and time, contamination may vary considerably according to its vertical position in the water column. Some pollutants concentrate at the very surface of the water-column

---

[3]A photosynthetic organism is one which uses sunlight to synthesize compounds.

[4]The precise conditions needed for the initiation, propagation, and maintenance of these blooms are not completely understood. It is known, however, that nutrients are required for a bloom to proceed. If nutrients from waste disposal have enriched marine waters, they may contribute to a bloom greater (and hence more damaging) than it otherwise would be.

Figure 15.—Extent of Oxygen Depletion in Bottom Waters of the Louisiana Shelf, July 1985

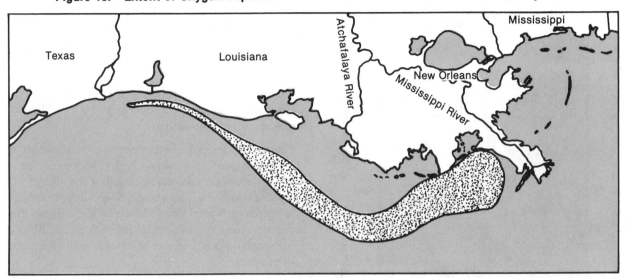

Stippled area depicts bottom waters with dissolved oxygen concentrations less than 2 milligrams per liter (a condition known as hypoxia). Most animal life cannot survive for long in water with such low oxygen concentrations.

SOURCE: N. Rabalais and D.F. Boesch, *Extensive Depletion of Oxygen in Bottom Waters of the Louisiana Shelf During 1985* (draft manuscript, 1986).

**Figure 16.—Volume of Water in Chesapeake Bay With Levels of Dissolved Oxygen
Lower Than 0.7 Milligrams Per Liter, 1950-80**

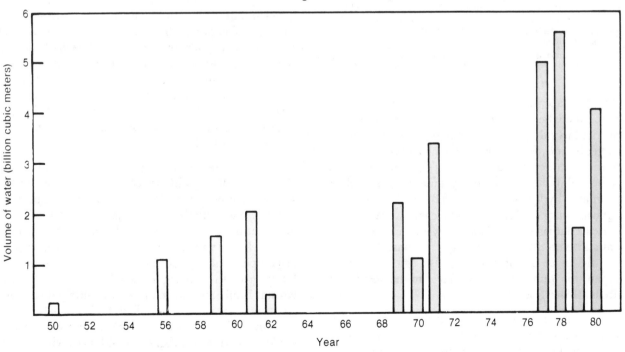

NOTE: Years lacking bars, no data are available.

SOURCE: Office of Technology Assessment, 1987; after U.S. Environmental Protection Agency, Region 3, Chesapeake Bay Program, *Chesapeake Bay: A Profile of Environmental Change*, PB 84119197 (Philadelphia, PA: 1983).

Figure 17.—Concentration of Nickel and Hydrocarbons in Narragansett Bay, Rhode Island, in Relation to Distance From Discharge Points

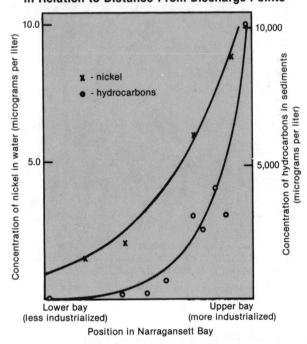

As in many other locations in the United States, concentrations of pollutants in Narragansett Bay tend to be greatest near discharge points, especially where many such discharges are concentrated in a highly industrialized or urbanized area. Note that one microgram equals one-millionth of a gram.

SOURCE: Save The Bay, Inc., *Down the Drain: Toxic Pollution and the Status of Pretreatment in Rhode Island* (Providence, RI: September 1986).

(the "surface microlayer"[5]), an ecologically important zone where the presence of pollutants may be particularly damaging. In the urbanized areas of Puget Sound, for example, the microlayer has been found to contain relatively high concentrations of some pollutants. One type of pollutant—polynuclear aromatic hydrocarbons (PAHs)—was present at concentrations which in many cases were acutely toxic to flatfish eggs in laboratory experiments. Scientists believe the pollutants are responsible in part for the lower quantities of flatfish eggs and other organisms found in the microlayer in the developed areas of the Sound (610).

The presence of pollutants in the water column is important in three respects. First, marine organisms may be affected by the direct exposure to con-

taminated water. In Puget Sound, for example, some samples of contaminated bottom water were found to cause sublethal toxic effects in some organisms (89). Evidence from the Chowan River in North Carolina suggests that herring have detected and avoided pulp mill effluent in the river, to the detriment of some of the river's fishermen (398). These direct impacts can also give rise to additional ecological repercussions.

Second, the pollutants may be transported to other locations and transferred to sediments or to the atmosphere, thereby increasing the chances of exposure to living organisms and further ecological impacts. Third, in addition to impacts on marine organisms, elevated levels of these pollutants may reach a point where human health is directly threatened.

Human Pathogens

Pathogens often are discharged from combined sewage overflows, municipal treatment plants, runoff, raw-sewage outfalls, and boats in marinas and elsewhere. As a result, high levels of fecal coliform bacteria in the water frequently create the need for government authorities around the country to restrict shellfish harvesting. High coliform levels also result in temporary or permanent beach closures, particularly along the north Atlantic coast (486). Beach contamination appears to be less common in other regions of the country, but complete information on the nationwide extent of beach closures is not readily available and trends are not clearly discernible.

In some areas—such as parts of Chesapeake Bay (205,335)—fecal coliform contamination is not as serious as it was 10 or 15 years ago. The improvements are usually the result of greater levels of sewage treatment. Conversely, such contamination has not declined and has actually worsened in other areas, particularly those experiencing high population growth and rapid development (221). In coastal Louisiana, for example, municipal sewage treatment capacity has failed to keep pace with growth and is unable to adequately treat wastes (315). Growing numbers of residences with septic systems and increasing numbers of small boats also pose problems in many coastal areas. These water quality threats are expected to increase in some areas over the next decade.

---

[5]The surface microlayer comprises a very thin layer at the upper portion of the water column (ranging in thickness from less than one-tenth of a millimeter to several centimeters) (see ch. 4).

Metals and Organic Chemicals

The contamination of waters with metals and organic chemicals also is common. Concentrations of some metals and organic chemicals (e.g., DDT) in the water column have declined in many areas over the past 15 to 20 years. This is often because emissions of these pollutants have been reduced substantially from specific point sources, usually because of curtailed production of the wastes or because of greater waste treatment by industrial and municipal entities. For example, discharges of most key metals and organic chemicals into New York Harbor declined during the 1970s and early 1980s (11).

But there also are areas where improvements have not occurred and where concentrations of specific pollutants have increased (220,394). This has been especially true where rapid residential, agricultural, and industrial growth has resulted in greater emissions from both point and nonpoint sources. For example, monitoring data from the lower St. Johns River, Florida, indicate that concentrations of waterborne toxic metals increased from 1970 through 1980 (164).

Impacts on Sediments

Sediments may be physically, chemically, or biologically altered by waste disposal activities and runoff. Physical alterations can occur when solids from pipeline discharges or dumping accumulate on the bottom. If this material differs substantially from the original sediment, then the substrate available to bottom-dwelling organisms can change significantly. In southern California, for example, the accumulation of solids discharged by ocean outfalls, in combination with other environmental changes associated with the discharges, has affected the distribution and abundance of benthic organisms over an area of approximately 170 km² (52,354).

Contamination of sediments with metals, organic chemicals (e.g., PCBs, other chlorinated hydrocarbons, and polynuclear aromatic hydrocarbons), and pathogens poses a particular problem. Contaminated sediments have been found around the country, and they are generally adjacent to industrial and urban areas where large volumes of contaminated material such as industrial wastes or municipal effluent have been discharged or dumped, or

in estuaries that receive substantial pollutant loads from upstream. Sediment contamination is most prevalent and severe in the estuaries and coastal areas of the Northeastern United States (figures 17 and 18). The character of sediment contamination varies widely, as do its origins and consequences.

In some cases, the consequences of such contamination are relatively apparent and serious, especially where there are extremely high concentrations of particularly toxic pollutants. Among such areas are portions of Puget Sound like Commencement Bay and Everett Harbor, the Southern California Bight, and several areas along the northern Atlantic coast like Buzzards Bay. Some of these areas have been classified as Superfund sites. Many other areas exhibit various mixes and concentra-

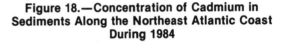

Figure 18.—Concentration of Cadmium in Sediments Along the Northeast Atlantic Coast During 1984

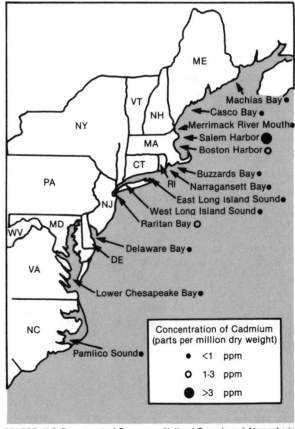

SOURCE: U.S. Department of Commerce, National Oceanic and Atmospheric Administration, Ocean Assessments Division, *Progress Report and Preliminary Assessment of the Findings of the 1984 Benthic Surveillance Project* (Rockville, MD: 1986).

tions of pollutants with specific subsequent effects on the biota.

The restrictions imposed on point sources over the past 15 to 20 years have reduced discharges of some metals, organic chemicals, and human pathogens, and helped limit sediment contamination. Once sediments are contaminated, however, the duration and consequences of contamination vary. For example, in time, the pollutants may break down into less harmful byproducts. Or subsequent sediment deposition may bury the pollutants and prevent further exposure to living organisms (unless the sediments are subsequently disturbed).

Despite some progress, serious problems from contaminated sediments will continue to persist. Although releases of some pollutants have been curtailed or reduced in some areas, in other instances a growing variety and quantity of pollutants continue to be released to the water column and make their way to the sediments. The sediments often act as a repository for such pollutants, holding them for days, years, decades, or even centuries. As long as the pollutants persist in a toxic form, contaminated sediments can continue to affect organisms.

IMPACTS ON ORGANISMS

Pollutants from waste disposal activities and other sources have affected marine organisms and ecosystems in many different ways. The impacts vary widely, from acute and lethal to minor, from extremely adverse to relatively beneficial. The geographic scale also varies, ranging from very small areas to many thousands of square miles (417).

Some organisms are especially vulnerable to waste disposal activities and pollutants. Among these are bottom-dwelling (benthic) organisms and those which spend all or part of their lives in coastal waters or estuaries. Organisms that inhabit polluted waters during sensitive life-stages are particularly susceptible to environmental perturbations.

Striped bass, for example, spend their early life-stages in or near estuaries and during that time are very sensitive to substances (e.g., copper, cadmium, and aluminum) contained in waste discharges. High mortality rates during these stages appear to be related to the presence of pollutants and other factors. The precipitous declines in striped bass stocks in recent years are thought to result in part from low survival rates during the first 60 days of life (218,219,625).

Birds and Mammals

Birds and mammals are affected by pollutants in several ways. For instance, they can be affected indirectly when pollutants alter their habitat or food supplies, as is the case with canvasback ducks in the Chesapeake Bay. Pollutants caused drastic declines in the Bay's seagrasses, including the ducks' preferred food—wild celery. This has contributed to a precipitous decline in the Bay's population of canvasback ducks (624) (figure 19).

The strongest evidence linking pollutants to impacts on birds and mammals occur when organisms ingest a pollutant or a metabolite of a pollutant. The most important pollutants are those—e.g., chlorinated hydrocarbons—that persist and tend to increase in concentration as they are transferred through the food web (i.e., "biomagnify") (ch. 4). Marine birds and mammals often feed at relatively high trophic levels and thus are particularly susceptible to biomagnification.

Some evidence exists to directly link pollutants discharged from point sources to elevated concentrations of pollutants (or body-burdens[6]) and adverse effects in birds and mammals (65,421,698). Although the full consequences of such contamination in birds and mammals are not always known, impacts such as reproductive impairments have been observed. A well-known example involved discharges of DDT into southern California waters, where elevated concentrations of organochlorine chemicals and subsequent population declines in brown pelicans and several other bird species were linked to DDT-contaminated fish (92,422,423,485).

---

[6]The body-burden of a contaminant is its concentration in an organism's body.

Figure 19.—Canvasback Duck Population on the Chesapeake Bay, 1954-85

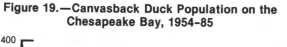
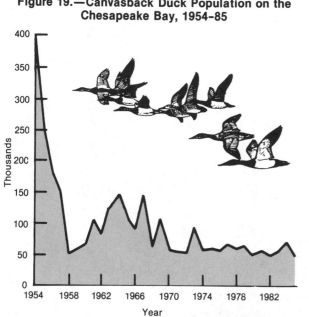

SOURCE: U.S. Department of the Interior, Fish and Wildlife Service, *The Canvasback* (Annapolis, MD: 1985).

Photo credit: Stephen C. Delaney

Photo credit: A.W. Smith, Naval Biomedical Research Laboratory

Reproductive problems in sea lions breeding in southern California have been associated with DDT, large quantities of which were dumped offshore or discharged from marine pipelines prior to the early 1970s. Pictured here is a female sea lion attempting to carry her prematurely delivered pup.

Photo credit: K.A. King, U.S. Fish and Wildlife Service

Elevated levels of selenium and othr metals have been found in waterbirds in Galveston Bay, Texas, sometimes at levels sufficient to impair reproduction. The birds pictured here are Laughing Gulls.

Reproductive problems in sea lions also have been associated with organochlorine pollutants (126,187). Other less dramatic, yet nevertheless significant, examples exist elsewhere in the country. For example, three waterbird species nesting in Galveston Bay, Texas, were found to contain elevated levels of several metals (including selenium), in some cases at levels associated with impaired reproduction (268).

Reduced emissions of some pollutants have led to noticeable improvements. In particular, the banning of DDT production and disposal helped reverse the decline in brown pelicans (5). Nevertheless, impacts related to pollutants continue to be documented in birds and mammals (162).

Acute Lethal Effects on Finfish and Shellfish

Given large enough quantities, some pollutants or combinations of pollutants will quickly kill finfish and shellfish. The mechanisms by which this occurs can vary from the depletion of oxygen (associated with discharges of nutrients and organic mat-

Fish kills are frequent and severe in the Gulf of Mexico and along the southern Atlantic Coast. They result from low oxygen levels (hypoxia) that are caused by various factors, including waste disposal and natural processes.

ter) to the crippling of an organism's nervous system by certain toxic organic chemicals. These effects are difficult to measure in the field because seriously debilitated or dead organisms quickly disappear.

One type of uniquely compelling evidence which does arise in the field occurs when large numbers of organisms are killed at once by pollutants. The occurrence of mass mortalities varies around the country. They are least frequent along the Pacific coast, but are more common and serious in the Northeast. However, the greatest problems exist in the Gulf of Mexico and along the southern Atlantic coast, where hypoxic conditions cause frequent fish kills (315,359,554).

The magnitude of the kills varies widely. The largest incidents can involve millions of fish. The species reported as killed are often commercially valued species. For example, 109 fish kills were reported in the State of Maryland in 1985; 97 percent of the estimated 4.6 million fish killed were menhaden, a very abundant and important commercial species. The majority of the kills investigated in Maryland—some involving hundreds of thousands of fish—occurred in estuarine waters in the Chesapeake Bay (452).

While the causes of fish kills are not always clearly understood, evidence suggests that waste disposal activities in many instances are often significant contributors. Most kills occur in estuaries and are caused by low levels of dissolved oxygen (hypoxia). Municipal sewage treatment plants appear to be important contributors to the hypoxic conditions that cause fish kills because they discharge nutrients or oxygen-demanding materials that lead to oxygen depletion. Industrial dischargers too are important, although to a lesser extent. In the Chesapeake Bay, discharges contribute large quantities of nutrients but the precise magnitude of their contribution to the fish kills is unknown (452). The extent and severity of fish kills have been reduced in many areas over the last 10 to 15 years, but information is not available to accurately judge the reduction (N. Harllee, U.S. EPA, pers. comm., March 1986).

High Levels of Pollutants in Finfish and Shellfish

Shellfish Contamination

The concentrations of coliform bacteria and natural marine biotoxins in shellfish have been periodically surveyed for many years. This information is now supplemented with information from "biomonitoring" surveys, which measure the concentrations of toxic chemicals (e.g., metals and organic chemicals) in shellfish.

Since 1966, the National Shellfish Register (ch. 7) has provided an important indicator of the extent to which shellfish in U.S. waters are contaminated with coliform bacteria. In 1985, the register showed that 58 percent of the "productive" shellfish areas in the United States were approved for harvest, while the rest were subject to some level of restriction (603) (table 7). **Commercial shellfish harvests from roughly one-third (27 to 42 percent) of the productive areas are limited because of actual or potential contamination.** Over 80 percent of the harvest-limited productive shellfish areas in the Nation are in the Gulf of Mexico and along the southern Atlantic coast. (603).

Although the register does not show a clear, overall national trend in shellfish contamination, it does, in combination with other evidence, indicate that bacterial contamination is a significant problem nationwide. Trends vary from one body of water to the next. In some areas, such as in the vicinity of Savannah, Georgia (9) or San Francisco Bay (69), shellfish contamination by fecal coliform has fallen to the point where shellfishing areas have been reopened for the first time in decades.

However, in other regions—particularly in rapidly developing areas such as the coastal portions

of the Gulf of Mexico and southern Atlantic States —the problem is growing (221,356,394). The contributing causes include both point sources (primarily municipal sewage treatment plants and combined sewage overflows, and growing numbers of recreational boats) and nonpoint sources (including runoff and groundwater seepage from increasingly developed and often unsewered coastal areas).

Shellfish contamination with metals and organic chemicals also has been surveyed by State and Federal authorities (71,512,537,597,621). These efforts have varied widely among programs and from year to year. In some areas—usually in marine waters adjacent to or downstream from urban, industrial,

N.J. DEPT. OF ENVIRONMENTAL PROTECTION

Photo credit: Office of Technology Assessment

Contamination of fish and shellfish with toxic pollutants sometimes compels local or State authorities to issue warnings or impose restrictions on the harvest, sale, or consumption of contaminated organisms. This sign has been posted on the tidal Passaic River in New Jersey, where dioxin contamination is so severe that the State forbids the sale or consumption of any fish or crabs from the waterway.

Table 7.—Classification of Shellfish Growing Waters (thousands of acres)

Region and State	Approved for harvest	Productive			Percent of total productive waters approved	Nonshellfish/ nonproductive	Total
		Harvest limited areas					
		Prohibited	Conditionally approved	Restricted			
Northern Atlantic:							
Maine	936	87	13	10	89	0	1,046
New Hampshire..	4	6	0	0	40	0	10
Massachusetts ..	255	41	1	5	84	500	802
Rhode Island	96	20	12	0	75	0	128
Connecticut	309	78	6	0	79	0	393
New York	828	192	1	0	81	0	1,021
New Jersey	236	118	20	21	60	0	395
Pennsylvania	0	0	0	0	—	6	6
Delaware........	209	19	3	0	90	44	275
Maryland........	1,369	64	0	0	96	97	1,530
Virginia	1,295	174	33	0	86	2	1,504
Subtotal	5,537	799	89	36	86	649	7,110
Southern Atlantic:							
North Carolina...	1,755	370	0	0	83	0	2,125
South Carolina...	200	72	9	0	71	0	281
Georgia.........	61	144	0	0	30	0	205
Florida..........	40	36	37	0	35	748	861
Subtotal	2,056	622	46	0	75	748	3,472
Gulf of Mexico:							
Florida..........	266	260	306	0	32	578	1,410
Alabama	74	103	195	0	20	2	374
Mississippi......	123	96	171	0	32	0	390
Louisiana	0	31	3,462	0	—	0	3,493
Texas...........	1,310	358	0	0	79	2	1,670
Subtotal	1,773	848	4,134	0	26	582	7,337
West coast:							
California	2	263	12	1	1	248	526
Oregon	14	14	0	12	35	44	84
Washington	147	49	45	0	61	1,795	2,036
Subtotal	163	326	57	13	29	2,087	2,646
U.S. total........	9,529	2,595	4,326	49	58	4,066	20,565

DEFINITIONS:

Productive: Any areas which are not classified "nonshellfish/nonproductive." At one time this category only contained areas which did or could produce shellfish (either naturally or aquaculturally) in quantities sufficient to justify commercial harvesting. As a result of changes in the classification system, however, there is an effort underway nationwide to classify all coastal waters within subcategories of this category; consequently, it includes areas which formerly were termed "nonshellfish/nonproductive."

Approved for harvest: Area surveyed and found free of hazardous concentrations of pathogenic organisms and/or pollution. Molluscan shellfish may be commercially harvested at any time.

Harvest limited:

 a) *Conditionally approved:* Area surveyed and shellfish are found to meet "approved" area requirements for only part of the year. Molluscan shellfish may be harvested only during periods when pollutant levels are deemed acceptable. The area may be closed for the balance of the year because of high pollutant levels or because the shellfish control authorities have failed to establish that "approved" area standards are being met during that period; such failure may result from various factors, including cutbacks in funding of classification activities.

 b) *Restricted:* Area surveyed and shellfish are found to be contaminated. Shellfish may be harvested but only can be marketed if they first are purified in a depuration facility or "relayed" to an approved area. In either case, the shellfish may be marketed once they are depurated (cleansed of pollutants).

 c) *Prohibited:* Area surveyed and closed due to hazardous levels of contamination; or area has not been surveyed at all. Molluscan shellfish may not be commercially harvested at any time.

Nonshellfish/nonproductive: At one time, if areas were determined to be inaccessible, or did not or could not produce shellfish (either naturally or aquaculturally) in quantities sufficient to justify commercial harvesting, waters were classified into this category. As a result of changes in the classification system, however, acreage in this category is now being transferred into subcategories of the "Productive" category. At present the "Nonshellfish/nonproductive" category accounts for less than 20 percent of total classified acreage.

SOURCE: U.S. Department of Commerce, National Oceanic and Atmospheric Administration; and Department of Health and Human Services, Food and Drug Administration, *1985 National Shellfish Register of Classified Estuarine Waters* (Washington, DC: December 1985).

or agricultural areas—elevated levels of metals and organic chemicals are frequently present in shellfish. These concentrations sometimes are high enough to adversely affect the shellfish and to threaten consuming organisms, including humans. In some cases this has prompted government warnings or restrictions on fishing or consumption (see ch. 6). The full national extent of contamination by metals and organic chemicals, and its consequences and trends, are not known.

Finfish Contamination

Only a limited number and variety of fish have been analyzed for specific pollutants. These data reveal that the level of measured contamination varies widely—geographically, among species, among individuals, and even in different tissues of a single contaminated specimen. Likewise, the origins of the contamination and its significance to the health of both humans and marine organisms varies. Finally, there are wide differences in trends; some contaminants are increasing in importance while others are declining (340).

Generally, contamination by metals and organic chemicals from point sources is most severe near urban and industrial centers and in estuaries downstream from such areas (figure 20). Contamination also has been detected at distant points in the open ocean, but little information is available on the level of contamination and its consequences (198).

Bottom-dwelling fish that spend a substantial portion of their lives in close proximity to contaminated sediments are the most seriously exposed and contaminated, as are other fish in the same food webs. Sole and other bottom-dwelling fish have been contaminated with metals and organic chemicals in many areas of the country, including Boston Harbor; Commencement Bay, Washington; Santa Monica Bay, California; and others. In most cases, contaminant levels do not pose a clear threat to the well-being of the fish or to consumers of such fish (other organisms or humans). The concentrations found thus far usually have been below the levels set by the Food and Drug Administration (FDA), which has set some standards to restrict consumption of contaminated fish by humans (178, 593).

There are, however, instances around the country where contamination levels have been sufficient to move officials to warn the public or restrict or prohibit the capture or sale of the fish. Many of the most serious and widely publicized problems have resulted from point source discharges that contain long-lasting toxic chemicals (e.g., DDT and PCBs) that accumulate in the tissues of fish (337, 463). For example, the capture and sale of striped bass in New York has been banned because of PCB contamination and signs have been posted warning against the eating of fish caught in Santa Monica Bay in southern California (194,340,577).

Other Effects on Finfish and Shellfish

In addition to acute lethality and elevated body-burdens of pollutants, individual finfish and shellfish also exhibit behavioral and physiological effects, and populations of these organisms exhibit changes in abundance and distribution. These effects may be negative, positive, or inconsequential from the human standpoint.[7] Although many effects are difficult to document, a growing body of evidence links these effects to exposure to pollutants that sometimes are present at very low concentrations or to environmental changes induced by pollutants (127). The effects are concentrated in estuaries and coastal waters, but detectable effects also have been found in fish far from shore in the open ocean (198,535).

The effects of pollutants on behavior are diverse. Some fish and shellfish will, if they can, avoid hypoxic bottom waters or waters containing various contaminants. Likewise, organisms living on or in sediments may avoid sediments that have been altered physically or chemically. Other aspects of their behavior may also change: for example, they may eat fewer or different organisms, be less active, or grow more slowly.

Pollution also has been linked with physiological and biochemical changes and diseases in fish.

---

[7]Positive impacts can result from a decline in pollutant inputs, or even from increases in the volumes of wastes. Pipeline discharges or dumping of organic matter and nutrients, for example, may increase productivity of marine waters. Some observers argue that this increased productivity has in some cases been beneficial (509,526).

Figure 20.—Concentrations of Polychlorinated Biphenyls (PCBs) in the Livers of Fish From Selected Sites

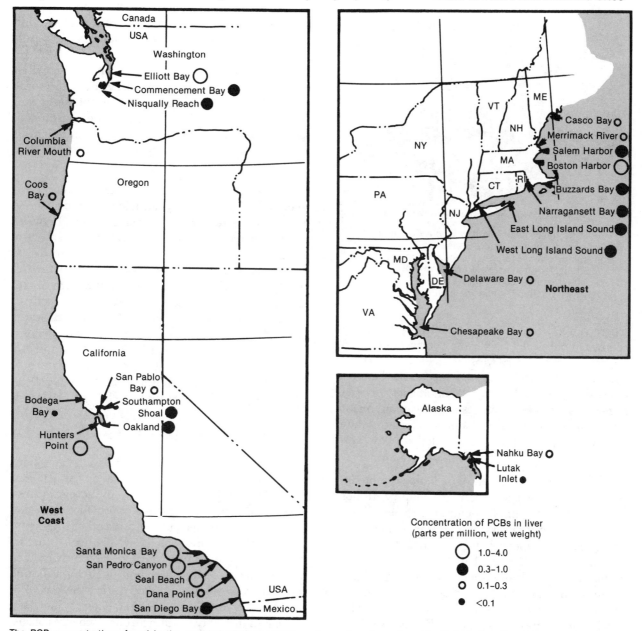

Concentration of PCBs in liver
(parts per million, wet weight)

○ 1.0–4.0
● 0.3–1.0
o 0.1–0.3
• <0.1

The PCB concentrations found in the sampled areas generally are lower than those found to biologically affect freshwater or saltwater organisms. In *some* situations, however, biological effects have been detected at the PCB concentrations found at these sites. PCBs may cause reproductive failures, birth defects, tumors, liver disorders, and skin lesions, and they may suppress the immune system.

SOURCE: U.S. Department of Commerce, National Oceanic and Atmospheric Administration, Ocean Assessments Division, *Progress Report and Preliminary Assessment of the Findings of the 1984 Benthic Surveillance Project* (Rockville, MD: 1986).

Photo credit: Southern California Coastal Water Research Project

Pollutants have been linked with physiological and biochemical changes and diseases in fish, ranging from minor effects to conspicuous pathological abnormalities. Pictured here are two Dover Sole from the coastal waters near Los Angeles; the top fish exhibits severe "fin erosion," while the lower one is normal.

These range from subtle and relatively minor responses to physically visible and conspicuous pathological abnormalities. Most noticeable are effects such as fin erosion (or fin rot), ulcers, shell disease or erosion, tumors, and skeletal anomalies. Affected organisms may be less resistant to infection, or suffer impaired growth or reproduction. Some of these effects, although not immediately lethal, may eventually precipitate an organism's death.

These kinds of effects have been documented in polluted marine waters around the country. In Boston Harbor, for example, pollution has been linked to fin erosion and cancerous lesions in winter flounder, a major commercial and recreational fish (373). In San Francisco Bay, evidence links pollutants and pathological problems—including impaired reproduction—in striped bass and starry flounder. In Puget Sound, various pathological conditions, most notably liver tumors, found in English sole and other fish are correlated with exposure to pollutants. Numerous other examples have been documented, as well (for example, see refs. 307,516, 544,621).

In some instances, especially in small areas or for limited periods of time, evidence links the disposal of wastes to changes in the abundance, distribution, or diversity of some fish and shellfish. These changes most frequently result from pollution-induced changes in various closely interrelated ecological parameters such as food supplies, water quality, and habitat. For example, pulp mill effluents discharged into the Fenholloway River estuary, located along Florida's Gulf coast, have been significant contributors to the decline in the extent and productivity of the area's seagrasses and some types of algae. Because those photosynthetic organisms are of central importance in the coastal ecosystem, their decline has had major repercussions on other populations and on community structure (307).

On longer time-scales and over larger areas, however, evidence is rarely sufficient to conclusively establish cause-and-effect relationships between changes in fish populations and waste disposal activities. Nor is evidence usually adequate to detect trends. Despite this, **considerable circumstantial evidence indicates that pollutants from waste disposal activities have contributed to declines of major fish populations in the United States** (529,610,640,691). For example, officials in eight Southeastern States along both the Gulf and Atlantic coasts believe that widespread declines of anadromous species in those States have been caused in part by pollutant discharges (492).

Submerged Aquatic Vegetation

Waste disposal activities have had substantial impacts on submerged aquatic vegetation (SAV). This is particularly important in view of the significance of SAV in marine ecosystems; it provides vital shelter and food sources and performs other important ecological functions such as stabilizing sediments. During the past century, the general trend has been toward decreases in the extent of SAV, although some increases have occurred during the last 10 or 15 years. A major cause for the declines has been increased turbidity resulting from discharges of suspended solids and from growing populations of plankton fostered by releases of nutrients.

Examples of vegetation loss exist around the country (17). In Florida, for example, seagrass meadows have suffered significant losses; indeed, seagrasses in some areas have been virtually wiped out since 1940 and the outlook for remaining seagrass beds in Florida is bleak (221,309). Perhaps the best known example in the United States is the

Chesapeake Bay, where the SAV has declined precipitously over the last 15 to 20 years (427,640).

Benthic Organisms

Changes in sediment or water quality induced by waste disposal often affect the benthic plants and animals that live on or near the bottom (including many fish, shellfish, and plants). Benthic communities have been affected by waste disposal in every region of the country. The impacts are most severe in and near estuaries that receive high inputs of pollutants from rivers, near developed coastal areas, and near dredged material disposal areas—particularly in the estuaries and coastal areas of the Northeast.

Problems related to waste disposal arise from sediment contamination, hypoxic bottom waters, increased turbidity, and physical changes in the sediment resulting from the settling of solids (e.g., from dredged material). The effects on benthic organisms vary from relatively rapid death to subtle effects on species diversity and numbers; the effects range from long term and permanent to short term and transitory. For example, a study of the disposal of fine-grained dredged material at a site in the Chesapeake Bay found that while many organisms were buried and consequently killed, the area apparently had recovered completely within 15 months (Harrison, 1976, cited in ref. 701). Conversely, far more serious impacts have been observed in the New York Bight. Among the effects caused at least in part by waste disposal have been mass mortalities of benthic organisms, large-scale and long-term contamination, diseases and abnormalities, changes in abundance and distribution of particular species, and changes in community structure (212,343,546,621).

As is the case with other organisms, trends pertaining to impacts of waste disposal on benthic communities vary from place to place. Some areas have improved since the early 1970s. For example, one study of the coastal shelf of Palos Verdes, California, between 1971 and 1981 showed that reduced emissions of pollutants (e.g., suspended solids, DDT, and PCBs) resulted in a reduction in the extent of observed benthic impacts (539). On the other hand, continued deterioration is being observed in other waters; for example, this is the case in the areas where shellfish contamination is increasingly prevalent.

GEOGRAPHICAL VARIATIONS IN IMPACTS

The extent of impacts on marine resources that are caused by waste disposal activities and nonpoint sources varies considerably among different waterbodies. Limited space precludes a detailed discussion of site-specific impacts, but generalizations can be made about the physical characteristics, degree of development, and types and extent of impacts that are exhibited in different regions of the country (figure 21).

Northern Pacific Coast

The Northern Pacific region includes the marine resources off the coasts of Alaska, Washington, and Oregon. This region contains more coastline and more stretches of relatively enclosed bodies of water (e.g., estuaries, bays, and sounds) than any other region, largely because of the size and shape of Alaska's coast. Much of the region, with some notable exceptions, is relatively free of conspicuous and serious marine impacts induced by waste disposal activities.

The region does have some major industrial development, including the forest products, seafood processing, petroleum refining, and chemical industries. The municipal and industrial effluent discharged into the region's coastal waters originates primarily from two areas: coastal areas around Puget Sound and inland areas along the rivers, particularly the Columbia River. The region's most severe impacts have occurred in Puget Sound.

The impacts of pollutants have been manifold in Puget Sound (313,463,483). The most severe problems occur in urban embayments. Many commercial shellfish beds have been closed because of fecal coliform contamination. In Commencement Bay, levels of toxic pollutants are high enough to

Figure 21.—The Five Coastal Regions of the United States Used in This Report

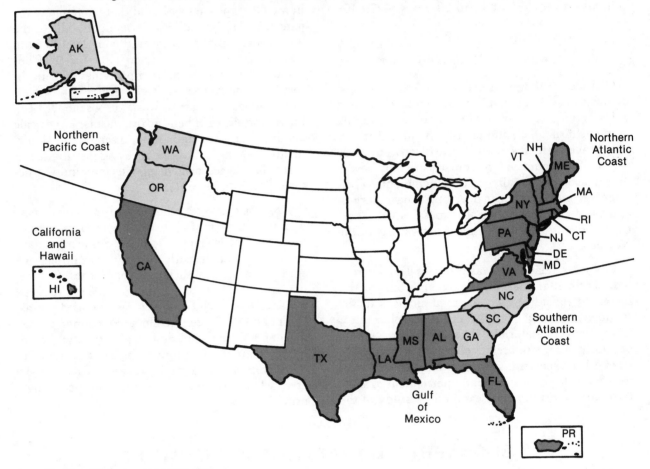

SOURCE: Office of Technology Assessment, 1987.

make it one of the most contaminated areas in the country and a "Superfund" site. Diseases and abnormalities, most notably liver tumors, have been detected in bottom-dwelling fish. These pathological conditions and other impacts have been linked with exposure to chemicals found in sediments, the water column, and food particles (89,326,351,376). The health of humans who consume large amounts of contaminated fish also may be endangered (293).

Outside of Puget Sound, waste disposal activities and pollutant inputs generally tend to be less intense and appear to cause less severe problems. Impacts are generally localized and poorly documented (425). They include contamination of shellfish with fecal coliform bacteria, decreased levels of oxygen near outfalls from mills and seafood processing plants, and effects from dumping of dredged material offshore of the Columbia River estuary.

California and Hawaii

The region that includes California and Hawaii differs from other regions in that relatively more of its coast are open, rather than enclosed in areas such as bays. In addition, its continental shelf is relatively narrow.

Relatively few impacts have been documented in Hawaii. Some problems, however, have been associated with the discharge of nutrients. These problems were alleviated during the 1970s and early 1980s as sewage treatment plants were built and upgraded, and as outfalls were extended into deeper, open waters. In Kaneohe Bay, for example, sewage discharges prior to the late 1970s had seriously degraded marine communities. In the late 1970s, the discharges were diverted to a deep-ocean outfall and conditions in the Bay improved considerably, with notable reductions in turbidity and a marked recovery in coral communities. In areas such as Pearl Harbor and Mamala Bay, however, treated sewage from dense urban populations continues to result in impacts (228,330,521).

Most of the urban and industrial development, and the associated waste discharges, are in California, most notably in the San Francisco Bay and along the coast of southern California. Municipal wastes from California's large and growing population are voluminous and contribute substantial quantities of many different pollutants. Industrial effluents, dominated in the coastal regions by the petroleum refining, metal finishing, and inorganic chemicals industries, also are sizable. Runoff from cities and from the State's extensive agricultural areas is also a major contributor of pollutants.

The State contains waters of widely varying quality, ranging from relatively pristine to some of the most polluted in North America (69,71,351). In particular, two areas are heavily affected by waste disposal activities—the San Francisco Bay and the southern California coast. Pollutant impacts also frequently occur in other localized areas.

San Francisco Bay is a large and enclosed estuary. Much of it is ringed by intense urban development, including San Francisco, Oakland, and San Jose, and industry, including major petroleum refineries. The Bay receives pollutants from these municipal and industrial sources and from the rivers that drain California's Central Valley. These pollutants, along with other factors, have significantly altered the Bay's ecosystem (395).

Some impacts have been markedly reduced since the early 1960s, largely because of the construction of waste treatment facilities. Serious problems persist, however, especially in shallow and poorly flushed portions of the Bay, because substantial volumes of pollutants continue to be discharged directly and indirectly into the Bay. Eutrophication and low concentrations of dissolved oxygen are localized problems. Large numbers of the organisms are exposed to elevated concentrations of various pollutants (e.g., pathogens, metals, PCBs, and DDT), and impacts to benthic organisms, fish, and birds continue to be documented (71,90,321,421, 493,529,691).

The marine waters of southern California support a wealth of marine resources that are of considerable value, including extensive commercial and recreational fisheries, numerous beaches, refuges, and sanctuaries. The great beds of giant kelp present along the open coast provide habitat for many valuable fish and shellfish and support a substantial kelp harvesting industry. Marine mammals and birds also are present and many breed in the area (24).

Juxtaposed to the waters and their resources is one of the continent's great urban concentrations. These urban areas discharge large volumes of wastes (mostly municipal) to marine waters, which exhibit elevated concentrations of many pollutants. Fecal coliform bacteria reach high concentrations near many ocean outfalls, but serious problems are confined primarily to an area near the Mexican border where high concentrations orginating from Mexico have compelled the closure of U.S. beaches and restrictions on shellfishing (356).[8] Elevated concentrations of chlorinated hydrocarbons and metals have been detected in various organisms (194, 423,499).

Shellfish, finfish, birds, mammals, and aquatic vegetation have all been affected (17,126,187,485). Fish, for instance, suffer liver abnormalities, fin erosion disease, and reproductive problems linked to pollutants (18,52,92,119,350,539). Kelp beds have undergone dramatic changes during the last 50 years (figure 22). In 1984, 108 km² of the benthic community around three of the area's major outfalls was changed or degraded. Although a substantial area, this is an improvement over the 163 km² that were changed or degraded in 1977 (53).

One issue of particular importance is the human consumption of fish and shellfish contaminated with toxic pollutants. Although DDT concentrations in fish and shellfish have fallen since discharges of DDT were curtailed in the early 1970s (340), residues are still high in some areas and some organ-

---

[8]The problem of bacterial contamination in U.S. waters close to the border should be greatly alleviated by a new sewage treatment plant in Tijuana, Mexico, which began operating in January 1987. Furthermore, under an international agreement, Tijuana will divert its sewage to treatment plants in San Diego when pollutants in discharges from the Tijuana plant would otherwise be moved by marine currents into U.S. waters.

Figure 22.—Relationship Between Size of Kelp Beds and Quantity of Solids From Municipal Discharges at a Southern California Site

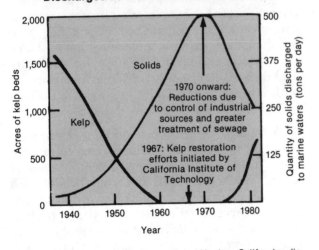

These kelp beds are located near Palos Verdes, California; discharges are from the Joint Water Pollution Control Plant, operated by the Los Angeles County Sanitation Districts.

SOURCE: After J.C. Meistrell and D.E. Montagne, "Waste Disposal in Southern California and Its Effects on the Rocky Subtidal Habitat," *The Effects of Waste Disposal on Kelp Communities*, W. Bascom (ed.) (La Jolla, CA: University of California Institute of Marine Resources, 1983).

isms. High concentrations of DDT and its metabolites, as well as PCBs, have been found in fish caught by southern California fishermen and sold in the area's fish markets; in shellfish sampled as part of California's Mussel Watch monitoring program (see ch. 7); and in the blood of recreational fishermen (577).

Because of apprehension over human exposure to these pollutants, especially DDT and its metabolites, commercial fishing has been prohibited around some outfalls. The State has established guidelines to reduce ingestion of contaminated fish and posted signs warning against consumption of fish caught in Santa Monica Bay.

Gulf of Mexico

The Gulf of Mexico, with its extremely productive habitats and wealth of sea life, is one of the most important marine environments in the United States (23). It also receives large amounts of pollutants. By far the largest volume of many pollutants is carried to the Gulf by the region's rivers, especially the Mississippi River. Many pollutants are generated from both point and nonpoint sources

in areas beyond the immediate coastal area. The Mississippi River, for example, carries wastes from the heavily developed Baton Rouge and New Orleans area, and from urban and rural areas deep in the Nation's interior (607).

Although small relative to the quantities of riverborne pollutants, considerable waste is discharged from municipal and industrial sources along the coast. Large amounts of dredged material also are dumped in the region's marine waters. The major industrial discharges along the coast are associated with refineries and the petrochemical industry, especially in Louisiana and Texas. The forest products and seafood processing industries are major contributors all along the Gulf coast. Despite the overall dominance of riverborne pollutants, in many areas these local sources substantially affect the quality of the marine environment.

Wastes from permitted discharges have been linked to a variety of ecological impacts (2,220,221, 315,359,554). The first problem is the depression of dissolved oxygen levels and accelerated eutrophication in areas close to shore (156,181,454,465,466, 703). Extensive hypoxia also has been documented in the waters further offshore, south of Louisiana, but the degree to which waste disposal contributes to the phenomenon has not been ascertained.

The second major problem is the contamination of waters with human fecal coliform. This happens in virtually every coastal State in the Gulf and appears to result primarily from nonpoint sources (e.g., contaminants from septic tanks are washed into estuaries and coastal waters by runoff). However, point sources, including municipal sewage plants, also contribute to the problem in some areas.

Other problems, often less evident, result from releases of metals, chlorinated hydrocarbons, and other chemicals. The impacts usually are localized, but they can be quite serious in highly developed areas and in waters where circulation is relatively poor—for example, in the Mississippi Sound (figure 23) and in Galveston Bay. Other affected areas occur throughout the Gulf (185,268,309,333,428).

Marine resources have been harmed by these types of pollutant discharges. Many of the region's shellfish beds are contaminated with fecal coliform. Vital beds of seagrasses are declining throughout the region (309). Fish and shellfish populations have

Figure 23.—Environmental Stress in the Pascagoula Area of the Mississippi Sound

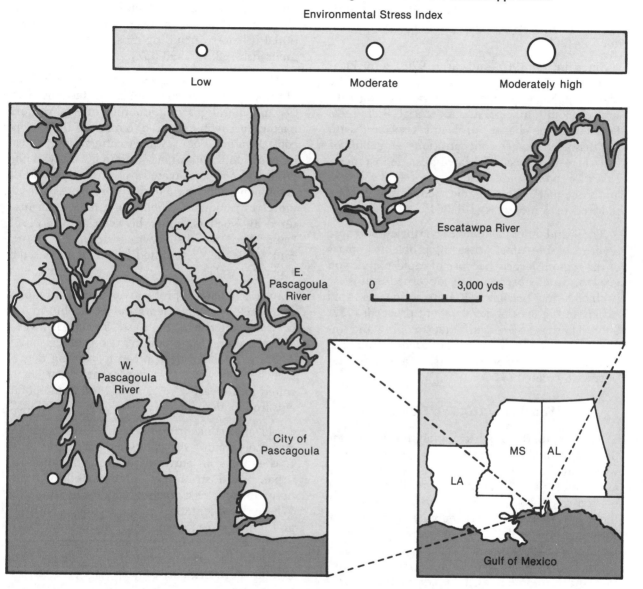

Environmental Stress Index

Low   Moderate   Moderately high

The Environmental Stress Index indicates the ecological risks associated with sediment contamination; a high value represents a higher risk of serious environmental stress. The index is a mathematical product of numerical ratings in four categories:

1. toxicity of sediments to selected organisms under laboratory conditions;
2. how readily sediment settles to the bottom after disturbance;
3. likelihood of sediment disturbance (e.g., from boat traffic or dredging); and
4. vulnerability of organisms to toxic substances (including factors such as ecological importance of indigenous species, life stages present, species diversity, mobility, and others).

SOURCE: Adapted from T.F. Lytle and J.S. Lytle, *Pollutant Transport in Mississippi Sound* (Ocean Springs, MS: Mississippi-Alabama Sea Grant Consortium, 1985).

been declining in some areas (476,492), and birds have been found with elevated levels of contaminants—sometimes at levels that may impair reproduction (268).

Since the late 1960s and early 1970s, some locations along the Gulf have significantly reduced the release of nutrients and oxygen-demanding substances from municipal and industrial discharges. Reflecting this change, problems associated with low dissolved oxygen concentrations—e.g., major fish kills—have been alleviated in some instances. Likewise, point source discharges of fecal coliform and some metals and organic chemicals have been reduced in some areas (137,364).

These and other pollutants nevertheless still pose severe and sometimes worsening problems in parts of the region, largely because of rapid population growth. Some observers are concerned that these problems may become still more serious as rapid urban and industrial development proceeds (137, 308). There is also concern that dredging and impacts associated with disposal of dredged material may increase dramatically as several deep-water ports are created (314).

Southern Atlantic Coast

The Southern Atlantic region bordering the Southeastern U.S. coast has an irregular shoreline, with many bays, drowned river valleys,[9] wetlands, and islands (25). Municipal waste discharges are concentrated along the relatively small portion of the coast that is densely populated, largely in Florida. Industrial effluent in the region is dominated by the forest products industry, which is scattered along the coast. Nonpoint sources are important and are sometimes the predominant sources of pollution. They contribute significant amounts of pollutants directly to marine waters and indirectly through rivers and streams.

Waste discharges have been linked to various impacts in the region (182,220,221,276,399), including increased levels of nutrients and fecal coliform and reduced concentrations of dissolved oxygen (13, 534,703). The resultant hypoxic and eutrophic con-

ditions have been associated with fish kills, depressed populations of benthic organisms, fish diseases, and the decline of commercial and recreational fisheries, including those based on anadromous fish (42,397,398,552).

Although some localized reductions in releases of fecal coliform in some instances have occurred (9), in general growing amounts of fecal coliform bacteria are being released to many of the region's coastal waters. Poor sewage treatment and increasingly serious contamination from a variety of nonpoint sources, aggravated by extremely rapid development, contribute to the problem. As a result, more restrictions on shellfishing have been instituted. In North Carolina, between 1980 and 1985, "approved" shellfishing areas declined by 1 percent, while the "prohibited" acreage increased by 4 percent (603).

In selected areas, problems associated with point source releases of nutrients and fecal coliform have been alleviated over the past 15 years as a result of Federal and State pollution control legislation. These gains have been offset to varying degrees elsewhere, however, by development that has resulted in increasing releases of these same pollutants from both point and nonpoint sources (13,552).

Elevated concentrations of metals and organic chemicals occur in the waters and sediments of some coastal areas, in particular those with substantial urban or industrial development. Both point and nonpoint sources contribute to the contamination. Where documentation exists, trends in emissions and impacts are mixed. Some areas show considerable reductions in the concentrations of metals and synthetic organic chemicals, while elsewhere increases are evident.

Pollutants may be linked to several important changes in fish populations, particularly anadromous fish populations in many of the region's river systems (492). In North Carolina over the last decade, for example, pollutants from point sources may be partly responsible for the decline of commercial fisheries relying on striped bass and herring. Generally, however, it is difficult to link specific pollutants and declines in fish populations. The relatively high incidences of "ulcerative mycosis" in some fish, a disease characterized by skin ulcers, may be linked to pollutants but a clear explana-

---

[9]These are valleys that cut through coastal lands when the shoreline extended further out than it does today. These river valleys have since been "drowned" by higher marine waters.

tion for the disorder has yet to be found (36, 135,396).

Northern Atlantic Coast

The Northern Atlantic coast, running from the North Carolina/Virginia border to Canada, contains many major bays, estuaries, and shallow coastal areas, and is graced with remarkably rich marine resources. It also is the location of extensive agricultural, urban, and industrial development which has occurred for several centuries. Consequently, marine ecosystems in many parts of the region are polluted and degraded, sometimes severely (621).

The problems in these estuaries and coastal waters (e.g., Chesapeake Bay, New York Bight, Long Island Sound, Narragansett Bay, Boston Harbor, and Buzzard's Bay) have been extensively studied (7,343,373,512,640,688). Impacts in the Chesapeake Bay, New York Bight, and Deepwater Disposal Sites are described in chapter 1, and details of impacts in other waterbodies are available from the U.S. Congress (588). This section describes the nature and extent of impacts in these waters in more general terms.

Municipal and industrial discharges, plus dumped sewage sludge and dredged material, are important sources of pollutants in this region; their quantity and composition vary from place to place. Pollutants from these waste disposal activities have been associated with various impacts. Some improvements have occurred since the late 1960s and early 1970s, but deterioration has occurred in other cases (108,335,336,347,394,480,486,597,609,681).

Eutrophication and hypoxia appear to be the most pervasive and serious consequences of pollution in the region. These problems occur in many estuaries and bays, and over wide, shallow areas of the continental shelf that are often quite distant from the original sources of the nutrients and organic material. Such impacts are most severe south of Connecticut. Natural changes in water quality along with inputs of nutrients and organic material from numerous sources (e.g., municipal and industrial effluents, runoff, raw sewage, dredged material, combined sewer overflows) all contribute to the problems (302,347,419,694).

Contaminated water and sediments are common throughout the region. Bacterial contamination of the water, particularly from raw sewage in combined sewer overflows, has sometimes closed beaches, in most cases temporarily but sometimes permanently (77,199,302). Sediments in many areas contain elevated concentrations of pathogens, metals, and organic chemicals (512). Among the most seriously contaminated sediments are those in the James River estuary, the Patapsco River around Baltimore, the Hudson River estuary, Raritan Bay and the New York Bight, New Bedford Harbor, and Boston Harbor (57,239,640,687).

Many impacts on marine organisms have been linked, with varying degrees of certainty, to waste discharges. These include major kills of fish and benthic organisms (452), increased incidence of disease and abnormalities, declines in major fisheries, and changes in community structure (407,495,711, 713). For example, in Boston Harbor, fin erosion and cancerous lesions have been found in winter flounder, a major commercial and recreational species.

Commercial harvesting is limited in 14 percent of productive shellfish areas, mostly because the shellfish contain high concentrations of bacteria. Over one-half of the shellfish beds in Boston Harbor are closed, at an estimated annual loss of $4 million. The size of the areas in the region in which harvesting is limited has been slowly increasing over the past 5 years, although there are localized exceptions to the trend (277,603).

In addition to being contaminated with pathogens, many fish and shellfish also contain elevated concentrations of other pollutants, especially metals and hydrocarbons. As with pathogens, these concentrations have sometimes been high enough to prompt officials to restrict fishing or harvesting.

Some of the most extensive and serious instances of contamination are associated with large releases of PCBs by industrial manufacturers into the Hudson River between 1950 and 1976 and into New Bedford Harbor (in Buzzard's Bay, Massachusetts) from 1947 to 1977 (271,339). This has caused widespread contamination of some fish and shellfish, and diseases and abnormalities in some organisms (76, 475). Fishing and the sale of contaminated organisms is restricted over wide areas. For instance, in

New Bedford Harbor, a total of 18,000 acres were closed to lobstering.

Impacts to submerged aquatic vegetation (SAV) and birds also have been severe in some cases. The most important impact to aquatic vegetation in the region apparently results from the large-scale introduction of nutrients in coastal estuaries. This has had major repercussions on aquatic organisms of all kinds, including valuable fish and waterfowl (427,640). Some birds have exhibited elevated concentrations of pollutants as a result of ingesting contaminated organisms (162). In some cases these high concentrations led to reproductive failures and population declines—most notably, in the fish-eating osprey that once was common throughout the region (698). Since restrictions on the production and use of DDT were imposed in the early 1970s, many parts of the region have witnessed a dramatic increase in osprey populations.

WARNING
Eating fish caught in Santa Monica Bay may be harmful to your health because of chemical contamination. You should not eat the fish called White Croaker, King Fish or Tom Cod.

FOR MORE INFORMATION, CONTACT THE SANTA MONICA HARBOR OFFICE 458-8694

ATENCION
Comer Pescado de la Bahia de Santa Monica puede ser dañino a su salud a causa de contaminacion quimica. Evite comer los pescados Whitecroaker, Kingfisher o Tomcod.

MAS INFORMACION SOBRE ESTOS RIESGOS SE ENCUENTRAN EN SANTA MONICA HARBOR OFFICE 458-8694

Photo credit: J. Jocoy, L.A. Weekly

Chapter 6
Impacts of Waste Pollutants on Human Health

CONTENTS

Tables

Figures

Boxes

Impacts of Waste Pollutants on Human Health

INTRODUCTION

Many substances found in wastes disposed of in marine environments have the potential to produce a variety of acute and chronic human health effects. The likelihood of human exposure to these pollutants depends on their physical, chemical, and biological form; concentration; and persistence or survival. The character of the disposal environment, method of disposal, and nature of environmental pathways leading to human exposure are also important variables.

There are two major contaminant pathways to humans. Contaminants can pass *directly* from contaminated media (water or air) to humans, typically by uptake through the skin or lungs, or through ingestion (e.g., swallowing water). Alternately, they can reach humans *indirectly* by ingestion of plants or animals that have taken up these substances directly or indirectly from contaminated environments. The relative significance of these two kinds of pathways varies for different pollutants.

Indirect exposure from water can be a significant route of exposure to toxic organic chemicals and metals because many of these substances have a capacity to persist in the environment, to concentrate in particular parts of the environment, and to bioaccumulate in certain plants and animals. Direct human exposure to toxic organic chemicals and metals from water is generally less significant because these substances are typically present in water at relatively low concentrations.[2] For pathogens,

both direct and indirect exposure can be significant because only small numbers of microorganisms are required to induce disease, and because microorganisms can reproduce in the environment or in infected animals.

Another general distinction can be made between different types of pollutants based on the extent to which they persist or are degraded in the environment. "Conservative" and "persistent" pollutants (e.g., toxic metals, PCBs) are broken down slowly or not at all, while "nonconservative" and "labile" pollutants (e.g., biodegradable or volatile organic chemicals) are rapidly rendered harmless or are removed from the system by environmental processes. Most of the problem substances in marine environments are in the former categories: their environmental persistence enhances their potential to reach and affect humans. Microbial contaminants of both types exist as well; some microbes are killed or inactivated by environmental processes, while others can actually proliferate.

Human health impacts can be associated with three major categories of pollutants that are present in wastes disposed of in marine environments, and that have the potential to reach and cause adverse impacts in humans. These categories are:

1. **toxic metals**: arsenic, cadmium, lead, and mercury;
2. **synthetic organic chemicals**: polycyclic aromatic hydrocarbons (PAHs), chlorinated hydrocarbons, and "specialized" chemicals (polychlorinated biphynyls (PCBs), pesticides, dioxins); and
3. **human pathogens**: viruses, bacteria, fungi, and parasites.

---

[1]Much of the information presented in this chapter is derived from extensive analyses contained in two contract reports prepared for OTA (205,409).

[2]Where high concentrations are present, direct exposure of humans to such substances in water can cause significant impacts. Such exposures might be experienced, for example, by bathers swimming at a beach in the immediate vicinity of an industrial discharge or by divers working in highly polluted waters.

LIMITATIONS OF HUMAN HEALTH EFFECTS DATA

Major impediments exist that limit a thorough evaluation of the human health impacts caused by waste-borne pollutants. Information is lacking on most of the individual substances detected in wastes and there is often uncertainty or controversy surrounding those for which there are data. In addition, it is difficult to accurately predict the behavior of a substance based only on data for chemically similar substances. An accurate assessment is also hindered by the complexity of most wastes, which are typically contaminated by a mixture of potentially harmful substances. Existing data typically allow at most an assessment of *acute, short-term* impacts associated with a particular waste; it is rarely possible to adequately assess *chronic, long-term* effects.

Moreover, studies of the environmental or ecological impacts of individual chemicals introduced into the environment through waste disposal rarely measure or sufficiently consider the potential *human health effects*. In fact, the human health risks associated with exposures to the vast majority—90 percent or more—of all chemicals found in different wastes are unknown (386). Knowledge about the human health effects of environmental pollutants generally comes from studies of only a few toxic compounds, such as mercury, cadmium, and PCBs (409); and these effects are often recognized only after large-scale occupational exposure, industrial accidents, or massive discharges of waste into the environment.

The accumulated knowledge in the field of toxicology is based primarily on experimental work with laboratory mammals. Thus, estimating human health risks from the disposal of chemical compounds in marine environments is generally based on extrapolating information on the toxicity of specific chemicals to laboratory mammals to the concentrations observed in marine organisms or humans. The estimates must also attempt to account for the persistence of such substances in the environment and their tendency to bioaccumulate or to biomagnify in marine organisms that might be consumed by humans. Public health information on human pathogens is largely derived from investigations of past incidents and, more rarely, on prospective epidemiological or clinical efforts.

TOXIC METALS

General Characteristics

Metals are chemical elements and as such cannot be destroyed or broken down through treatment or environmental degradation. However, a number of environmental processes—both chemical and biological—can alter the mobility and bioavailability of metals.

In general, toxic metals (including arsenic, beryllium, cadmium, chromium, cobalt, copper, lead, mercury, molybdenum, nickel, selenium, and zinc) are of potential concern whether they are found in wastes that will be disposed of on land or in the ocean. With respect to human health impacts arising from disposal of wastes in the marine environment, four metals are of primary concern: arsenic, cadmium, lead, and mercury. These are particularly important because of their known toxicity to humans and their presence in relatively high concentrations in wastes disposed of in estuaries and coastal waters. Metals of secondary concern include chromium, copper, and selenium. Other toxic metals are present in much lower concentrations both in wastes and in regions of the marine environment that are likely to lead to human exposure (409).

In marine environments, most dissolved metals are rapidly adsorbed and remain bound to particulate material that eventually settles out of the water column and is incorporated into sediments. However, some metals (e.g., cadmium, copper, nickel, and zinc) can be slowly—over a period of months or years—released from the sediment's oxidized surface layer and sublayers (124). In addition, metals can be released through other chemical and biological changes or processes, for example, by changes in salinity that commonly occur in estuarine waters. The action of microorganisms can also mobilize metals: for example, bacteria in sediments can convert slightly toxic inorganic mercury to the highly toxic and volatile methyl mercury (26). Finally, the burrowing of animals into sediments (bioturbation) and physical processes such as storms can release metals from sediments through oxidation and direct physical resuspension (709).

Pathways to Humans

Indirect pathways to humans vary with the particular environment and method used for disposal. **In marine environments, consumption of contaminated seafood is generally the major route of human exposure to metals.** *Direct* human exposure to metals is usually less important because they generally attain very low concentrations in the water column (409). However direct exposure resulting from volatization of certain metals (e.g., mercury, arsenic, and lead) due to their methylation by microorganisms, or direct exposure to marine waters with high concentrations of metals, would be a concern.

Bioaccumulation and biomagnification[3] are important processes that largely determine the potential for indirect human exposure to toxic metals and organic chemicals that result from marine waste disposal. Marine organisms, especially benthic organisms, can bioaccumulate metals by filtering water during feeding or swimming, ingesting particulate matter onto which such substances are adsorbed, or ingesting other contaminated organisms (46,232,260). Bioaccumulation generally increases as the degree of water or sediment contamination increases, but it varies considerably among metals, species of marine organisms, and types of sediment.

Biomagnification of a metal can result in the stepwise increase in an organism's tissue concentration of several orders of magnitude or more, and hence represents a major potential pathway for human exposure. According to available evidence, most toxic metals do not biomagnify into higher trophic levels of the marine food chain (33,260). However,

methyl mercury, and perhaps selenium[4] and zinc, are important exceptions to this rule.

Of the metals of primary concern, cadmium and mercury have significant potential for transport to humans through consumption of contaminated seafood (table 8). Arsenic is largely converted to nontoxic forms by marine organisms (129), and neither arsenic nor lead have been shown to accumulate significantly in seafood (409).

Even when bioaccumulation is not a factor, significant quantities of metals can concentrate in the gut or gills of marine organisms without actual absorption into the tissues. This is especially true for shellfish that filter large quantities of seawater and ingest solid matter during feeding (e.g., oysters, clams, mussels). Because people generally eat these organisms in their entirety, toxic substances can be passed to humans even in the absence of any actual bioaccumulation. This mechanism probably accounts for most instances of shellfish contamination involving metals that do not bioaccumulate.[5]

Potential and Actual Human Health Impacts

Toxic metals are capable of inducing a variety of human health effects—lethal and sublethal, acute and chronic. Some of the known properties and effects of exposure to the metals of primary concern in marine environments are summarized in table 8.

Acute environmental effects attributable to toxic metals may occur if concentrations are sufficiently high. Although documentation of human poisoning from consuming seafood contaminated with toxic metals is uncommon, there have been several well-known, catastrophic events. In Minamata, Japan, for example, over 100 people died and 700 others suffered severe, permanent neurological damage after consuming shellfish contaminated by industrial discharges of methyl mercury into Minamata Bay (128). A similar, though less severe, event occurred in Niigata, Japan in 1965 (550).

---

[3]In this context, the term bioaccumulation refers to the process whereby a substance enters an aquatic organism, either directly from the water through gills or epithelial tissue, or indirectly through consumption of other organisms. Biomagnification refers to the resultant process whereby tissue concentrations of bioaccumulated substances increase as the material is passed up through more than one trophic level in the food chain (260).

---

[4]High levels of selenium have been detected in the tissues of at least two species of marine birds: Texas laughing gulls in Galveston Bay (268), and California scoters in San Francisco Bay (424).

[5]This phenomenon applies equally to organic chemicals and pathogens, as discussed below.

Table 8.—Properties and Effects of Metals of Primary Concern in Marine Environments

	Arsenic	Cadmium	Lead	Mercury
Bioaccumulation	Low except in some fish species	Moderate	Low or none	Significant (methylated form)
Biomagnification	Low or none	Low or none	Low or none	Significant (methylated form)
Properties	Metallic form: insoluble Readily methylated by sediment bacteria to become highly soluble, but low in toxicity	Metallic form: relatively soluble Not subject to biomethylation Less bioavailable in marine than in fresh water Long biological residence time Synergistic effects with lead	Generally insoluble Adsorption rate age-dependent, 4 to 5 times higher in children than adults Synergistic effects with cadmium	Metallic form: relatively insoluble Readily methylated by sediment bacteria to become more soluble, bioavailable, persistent, and highly toxic
Major environmental sink	Sediments	Sediments	Sediments	Sediments
Major routes of human exposure:				
Marine environments	Seafood: very minor route, except for some fish species	Seafood contributes ≈10% of total for general population	Seafood comparable to other food sources	Seafood is primary source of human exposure
Other environments	Inhalation: the major route	Food, primarily grains	Diet and drinking water	Terrestrial pathways are minor sources in comparison
Health effects	Acute: gastrointestinal hemorrhage; loss of blood pressure; coma and death in extreme cases Chronic: liver and peripheral nerve damage; possibly skin and lung cancer	Emphysema and other lung damage; anemia; kidney, pancreatic, and liver impairment; bone damage; animal (and suspected human) carcinogen and mutagen	Acute: gastrointestinal disorders Chronic: anemia; neurological and blood disorders; kidney dysfunction; joint impairments; male/female reproductive effects; teratogenic	Kidney dysfunction; neurological disease; skin lesions; respiratory impairment; eye damage; animal teratogen and carcinogen
References[a]	Doull, et al., 1980 Harrington, et al., 1978 O'Connor and Kneip, 1986 Woolson, 1983	Chapman, et al., 1968 Nriagu, 1981 O'Connor and Kneip, 1986 Wiedow, et al., 1982	Callahan, et al., 1979 Heltz, et al., 1975 Kneip, 1983 NAS, 1980 O'Connor and Kneip, 1986 O'Connor and Rachlin, 1982	Grieg, et al., 1979 Kay, 1984 Nriagu, 1979 Windom and Kendall, 1979

[a]See list of references at end of report.

SOURCE: Office of Technology Assessment, 1987.

Other events involving metal contamination of food sources (though not seafood) have occurred. These include dietary transport of lead and cadmium from the application of sewage sludge to agricultural lands (470,471); and the pollution of Japanese rice paddies by industrial cadmium discharges, which resulted in 60 deaths (404).

Substantial laboratory evidence documents the potential for lethal and sublethal *chronic* effects to result from exposure to metals contributed by waste disposal activities. However, our capacity to detect chronic impacts in the field is limited, and this is in part responsible for the paucity of data documenting such human health effects.

Mercury is of special concern because it is easily taken up by humans through a diversity of exposure routes, including inhalation, ingestion, and through the skin. The source of mercury responsible for contamination of marine organisms differs among the various marine waters. In the open ocean, natural sources of mercury predominate, whereas in some estuaries and coastal waters, inputs from land-based sources are commonly the major source of contamination. In fact, marine organisms from polluted estuaries typically contain five or more times as much mercury as marine organisms from relatively clean estuaries (702).

Ingestion of contaminated shellfish and finfish, especially long-lived predatory fish such as tuna, is the primary route of human exposure to mercury (202,702). Again, however, in open-ocean fish mercury is generally attributable to natural sources, while shellfish contamination is typically more closely correlated with waste disposal.

Cadmium is used extensively in metal electroplating and found in a wide variety of wastes. It is more mobile than most other toxic metals. Large-scale cadmium contamination of estuarine waters by industrial and municipal wastewater discharges has occurred in some urban areas, for instance in the Hudson River estuary, and this has caused significant contamination of marine food chains (272,409). For example, studies show that

shellfish harvested near an industrial outfall formerly used to discharge cadmium wastes contained levels high enough to induce acute cadmium poisoning in an unwary consumer or urban fisherman (refs. 229,696; also see box N later in this chapter). Other studies of the entire Hudson estuary found that moderate consumption of shellfish could lead to exposure exceeding recommended safe levels (411).

The actual mechanism by which cadmium is transported through the marine food chain is controversial. Some investigators have found that cadmium present in sediments of some urban embayments is readily taken up by organisms, and that contaminated sediments represent the most likely source of human exposure (229). However, others suggest that sediments are not a significant contributor of cadmium even to contaminated organisms harvested from marine waters near urban-industrial areas (491).

TOXIC ORGANIC CHEMICALS

General Considerations

Some 65,000 chemical compounds are used in industry worldwide (386). Of these, approximately 10,000 are used regularly in one or more industrial processes; about 1,000 new chemicals are introduced into commerce each year (231). Clearly, individual evaluation of the health effects of each of these compounds is a hopeless enterprise. For several broad categories of industrial chemicals, figure 24 shows the paucity of data available on which to base health hazard evaluations.

Organic chemicals vary considerably with respect to their behavior in natural environments. Some chemicals accumulate in organisms; others do not. Some decompose rapidly if exposed to light, heat, or water; others are highly persistent. Some compounds may be metabolized by organisms into other compounds that may be more or less toxic

than the original compound; others resist biodegradation. Toxicities may vary from one organism to another, and toxic levels may also be affected by the presence of other compounds, producing synergistic or antagonistic effects.

Given this complexity, it is essential to invoke some system to simplify classification if a health hazard evaluation is to become manageable. One approach is to classify compounds according to how they behave in the environment, that is, on the basis of environmental fate. This approach can be used to reduce the size of the chemical universe down to those substances that have a potential to *reach* humans; information on human health effects would then only need to be developed for this subset.

Several relatively well-understood relationships exist between chemical properties of organic compounds and their fate in natural environments:

---

[6]Organic chemicals possess a molecular skeleton made of carbon and hydrogen and generally contain relatively few other elements, such as oxygen, nitrogen, or chlorine. Chlorine-containing organic compounds are referred to as ''chlorinated.'' Metals or metal-containing substances are usually termed ''inorganic,'' although in some cases metals can be chemically bound to organic compounds (e.g., methyl mercury).

Chemical Property:	*Related to:*
Solubility in water	Adsorption, route of absorption, mobility
Vapor pressure (volatility)	Atmospheric mobility, environmental persistence
Solubility in animal tissue	Bioaccumulation potential, adsorption by organic matter or sediments, persistence in organisms

Figure 24.—Ability To Conduct Health-Hazard Assessment of Substances in Seven Categories of Chemicals in Use

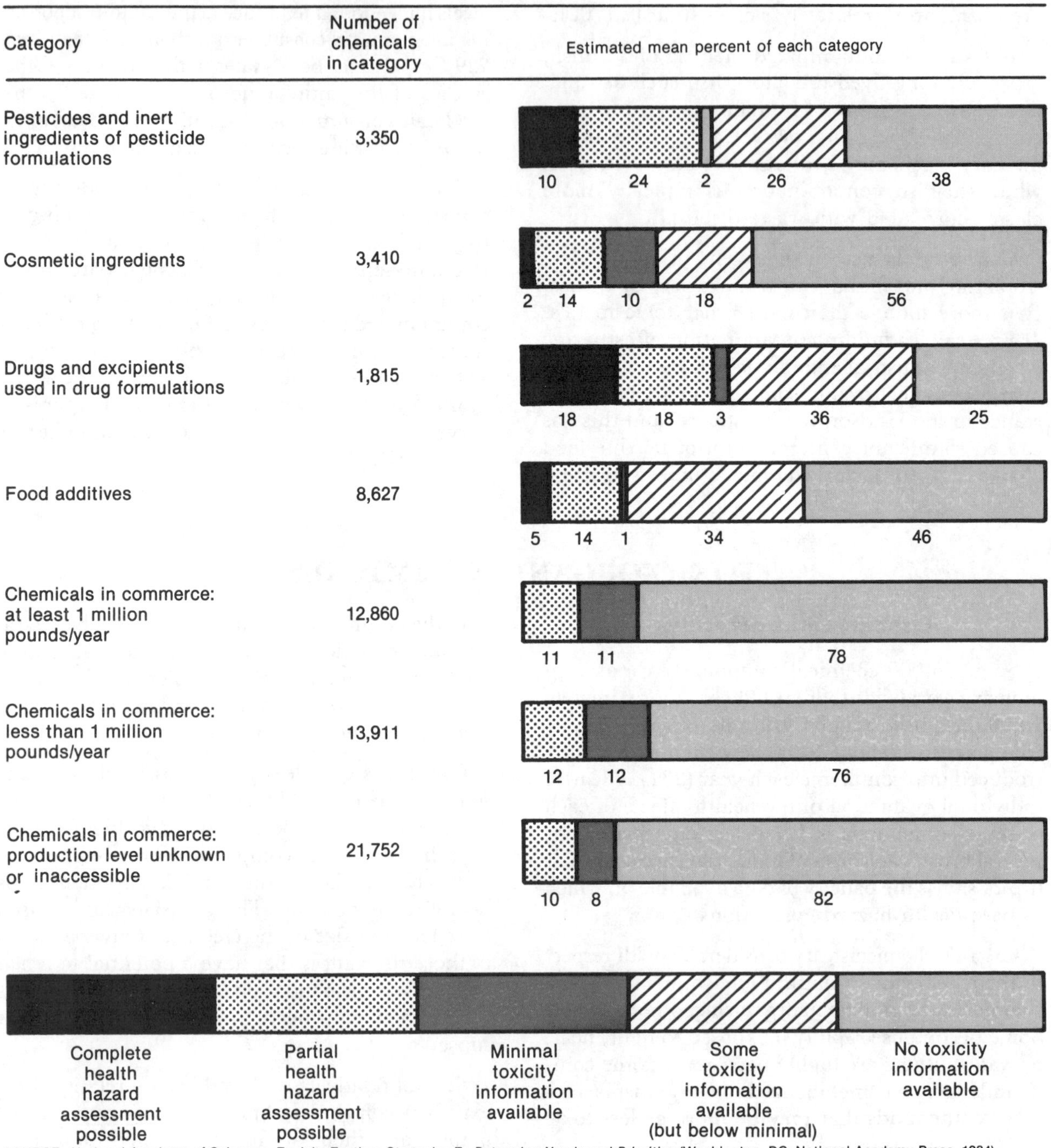

Category	Number of chemicals in category	Estimated mean percent of each category
Pesticides and inert ingredients of pesticide formulations	3,350	10 24 2 26 38
Cosmetic ingredients	3,410	2 14 10 18 56
Drugs and excipients used in drug formulations	1,815	18 18 3 36 25
Food additives	8,627	5 14 1 34 46
Chemicals in commerce: at least 1 million pounds/year	12,860	11 11 78
Chemicals in commerce: less than 1 million pounds/year	13,911	12 12 76
Chemicals in commerce: production level unknown or inaccessible	21,752	10 8 82

Complete health hazard assessment possible   Partial health hazard assessment possible   Minimal toxicity information available   Some toxicity information available (but below minimal)   No toxicity information available

SOURCE: National Academy of Sciences, *Toxicity Testing: Strategies To Determine Needs and Priorities* (Washington, DC: National Academy Press, 1984).

Using a scheme based on determinants of environmental fate such as these, it is possible to identify, for example, a class of compounds with low water solubility and high tissue solubility (e.g., PCBs, dioxins) that is likely to persist for long periods in the environment and bioaccumulate in organisms. Another class with the characteristics of high solubility and high volatility (e.g., chloroform, toluene) is likely to have low potential for accumulation in the food chain, but may pose a hazard to humans from inhalation or ingestion in drinking water. These considerations form the basis of the classification of organic chemicals used in this section (table 9).

Pathways to Humans

The most serious risks to human health are generally posed by those organic chemicals that are toxic, persistent, and have some means of reaching humans (i.e., present in significant concentrations in what has been termed the human exposure zone). As was the case for metals, humans can be exposed to organic chemicals either *directly* from the water (by absorption through the skin, by drinking contaminated water, or by inhaling contaminated air), or *indirectly* through ingestion of contaminated plants or animals.

In marine environments, many organic chemicals partition into sediments or onto the water's upper surface (the surface microlayer), where they are potentially available to marine plants and animals; this can provide an indirect pathway to humans through ingestion of contaminated seafood. Direct human exposure to organic chemicals is less common because these substances generally can be present in the water at only very low concentrations, although in some cases they may be of substantial concern.

As was the case for metals, the consumption of contaminated seafood is the primary pathway for human exposure to most organic chemicals in wastes disposed of in marine environments. Indeed, compounds such as PCBs and DDT have been shown to accumulate in humans through consumption of contaminated seafood (249,409).

Not surprisingly, the importance of **bioaccumulation and biomagnification** varies greatly for different organic chemicals and for different organisms. Several classes of organic chemicals, particularly those that are relatively insoluble in water, have a high bioaccumulation potential because of their solubility in animal tissue. Some of these organic compounds, including PCBs, benzo[a]pyrene, naphthalenes, and chlorinated pesticides (e.g., kepone, mirex, and possibly DDT), also appear to have a potential for biomagnification by marine organisms.

Other compounds that bioaccumulate but probably do not biomagnify include chlorinated phenols and benzenes, and most PAHs. However, there is relatively little information on the long-term fate and behavior of most organic compounds in aquatic environments (260).

Potential and Actual Human Health Impacts

Few documented cases of human health impacts from waste-derived organic chemicals in marine environments exist.[7] However, the potential for such exposure and effects clearly exists in the United States. Numerous estuarine and coastal areas—e.g., New Bedford Harbor, New York Harbor, and portions of Puget Sound—are sufficiently contaminated with toxic chemicals to preclude the harvest of fish and/or shellfish (box L; refs. 211,507). Commercial and recreational fishing for striped bass, bluefish, tautog, and eels has been curtailed in large portions of the New York Bight's apex due to high concentrations of PCBs and other organic compounds (28,30,237).

PCBs and PAHs are widely distributed in inland and coastal sediments and have been found in deep ocean sediments as well (62,325,484). PCBs and several other highly persistent and toxic chemicals (e.g., DDT) have been banned from further production or use in commerce in the United States. It can be expected, therefore, that the sediment concentrations and body burdens of these toxic compounds will probably decrease gradually over time

---

[7]Several cases involving land-based food sources have been reported. The most infamous is a disease outbreak in Japan, known as the Yusho incident, that was caused by the leakage of PCBs (in combination with dibenzofurans) from an industrial heat exchanger into a rice oil manufacturing process. Yusho symptoms included reduced birthweights and skin disorders such as chloracne and hyperpigmentation (287).

Table 9.—Properties and Effects of Major Classes of Organic Chemicals in Wastes Disposed of in Marine Environments

Chemical class	Major examples	Properties	Primary routes to humans	Health effects	References[a]
Low molecular weight hydrocarbons	Benzene Toluene Xylene	Volatile Biodegradable Low bioaccumulation potential	Inhalation Drinking water	Benzene: central nervous system (CNS) effects, blood disease, leukemia Toluene: possible CNS effects, low toxicity Xylene: irritant; teratogen	Callahan, et al., 1979 Doull, et al., 1980 NAS, 1977 O'Connor and Kneip, 1986 Snyder, et al., 1984
Low molecular weight chlorinated hydrocarbons	Chloromethanes: carbon tetrachloride (CTET) chloroform methylene chloride	Volatile Lipid-insoluble Low bioaccumulation potential Some (e.g., CTET and chloroform) are persistent; others (e.g., methylene chloride) are readily biodegraded	Inhalation Drinking water	CTET and chloroform: liver, kidney, blood, and gastrointestinal disorders; liver and kidney cancer Methylene chloride: possible CNS effects	Callahan, et al., 1979 Doull, et al., 1980 O'Connor and Kneip, 1986 Thom and Agg, 1975
	Trichloroethylene, tetrachloroethylene, tetrachloroethane	Volatile Lipid-insoluble Low bioaccumulation potential	Inhalation Drinking water	All: CNS effects, liver toxicity Tri- and tetrachloroethylene: liver cancer	Callahan, et al., 1979 Doull, et al., 1980 Sittig, 1985
	Vinyl chloride	Volatile Low bioaccumulation potential	Inhalation	Animal and human carcinogen; liver toxicity	Doull, et al., 1980
	Chlorobenzenes	Range of volatilities Lipid-soluble Significant bioaccumulation potential	Food (including seafood)	Hexachlorobenzene: carcinogen	Doull, et al., 1980 O'Connor and Kneip, 1986
Chlorinated pesticides	Cyclodiene pesticides: aldrin, dieldrin, heptachlor, chlordane DDT and metabolites Chlorinated phenoxyacetic compounds (2,4,5-T; 2,4-D) Hexachlorocyclohexanes: lindane, BHC	Nonvolatile High bioaccumulation potential Moderate to high toxicity Most are highly persistent	Food (including seafood)	Known or suspected human carcinogens; neurotoxic effects; chloracne and other skin diseases	Doull, et al., 1980 Mrak, 1969 NAS, 1977 Sittig, 1985 Walker, et al., 1969
High molecular weight chlorinated hydrocarbons	Polychlorinated biphenyls (PCBs) Chlorinated dioxins (TCDD) Chlorinated dibenzofurans	Nonvolatile High bioaccumulation potential Moderate to high toxicity Highly persistent	Food (including seafood)	All: neurological, liver, and skin disorders PCBs: tumor promoters or carcinogens TCDD: highly carcinogenic	Kimbrough, et al., 1975 Kolbye and Carr, 1984 Murai and Juroiwa, 1971 Poiger and Schlatter, 1983
Aromatic hydrocarbons	Phthalate esters (e.g., DEHP) Polycyclic aromatic hydrocarbons (PAHs)	Low to moderate volatility Highly insoluble Range of bioaccumulation potential Low to moderate toxicity	Food (including seafood)	Phthalates: many are teratogens DEHP: possible carcinogen PAHs: many (e.g., benzo(a)pyrene) are carcinogens; some are teratogens	FDA, 1974 Giam, et al., 1978 MacLeod, et al., 1981 NAS, 1977

aSee list of references at end of report.
SOURCE: Office of Technology Assessment, 1987.

Box L.—Public Health Advisories

One measure of the extent of contamination of seafood with potentially harmful levels of toxic substances (e.g., metals or organic chemicals) or microorganisms is the number of public health advisories that have been issued warning against consumption of seafood harvested from particular areas.* Unfortunately, there is no readily available national registry of such advisories (which are typically issued at the State level). However, a survey of coastal States to obtain this information is being undertaken by the National Marine Pollution Progam Office at the National Oceanic and Atmospheric Administration (R. Landy, NOAA, pers. comm., 1986).

While not yet complete, the survey has compiled information on about 40 public health advisories specifically involving toxic metals or organic chemicals (*not* contamination with microorganisms) that have been issued by coastal States. (Unfortunately, the data do not distinguish between fresh and marine waters or organisms.) Shown below is a preliminary listing of these advisories, broken down by the particular metal or organic chemical involved. Also listed are some of the waterbodies for which these advisories have been issued, most (but not all) of which are in estuaries or coastal waters (figure 25 illustrates the extent of advisories in the New York-New Jersey area). These advisories also demonstrate the importance of seafood consumption as a potential source of human exposure to particular classes of chemicals.

This type of effort represents an essential link in documenting the potential impacts associated with the input of pollutants that are of public health concern into estuarine and coastal marine waters. Further compilation and centralization of such data at both national and regional levels would play a useful role in furthering our understanding of human health risks posed by waste disposal and other activities affecting marine environments.

Substance	*Number of health advisories*
Mercury	7
Selenium	2
Heavy metals**	1
Chlordane	2
DDT	6
Pesticides**	2
PCBs	16
Dichlorobenzene	1
Tetrachlorobenzene	1
Dioxin	1
PAHs	1

Some waterbodies in which public health advisories have been issued:

- San Francisco Bay Delta Region and Santa Monica Bay, CA
- Baltimore Harbor, MD
- Hudson River, New York Harbor, Newark Bay, Lower Passaic and Hackensack Rivers, and Arthur Kill, NJ
- Inner New York Bight and most of Long Island Sound
- Duwamish River and Puget Sound, WA

*Advisories might be issued on any of several bases, for example, if tissue samples of shellfish were found to contain levels of toxic metals or organic chemicals that exceeded U.S. Food and Drug Administration action limits (see box N), or if levels of fecal coliform bacteria exceeded State water quality criteria.

**Unspecified or multiple substances.

(57,519).[8] Indeed, recent evidence shows such a decline along the west coast of the United States (340).

Direct exposure of humans to organic chemicals present in marine waters is possible in places where industrial discharges are located near bathing beaches. For example, a wastewater discharge from a New Jersey pharmaceutical firm, which enters marine waters about 2,500 feet from shore, has been identified in bioassays as the most mutagenic

industrial discharge in the State (T. Burke, New Jersey Department of Health, pers. comm., 1986). The wastewater from a Florida paper mill discharged into the Amelia River estuary has been found to be the most toxic among the State's nine mills (8). In neither case, however, has an actual effect on the health of bathers been demonstrated.

Classes of Organic Chemicals

To simplify the evaluation of organic compounds, they can be classified on the basis of similarities in their chemical structure, physicochemi-

[8]However, the manufacture and use of many of these chemicals have not been curtailed in other countries; in addition, existing stocks of some of these chemicals continue to pose major disposal problems.

Figure 25.—New York Bay-Newark Bay Fishing Sites and Advisory Areas

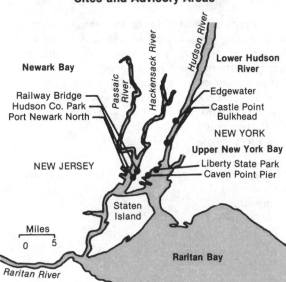

The PCB advisory for limited consumption of striped bass and bluefish applies to the coast and all the rivers and tributaries shown. Sale for human consumption of striped bass and American eel from most of these rivers is also prohibited.

SOURCE: T. Belton, et al., "Urban Fishermen: Managing the Risks of Toxic Exposure," *Environment* 28:19-20,30-36, November 1986.

cal properties (e.g., volatility, tissue solubility), and molecular weight. Chemicals having similar characteristics often tend to exhibit similar environmental fates or behavior, and therefore may follow similar pathways to reach humans. In some cases (though less predictably), compounds in the same group may also have similar toxicity characteristics. It should again be emphasized, however, that although such grouping of organic compounds helps determine their potential for significant human exposure, each compound must still be examined individually for specific environmental and health effects.

Table 9 lists the primary classes of organic chemicals that have some potential for reaching humans and inducing adverse health effects. Chlorobenzenes, chlorinated pesticides, high molecular weight chlorinated organic compounds, and aromatic hydrocarbons are of major concern in marine environments due to their ability to bioaccumulate in organisms. While wastestreams containing many of these compounds are specifically prohibited from ocean dumping, the major pathway by which they enter marine environments is as trace contaminants in industrial wastewater effluents, sewage sludge and effluent, and dredging spoils.

Chlorobenzenes exhibit a range of volatilities and tissue solubilities (and therefore bioaccumulation potential), dependent primarily on the degree of chlorination. Of all the low molecular weight chlorinated hydrocarbons, chlorobenzenes pose the greatest risk of transport to humans through marine or terrestrial food chains.

Chlorinated pesticides are significant because of the high risk of human exposure through consumption of contaminated seafood. They strongly associate with organic matter and sediments, where they are readily available to marine organisms. Their high molecular weight and complex, chlorine-containing structures make these compounds very resistant to degradation by bacteria or the metabolic systems of higher organisms. This class of substances exhibits a range of toxicities, from several cyclodiene pesticides that show adverse effects at the lowest doses tested (682) to DDT and related compounds that have moderate toxicity to humans and laboratory animals. Most of these compounds are known or suspected carcinogens.

Chlorinated organic compounds include PCBs dioxins, and dibenzofurans. These compounds generally share the same properties described for the chlorinated pesticides: low volatility, generally high bioaccumulation potential, strong ability to bind to organic matter and sediments, and high resistance to degradation by bacteria or the metabolic systems of higher organisms.

PCBs are actually a group of over 200 individual compounds possessing unique chemical and toxicological properties (494). PCBs are accumulated by marine organisms with very high efficiency. For example, one study found 85 to 95 percent assimilation of PCBs across the gut of striped bass (447). The ubiquity of PCBs in the environment and their extreme persistence and toxicity to marine and terrestrial organisms, led to a ban on their production in the United States in 1979.

Dioxins and dibenzofurans (the most well-studied of which are 2,3,7,8-tetrachlorodioxin (TCDD) and 2,3,7,8-tetrachlorodibenzofuran (TCDF)) are found as contaminants of numerous chemical preparations (467). Both sets of compounds are quite tissue-soluble, and are therefore easily bioaccumulated to high levels. TCDD and TCDF are among the most toxic compounds known and can produce lethal effects at low doses in aquatic organisms and birds (191,365).

Aromatic hydrocarbons include phthalate esters, which are common contaminants in water, sediment, and marine organisms (72) as well as in various types of wastes, especially sewage (73). The most common are diethylhexyl phthalate (DEHP) and dibutyl phthalate (DBP). Phthalate esters have low water solubilities, tightly adsorb to sediment particles, and are not readily degraded through biological activity (72,541). Chronic effects from exposure to phthalate esters include reduced weight gain, enlarged livers and kidneys (381), and an increased incidence of liver cell cancers (274).

Polycyclic aromatic hydrocarbons (PAHs) are derived from petroleum and other chemical processes and constitute most of the ''oil and grease'' regulated as a conventional pollutant under the Clean Water Act. Major sources to marine environments include spills and routine releases from ships, natural seepage, and municipal wastewater, sewage

sludge, dredged material, and runoff from urban areas (409). PAHs are readily adsorbed by suspended particulates and bottom sediments, where they persist and can accumulate to levels as high as 1 percent (10,000 parts per million) in highly industrialized areas such as New York Harbor (325).

Marine organisms, particularly mollusks, have a high potential for bioaccumulating PAHs, and they can achieve tissue levels far in excess of water concentrations and roughly comparable to sediment concentrations (409). Certain PAHs (e.g., benzo-[a]pyrene and naphthalenes) can also biomagnify in higher level predators of the marine food chain (260). Disposal of wastes at sea is considered the most likely source of PAHs to marine waters; subsequent transport of PAHs to humans can then occur through the food chain (409). How seafood consumption compares to other sources of PAH exposure is not well-understood, however, so the significance of PAH contamination of seafood as a source of exposure to humans is uncertain.

HUMAN PATHOGENS

General Considerations

Essentially all wastes that are disposed of in the marine environment either contain microorganisms or have the potential to modify the microbial community at the disposal site. Among the microorganisms entering the marine environment through waste disposal, or induced to proliferate as a result of such activity, are a variety of *human pathogens,* microorganisms that are capable of inducing human disease (box M).

Human pathogens in the marine environment come primarily from discharges of raw sewage and from sewage sludge and wastewater effluent from sewage treatment plants. For example, it is estimated that about 40 million gallons of raw sewage are discharged daily into the Hudson and East Rivers of New York City.[9] Pathogens are also found

in domestic and commercial food wastes, animal wastes, and biological wastes from hospitals and laboratories, many of which are discharged to surface waters or sewage treatment plants. Where combined sewer systems are employed, overflows during times of heavy precipitation can also be a significant source of pathogens. In rural areas, most bacterial contamination comes from non-urban runoff, animal wastes, poor septic systems, and poorly treated sewage discharges. Finally, the marine environment itself is a source of pathogens, because some pathogens naturally occur and propagate in marine waters.

In the United States, the most important waste-borne agents of human disease are viruses and bacteria, with respect to both their concentrations in wastes and the environment, and the incidence of disease attributable to them.[10] Enteric

---

[9]This discharge is scheduled to be halted in 1987 when a new secondary treatment plant is completed (G. Lutzic, New York City Department of Environmental Protection, pers. comm., 1986).

[10]However, in some parts of the world pathogens from other classes can be very important. The World Health Organization has indicated that the parasitic helminth *Schistosoma* is second only to malaria in causing disease and death in the tropics (252).

Box M.—Classes of Human Pathogens

Viruses.—Over 100 different types of human intestinal viruses are present in sewage and have the potential to be spread through contamination of water or food sources. Enteric Hepatitis A virus, which induces the potentially life-threatening liver disease infectious hepatitis, probably is the greatest threat to public health; this virus has been demonstrated to have been spread as a result of sewage contamination (197). Many outbreaks of viral gastroenteritis linked to contaminated coastal waters or shellfish are reported each year in the United States.

Bacteria.—A wide range of bacteria are found in sewage, including many known human pathogens that can cause food poisoning, typhoid fever, strep throat, dysentery, and other intestinal fevers. A number of other human pathogens have been identified as natural members of the marine bacterial community as well as constituents of sewage, including the causative agents of cholera and acute bacterial gastroenteritis. Several commonly known pathogenic bacteria found in sewage are *Salmonella*, *Streptococcus*, and *Campylobacter*.

Bacteria that are resistant to antibiotics are readily identified in sewage-contaminated waters. Because such bacteria can pass their resistance characteristics to other bacteria, the potential exists for humans to contract serious bacterial infections that would be difficult to treat. Pathogenic antibiotic-resistant bacteria can survive and have been detected in marine environments, but there is as yet no strong correlation between their occurrence and the disposal of sewage wastes. Nor has their actual public health significance been established.

Protozoa.—*Protozoa* are single-celled organisms, including such species as *Giardia* and *Entamoeba* (causes of giardiasis and amoebic dysentery, respectively), that are typically found in sewage in the form of cysts. Cysts are inactive and extremely resistant to environmental damage. Upon ingestion by a host organism, the cysts are activated and can undergo maturation and reproduction to form additional cysts that are excreted in the feces.

Helminths.—*Helminths* include various parasitic flatworms and are transmitted through ingestion of cysts present in inadequately cooked meat; infection leads to disease and excretion of eggs in the feces. Helminths are responsible for trichinosis, schistosomiasis, diarrhea, and various liver disorders.

Nematodes.—*Nematodes* are parasitic roundworms and hookworms, the most common of which is *Ascaris*. They are capable of forming cysts or eggs, which are excreted in the feces and require activation outside of an organism in order to initiate infection; hence, they are not capable of direct human-to-human spreading.

Fungi.—These are generally present in significant quantity in sewage material only when given the opportunity to grow during treatment or storage. Large numbers are commonly found in composting sludge. Reproduction involves the formation of spores, the inhalation of which provides the most direct pathway to human infection. Pathogenic fungi include *Actinomyces* and *Candida*.

Cyanobacteria and Eucaryotic Algae.—Certain types of marine bacteria and algae, whose growth appears to be generally stimulated by nutrients, produce toxins that can be subsequently concentrated by shellfish. Eating toxin-contaminated shellfish can lead to diarrhea, muscular paralysis, neurological damage, and liver disorders. Shellfishing is usually restricted during "blooms" of toxin-producing algae, called red or green tides (4). In addition, skin and respiratory problems have been reported from swimming in such waters (545). Limited evidence exists that links blooms in the New York Bight to nutrients or metals introduced through waste disposal (352).

viruses are especially significant with respect to disposal of wastes in water, since they are ideally suited to be spread by contact with contaminated water (205).

Pathways to Humans

Both direct and indirect exposure pathways can be significant for pathogens because only a small number of organisms are required to induce disease, and because they can reproduce in wastes, contaminated media, or infected organisms. Several properties of pathogens are important in determining their potential to pose risks to human health.

Survival

A growing body of evidence indicates that some bacteria, including a number of known human pathogens, may persist in the marine environment for periods of many months or longer in a nonculturable, but virulent form (205,366). For example, the agent responsible for an outbreak of cholera along the Gulf coast of Texas appears to have persisted for at least 5 years in coastal waters (35). In addition, many viruses and parasites are extremely resistant to environmental inactivation or destruction.

Viruses and bacteria strongly adhere to particulate matter, which provides a degree of protection and increases their survival in sediments (183). Concentrations of enteroviruses may be 10 to 10,000 times greater in coastal sediments than in the overlying water; most pathogens are concentrated in the surface layers of bottom sediments (197). Subsequent dredging of these sediments may increase the concentrations and availability of human pathogens in the areas where dredging or disposal of dredged materials takes place (203,204).[11]

A number of human pathogens appear to survive better in estuarine and other coastal environments than in the open ocean (253). However, in some cases, the colder temperature of the open ocean, especially bottom waters beyond the con-

tinental shelf, can actually enhance the survival of some pathogens, although it also retards growth (16).

Propagation

Different classes of pathogens have different requirements for propagation. In the absence of an appropriate host, viral or parasitic propagation generally cannot occur, so there are no mechanisms available for increasing the number of viruses in sewage material or the marine environment; they can, however, become concentrated in sludge or sediment. Bacteria introduced through wastes have the potential to replicate and increase their numbers, but this potential has not been well-studied for most organisms. Some pathogenic bacteria that naturally occur in the marine environment are fully capable of propagation (e.g., certain species of *Vibrio*).

Exposure and Infection

For viruses and microorganisms present in the marine environment to exert an impact on human health, they must both reach and infect humans. The ability of microorganisms to infect humans depends on numerous factors, including the minimum infective dose. As few as 10 to 100 bacteria, or a single virus, are capable of inducing infection and disease under the appropriate conditions (205). Moreover, the tendency for viruses and bacteria to adhere to particulate matter increases the risk of exposure and infection in two ways: 1) survival is enhanced through the protection the particles provide; and 2) because particles literally serve to "collect" viruses and bacteria on their surfaces, a single ingested particle can contain a large dose of microorganisms. The concentration of viruses and bacteria in sediments also increases the potential for their uptake by shellfish.

Potential and Actual Human Health Impacts

Shellfish-Borne Disease

Large areas of estuarine and coastal waters have been closed to shellfishing and/or finfishing because they are contaminated with sewage-derived microorganisms in excess of Federal standards (see box N). While such closures have largely eliminated

---

[11]For example, a total restriction on shellfishing has been imposed in estuarine waters around a dredging project near Cape May, NJ, because disturbance of the sediments was shown to cause elevated densities of coliform bacteria in the water (J. Staples, New Jersey Department of Environmental Protection, cited in ref. 10).

Box N.—Exposure to Toxic Chemicals and Pathogens Through Seafood

The risk of human exposure to toxic chemicals or pathogens through consumption of tainted seafood is growing, and increasing numbers of experts are questioning the adequacy of current programs designed to address this risk (29,31,174,293,367,430, 455,457). The risk extends to both commercially and recreationally harvested seafood and to coastal areas in all parts of the country. Of particular concern are urban recreational fishing and illegal commercial harvesting, which are widespread practices even in areas for which health advisories or bans on fishing have been issued.

The U.S. Food and Drug Administration (FDA) has authority to set and enforce allowable levels of toxic substances in food or food products, including seafood. Table 10 lists all of the substances for which

Table 10.—FDA's Criteria for Initiating Enforcement Actions Against Seafood

Substance	Action level	Type of food (edible portion only)
Methyl mercury	1.0 ppm	Fish, shellfish, crustaceans, other aquatic animals
PCBs[a].................	2.0 ppm	Fish and shellfish
Aldrin	0.3 ppm	Fish and shellfish
Chlordane	0.3 ppm	Fish
Dieldrin	0.3 ppm	Fish and shellfish
DDT, DDE, & TDE[b]........	5.0 ppm	Fish
Endrin	0.3 ppm	Fish and shellfish
Heptachlor and heptachlor epoxide...............	0.3 ppm	Fish and shellfish
Kepone.................	0.3 ppm	Fish and shellfish
	0.4 ppm	Crabmeat
Mirex	0.1 ppm	Fish
Toxaphene..............	5.0 ppm	Fish

[a]The value for PCBs is a tolerance for unavoidable poisonous or deleterious substances.
[b]DDE and TDE are toxic metabolic products of DDT.

SOURCES: Office of Technology Assessment, 1987; based on U.S. Food and Drug Administration, Industry Programs Branch, Center for Food Safety and Applied Nutrition, *Compliance Policy Guides Manual* (Washington, DC: October 1986); except for PCBs: 21 CFR Subpart B, Section 109.30.

FDA has developed limits in fish or shellfish, and indicates how much of each substance must be present in an organism's edible tissues before an enforcement action or seizure is authorized (termed the "action level"). While the list includes a fairly comprehensive list of pesticides, it includes only one metal (methyl mercury) and one other toxic organic chemical (PCBs). No action levels have been developed for other substances of concern, such as PAHs, cadmium, or chlorobenzenes.

The National Marine Fisheries Service of the U.S. Department of Commerce is the primary Federal agency authorized to inspect seafood harvested for commercial use. The activity of this largely voluntary inspection program has declined in recent years, despite increases in both total and per capita seafood consumption in the United States. In 1983, the service inspected 567 million pounds of fish, almost 19 percent of all that was consumed that year; in 1985, however, inspection had dropped to 443 million pounds, 13 percent of the total consumed (457).*

Some sampling of pollutant levels in fish and shellfish for research purposes also takes place (see ch. 5; ref. 109). These studies have in numerous cases revealed tissue levels of certain pollutants in excess of FDA action levels (29,31). They also reveal the lack of a concerted and coordinated program of routine monitoring that involves a sufficient number of species and individual samples to provide meaningful information on the nature of risks posed by seafood consumption.

---

*The extent of such inspections (e.g., the number of individual substances or the tissues of organisms tested) was not specified. Testing is performed routinely for fecal coliform bacteria, for example, but not for viruses.

outbreaks of serious shellfish-borne, bacterial disease, including epidemics of typhoid and paratyphoid fever, they cause major economic impacts. In Washington, 21 percent of the shellfish-growing areas are closed and another 11 percent are only provisionally open because of bacterial contamination. Moreover, contamination appears to be increasing: six previously pristine areas in Puget Sound have been closed in the last 3 years (463). Fishing in many other areas has been restricted

periodically due to sewage contamination (e.g., Boston Harbor, New York Harbor, and portions of Narragansett Bay, the Delaware River estuary, Chesapeake Bay, Mobile Bay, and San Francisco Bay).

Overall, the incidence of shellfish-borne disease is not decreasing in the United States and may be increasing (197). This trend largely involves increases in the number of outbreaks of *viral* disease

Photo credit: *Northeast Technical Services Administration, U.S. Food and Drug Administration*

Contamination of shellfish is a significant problem nationwide. The problem is growing, particularly in rapidly developing areas such as coastal portions of the Gulf of Mexico and southern Atlantic States.

(e.g., 80,367). For example, in the State of New York in 1982 alone, consumption of contaminated shellfish was identified as the cause of 103 different reported outbreaks of viral gastroenteritis involving over 1,000 people (367). Smaller outbreaks of more serious bacterial diseases also have been reported. In a series of apparently related cases stretching back over the last 14 years, several dozen people contracted cholera after consuming shellfish harvested from coastal marshlands in southwestern Louisiana (81,82). These represent the first indigenous outbreaks of cholera in the United States since 1911 (78).[12]

---

[12]While the bacteria responsible for cholera are natural constituents of marine waters, their survival in this case is thought to be enhanced through a cycle of: 1) sewage contamination of marsh water, 2) shellfish contamination, 3) human infection from consumption of shellfish (in some cases inadequately cooked), and 4) human fecal excretion into sewage (35). The presence of the infective agent in marsh water, shellfish, and sewage from several towns in the area has been documented repeatedly (35,82).

Water-Borne Disease

While the implementation of water quality guidelines and sewage treatment requirements has substantially reduced the outbreaks of serious human diseases attributable to direct contact with polluted waters, bathing in sewage-impacted waters is responsible for relatively high rates of gastrointestinal illness in the United States. In fact, the number of outbreaks of water-borne disease, particularly nonbacterial diseases such as viral gastroenteritis and hepatitis, has been steadily increasing in recent decades (79,197).[13] Recent epidemiologic evidence has shown that the incidence of gastrointestinal illness is significantly elevated in people

---

[13]Much of this type of illness goes unreported because of its relatively benign nature and the underreporting significantly hinders accurate measurement of true incidence. Data on the incidence of water-borne disease do not generally distinguish between fresh and marine waters.

swimming at several heavily used New York City beaches (63,641) and in Lake Pontchartrain in New Orleans, Louisiana (285).[14]

Epidemics of serious bacterial disease, while rare, have been caused by swimming in sewage-contaminated water. For example, in 1974 an outbreak of shigellosis was traced to swimming in a stretch of the Mississippi River downstream of a secondary treatment plant. Fecal coliform counts in the river were almost 90 times higher than the Federal standard (489). Similarly, outbreaks of typhoid fever in Australia and Egypt have been caused by swimming in sewage-contaminated marine waters (205).

Scuba diving in contaminated marine waters can also lead to increased incidence of waterborne disease (113,192,205,435). These diseases include dermatitis, wound infections, and other skin-related ailments, as well as enteric illness. Under-reporting of such diseases is judged to be considerable (205).

Shortcomings of Current Microbiological Standards

Many observers have raised concerns about the adequacy of current efforts to control and monitor microbiological contamination of marine waters and resources. Three major shortcomings need to be addressed:

1. current standards designed to protect humans against microbiological agents in marine waters or seafood may be too lenient,
2. monitoring protocols are inadequate to detect periodic violation of the standards, and
3. standards based on use of fecal coliform indicators do not adequately measure pathogen survival.

Current techniques used to measure marine water quality are probably significantly underestimating the true number of viable pathogens that are entering the marine environment, for at least four reasons (205). *First,* coliform bacteria, which have been used to indicate sewage contamination of water for 75 years, generally are not pathogenic, and do not survive as well as other pathogenic bacteria or viruses (481). In fact, studies show that gastroenteritis associated with swimming (at least in marine waters) is better correlated with enterococcal bacteria than with coliforms (63,285,641). In addition, outbreaks of gastroenteritis have been associated with shellfish harvested from waters that were deemed acceptable using traditional indicators (451,700).[15]

Second, existing standards use bacteria as indicators of contamination, while *viruses* appear to be the major cause of diseases resulting from exposure through both direct (swimming, diving) and indirect (seafood consumption) pathways (197). *Third,* current standards are based solely on water quality, while levels in sediments and shellfish are neither regulated nor routinely monitored. Yet sediments are probably an equal or more likely source of pathogens in shellfish (197).

Finally, increasing evidence suggests that bacteria (including certain human pathogens) introduced into the marine environment do not die off as rapidly as once believed, but remain viable for extended periods of time (e.g., months to years). These pathogens cannot be cultured in the laboratory and their presence cannot be detected using traditional tests, but they can be reactivated within a host organism (106,206). Thus, the apparent lack of human pathogens in the open ocean, especially at or near past sewage sludge disposal sites, may simply reflect our inability to detect these apparently viable, non-culturable pathogens (205).

The public health significance of the viable-but-not-culturable phenomenon is far from clear, however, and remains controversial. No definitive link has been established between the survival of bacterial pathogens through this mechanism and the occurrence of human disease. Moreover, because most disease related to exposure to marine waters or fish is caused by viruses rather than bacteria, the role that pathogenic bacteria in marine envi-

---

[14]These studies are significant because they demonstrate that not only can the presence of sewage-derived material in the marine environment result in human disease, but also that disease rates that are significant from a public health perspective can be difficult or impossible to discern in the absence of carefully performed, thorough (and expensive), epidemiologic studies.

[15]EPA has recently adopted enterococci as an indicator of microbiological water quality for marine *recreational* waters (664). It is not yet clear what influence adoption of the new standard and indicator will have on consideration of alternative indicators and standards for monitoring of the quality of shellfish and shellfish-harvesting waters (205).

ronments play in human disease is probably of secondary importance.

The Effect of Treatment on Pathogenic Microorganisms

Sewage Treatment

Wastewater treatment results in the partitioning of waste constituents into sewage effluent and sludge. Wastewater treatment and subsequent sludge treatment processes can in many cases significantly reduce the numbers of some types of sewage microorganisms. However, the actual extent of reduction varies considerably from operation to operation, and among classes of microorganisms. A full discussion of this topic is included in chapter 9.

Chemical Disinfection of Wastewater Effluents

In most instances, bacteriological water quality standards for recreational and shellfish-growing waters are met by chemically disinfecting sewage effluent prior to discharge. Chlorination traditionally has been viewed as an effective and economical means to reduce levels of microorganisms in effluent. However, concerns have been raised that chlorinated hydrocarbons (e.g., chloroform) formed as byproducts of chlorination pose significant risks to organisms in the immediate vicinity of treatment plant discharges. One alternative to chlorination is the use of long deep-ocean outfalls such as those employed in southern California. These achieve water quality standards through dilution.

The effectiveness of either approach in reducing pathogen exposure risks is questionable, however, given the traditional use of coliforms as the standard indicator species and the growing evidence that bacteria discharged in sewage effluent may persist in marine waters in a viable but nonculturable form. Substantial data indicate that:

1. chlorination is more effective against coliforms than against pathogenic viruses or even numerous pathogenic bacteria;

2. pathogenic viruses and bacteria can survive significantly longer in the marine environment than can coliforms; and

3. chlorination may only temporarily inactivate, rather than destroy, microorganisms present in effluent (205,297).

These findings suggest that the routine discharge of sewage effluent and the dumping of sewage sludge into estuaries, coastal waters, and the open ocean may be introducing large numbers of viable microorganisms, including pathogens, and that their densities in both the water and sediments may be increasing. Further study of the public health consequences of these practices is needed, particularly in light of the increasing incidence of shellfish- and water-borne disease.

Land Application of Sewage Sludge

Because sewage sludge is applied to land as fertilizer and for reclamation purposes, the survival and availability of pathogens in the sludge is of considerable public health significance. Viruses, bacteria, and parasites can survive in soil for many days or months depending on soil temperature, pH, clay content, cation exchange capacity, surface area, moisture content, and organic content. Viable pathogens have been found, for example, in surface runoff from sludge-amended fields. However, there are no documented cases of human disease resulting from land application of treated sewage sludge, although untreated sewage-derived wastes have often been implicated in disease outbreaks (60,205,379). Land application and landfilling of treated sewage sludge appear to pose less potential health risks to humans than disposal in freshwater or estuarine environments.

Depuration of Shellfish

Some countries (e.g., Japan) allow or even encourage the culturing and harvesting of shellfish in sewage-contaminated water to take advantage of the nutritive content of such wastes. Prior to marketing, these shellfish are depurated (i.e., placed in clean water for several days) to allow the shellfish to purge themselves of pathogens. This practice is controversial, however, because evidence indicates that depuration does not eliminate all pathogens, especially small bacteria or viruses such as the Hepatitis A virus (186,209,298,368,523).

These and other studies suggest that further systematic study of the health risks associated with various forms of depuration should be conducted prior to its use as an accepted means of decontaminating shellfish (205).

Chapter 7
Statutes and Programs Relating To Marine Waste Disposal

CONTENTS

Statutes and Programs Relating To Marine Waste Disposal

INTRODUCTION

Federal efforts to control and manage marine waste disposal are relatively recent in origin, with most programs being less than two decades old. In 1970, three major government reports recommended that a national policy for controlling ocean waste disposal be developed (115,382,623). In response to these and to the general environmental concerns of the 1960s and early 1970s, Congress passed a suite of major statutes that provide the general legal structure currently used to regulate all waste disposal activities. One of the reports, the Council on Environmental Quality's "Ocean Dumping—A National Policy" (115), became the primary basis for the Marine Protection, Research, and Sanctuaries Act (MPRSA) of 1972 and for much of the policy developed throughout the decade for regulating marine disposal (377,420).

MPRSA and the Federal Water Pollution Control Act (commonly referred to as the Clean Water Act, or CWA) are the two major statutes controlling waste disposal in marine environments. In general, the open ocean is reasonably well-protected

as a result of MPRSA, but other areas of the marine environment remain more vulnerable. In particular, estuaries and other coastal waters, primarily regulated under CWA, have received less protection. In fact, a 1981 study by the National Advisory Committee on Oceans and Atmosphere (NACOA) concluded that the Council on Environmental Quality (CEQ) report was responsible for the near total restriction of *open* ocean waste disposal (377). The NACOA report disagreed with this approach, and proposed that some wastes could be disposed of in marine waters under certain conditions. It recommended that a more comprehensive waste management strategy include greater use of the open ocean. This recommendation influenced an important 1981 court decision, *City of New York* v. *United States Environmental Protection Agency* (543 F. Supp. 1084) (155). The NACOA report and the court decision signaled a changing attitude toward the ocean, from relatively strict protection to carefully managed use (12,291,531).

OVERVIEW OF THE EXISTING LEGISLATIVE AND REGULATORY FRAMEWORK

The major provisions of the two major statutes, MPRSA and CWA, are summarized in table 11. (A number of other statutes that also have some affect on marine waters are described briefly in box O.) MPRSA regulates the *dumping* of any material in the territorial sea (0 to 3 nautical miles), the contiguous zone (3 to 12 nautical miles), and beyond in the open ocean. It applies to dumping of U.S.-origin materials from all U.S. vessels, but it only applies to foreign vessels dumping foreign-origin materials within 12 miles of the U.S. coast. CWA regulates *discharges* from all point sources into all U.S. waters, including the territorial sea,

the contiguous zone, and beyond.[1] Although both laws establish procedures to administer regulatory permit programs, there are basic differences in their regulatory approaches to marine waste disposal. MPRSA requires the balancing of all relevant factors (e.g., socioeconomic factors, land-based alternatives, etc.), while CWA primarily relies on technological considerations, giving some attention to economic feasibility.

---

[1]Except discharges from vessels beyond the 3-mile boundary.

Table 11.—Major Legislative Provisions Affecting Waste Disposal in Marine Waters

Statute and section	Purpose
Marine Protection Research, and Sanctuaries Act:	
Sec. 101	Prohibits, unless authorized by permit, the transportation of wastes for dumping and/or the dumping of wastes into the territorial seas or the contiguous zones.
Sec. 102	Authorizes EPA[a] to issue permits for dumping of nondredged materials into the contiguous zone and beyond as long as the materials will not "unreasonably degrade" public health or the marine environment, following criteria specified in statute or established by the Administrator.
Sec. 103	Authorizes Corps of Engineers to issue permits for dumping dredged material, applying EPA's environmental impact criteria to ensure action will not unreasonably degrade human health or the marine environment.
Sec. 104	Specifies permit conditions for waste transported for dumping or to be dumped, issued by EPA or the Coast Guard.
Sec. 107	Authorizes EPA and Corps of Engineers to use the resources of other agencies, and instructs the Coast Guard to conduct surveillance and other appropriate enforcement activities as necessary to prevent unlawful transportation of material for dumping or unlawful dumping.
Clean Water Act:[b]	
Sec. 104(n)	Directs EPA to establish national estuaries programs to prevent and control pollution; to conduct and promote studies of health effects of estuarine pollution.
Sec. 104(q)	Establishes a national clearinghouse for the collection and dissemination of information developed on small sewage flows and alternative treatment technologies.
Sec. 201, 202, 204	Specifies sewage treatment construction grants program eligibility and Federal share of cost.
Sec. 208	Authorizes a process for States and regional agencies to establish comprehensive planning for point and nonpoint source pollution.
Sec. 301	Directs States to establish and periodically revise water quality standards[c] for all navigable waters; effluent limitations for point sources requiring BPT should be achieved by July 1, 1977; timetable for achievement of BAT and other standards set. Compliance deadlines for publicly owned treatment works (POTWs) to achieve secondary treatment also set.
Sec. 301(h)	Authorizes waivers for POTWs in coastal municipalities from secondary treatment for effluent discharged into marine waters if criteria to protect the marine ecosystem can be met.
Sec. 301(k)	Allows industrial dischargers to receive a compliance extension from BAT requirements until July 1, 1987, for installation of an innovative technology, if it will achieve the same or greater effluent reduction than BAT at a significantly lower cost.
Sec. 302	Allows EPA to establish additional water quality-based limitations once BAT is established, if necessary to attain or maintain fishable/swimmable water quality (for toxics, the NRDC v. EPA consent decree sets terms).
Sec. 303	Requires States to adopt and periodically revise water quality standards; if they determine that technology-based standards are not sufficient to meet water quality standards, they must establish total maximum daily loads and waste load allocations, and incorporate more stringent effluent limitations into Sec. 402 permits.
Sec. 303(e)	Requires States to establish water quality management plans for watershed basins, to provide for adequate implementation of water quality standards by basin to control nonpoint pollution; Section 208 areawide plans must be consistent with these plans.
Sec. 304	Requires EPA to establish and periodically revise water quality criteria to reflect the most recent scientific knowledge about the effects and fate of pollutants, and to maintain the chemical, physical, and biological integrity of navigable waters, groundwater, and ocean waters and establish guidelines for effluent limitations.
Sec. 304(b)	Outlines factors to be considered when assessing BPT and BAT to set effluent limitation guidelines, including accounting for "non-water quality impact," age of equipment, etc.
Sec. 305(b)	Sets State water quality reporting requirements.
Sec. 306	Sets new source performance standards for a list of categories of sources.
Sec. 307	Requires EPA to issue categorical pretreatment standards for new and existing indirect sources; POTWs required to adopt and implement local pretreatment programs; toxic effluent limitation standards must be set according to the best available technology economically achievable.
Sec. 308	Requires owners or operators of point sources to maintain records and monitoring equipment, do sampling, and provide such information or any additional information.
Sec. 309	Gives enforcement powers primarily to State authorities. Civil penalties, however, and misdemeanor sanctions can be issued by EPA in U.S. district courts for violation of the act, including permit conditions or limitations; EPA also is authorized to issue criminal penalties for violations of Sections 301, 302, 306, 307, and 308. EPA may take enforcement action for violations of Section 307(d) which introduce toxic pollutants into POTWs.
Sec. 402	Establishes National Pollutant Discharge Elimination System (NPDES), authorizing EPA Administrator to issue a permit for the discharge of any pollutant(s) to navigable waters that will meet requirements of Sections 301, 302, 306, 307 and other relevant sections; States can assume administrative responsibility of the permit program.

Table 11.—Major Legislative Provisions Affecting Waste Disposal in Marine Waters—Continued

Statute and section	Purpose
Sec. 403	Directs EPA to establish Ocean Discharge Criteria as guidelines for permit issuance for discharge into territorial seas, the contiguous zone, and open ocean.
Sec. 404	Directs Secretary of the Army to issue permits for dredged or fill material; EPA must establish criteria comparable to Section 403(c) criteria for dredged and fill material discharges into navigable waters at specified disposal sites.
Sec. 405	Requires EPA to issue sludge use and disposal regulations for POTWs.
Sec. 504	Grants emergency powers to Administrator to assist in abating pollutant releases; establishes a contingency fund, and requires Administrator to prepare and publish a contingency plan to respond to such emergencies.
Sec. 505	Citizen suit provision allows citizens to bring civil action in district court against any person in violation of an effluent standard or limitation of an order by the Administrator for failing to perform a nondiscretionary act.

aUnless otherwise noted, the Environmental Protection Agency (EPA) is responsible for implementing provision(s).

bRelevant provisions of the recently-enacted Water Quality Act of 1987, which reauthorized and amended the Clean Water Act, are discussed in ch. 1, box C and corresponding text.

cWater quality standards are ambient standards designed to achieve certain uses of water; these now play a secondary role. Technology-based effluent standards are given the primary role and are designed to reduce pollutants so that ultimately all water is "fishable, swimmable." Effluent standards are performance standards and specify the maximum permissible discharge of a pollutant from a type of source and usually specify the degree of technology to be used ("best available," "best practicable," "reasonably available," etc.), but not the particular method needed to comply. Effluent limitation guidelines, on the other hand, apply to individual sources and specify their particular performance levels. Water quality standards (Sec. 303) are now the benchmarks by which to measure the success of the effluent standards in meeting clean water goals.

SOURCE: Office of Technology Assessment, 1987.

Box O.—Additional Federal Laws Affecting Marine Waste Disposal

Although MPRSA and CWA are the statutes having the most profound effects on marine waste disposal, a number of other laws are also relevant, whether by increasing the amount of waste needing disposal or by regulating certain uses. The Resource Conservation and Recovery Act is discussed in box P, while other statutes are briefly described here.

The *Clean Air Act Amendments* (CAA) of 1977 (42 U.S.C. 7401 et seq.) have indirectly resulted in the generation of large amounts of air pollution control wastes (fly ash, flue-gas desulfurization sludges, and other air pollution control sludges) which have been proposed for marine disposal at various times. These wastes are generated by the air pollution control equipment installed to comply with national emission and air quality standards for stationary sources of air pollutants.

The *Coastal Zone Management Act* (CZMA) of 1972 (16 U.S.C. 1451 et seq.) provides Federal grants to States to develop Coastal Zone Management Plans that balance the pressure for economic development and the need for environmental protection. EPA cannot issue a permit for an activity affecting land or water use in a coastal zone until it has certified that the activity does not violate a State's management plan. Through the National Estuarine Sanctuary Program, the act authorizes 50 percent matching grants to States to acquire and manage estuaries for research and educational purposes. Amendments to CZMA in 1980 state that management policies should protect coastal natural resources (including estuaries, beaches, and fish and wildlife and their habitat) and encourage area management plans for estuaries, bays, and harbors.

The *Comprehensive Environmental Response, Compensation, and Liability Act* (CERCLA) of 1980 (42 U.S.C. 9601 et seq.), better known as Superfund, was enacted to provide emergency response and cleanup capabilities for chemical spills and releases from hazardous waste treatment, storage, and disposal facilities. Its primary impact on marine waste disposal involves: 1) the identification of large numbers of hazardous waste sites in the coastal zone, with potential for movement of waste pollutants into marine waters; 2) the suggestion that some wastes generated by remedial action at Superfund sites possibly be disposed in the ocean; and 3) provisions regarding the liability of ocean incineration vessels (586).

The *Endangered Species Act* (ESA) of 1973 (16 U.S.C. 1531 et seq.) requires all Federal agencies and their permittees and licensees to ensure that their actions are not likely to jeopardize the existence of an endangered or threatened species or result in the destruction or adverse modification of critical habitats of such species. If an activity might affect an endangered or threatened species, the Federal agency must obtain a biological opinion from the U.S. Fish and Wildlife Service or the National Marine Fisheries Service about the potential effects. These opinions can be an integral part, for example, of the site designation process for marine dumping activities.

The *National Environmental Policy Act* (NEPA) of 1970 (42 U.S.C. 4321 et seq.) requires that an Environmental Impact Statement (EIS) be prepared for all proposed legislation and all major Federal actions that could significantly affect the quality of the human environment. Under NEPA, EPA is exempted from this provision, but it voluntarily prepared an EIS when it proposed revisions to the ocean dumping regulations and criteria in 1977 and it prepares EISs for site designations.

The *National Ocean Pollution Planning Act* (NOPPA) of 1978 (33 U.S.C. 1701 et seq.) directs the National Oceanic and Atmospheric Administration to coordinate the ocean pollution research and monitoring that is conducted by various Federal agencies and to establish Federal priorities in marine research.

The *Port and Tanker Safety Act* of 1978 (33 U.S.C. 1221 et seq.) regulates the operation of ships within U.S. ports and waterways to promote navigational and vessel safety and protect the marine environment. Regulations for operations at waterfront facilities include requirements for handling, storage, loading, and movement of dangerous materials.

The *Safe Drinking Water Act* (SDWA) of 1974 (42 U.S.C. 300(f) et seq.) mandated the development of primary and secondary drinking water standards. Compliance with these standards has resulted in the use of water treatment processes that generate sludges requiring disposal. Some of these wastes could be disposed of in marine waters.

The *Toxic Substances Control Act* (TSCA) of 1976 (15 U.S.C. 2601 et seq.) regulates the manufacture, processing, distribution, use, and disposal of chemical substances that present significant risks to human health or the environment. EPA is authorized to gather information concerning the toxicity of chemicals, mandate additional testing and research where necessary, and assess the extent of risk. TSCA affects marine waste disposal primarily through its regulations on the disposal of wastes contaminated with PCBs.

The *Low-Level Radioactive Waste Policy Amendments Act* of 1986 (42 U.S.C. 2021(b) et seq.) places responsibility with the States for managing commercial low-level radioactive waste (LLW) generated within their borders. States may enter into compacts with other States to establish and operate regional disposal sites. The original deadline for establishing operational sites was January 1, 1986, but it has been extended to December 31, 1992. Currently, there are only three operational disposal sites in the country. Although MPRSA effectively bans marine disposal of LLW, the difficulty of establishing future land-based disposal sites could lead to reconsideration of the marine disposal option.

The Marine Protection, Research, and Sanctuaries Act

Introduction

When passed in 1972, MPRSA became the first comprehensive legislation to regulate ocean dumping of all types of material that may adversely affect human health, the marine environment, or the economic potential of the ocean. MPRSA (33 U.S.C. 1401 et seq.) is the only pollution law exclusively devoted to the ocean and is the only law that explicitly requires consideration of alternative land-based disposal methods (Sec. 102(a)).[2] In contrast, other statutes such as the Resource Conservation and Recovery Act (RCRA) typically regulate disposal in one environment without explicitly considering the consequences in other environments.[3]

---

[2]One major finding in the *City of New York* v. *United States Environmental Protection Agency* decision (543 F. Supp. 1084) was the court's interpretation that the Act requires EPA to balance the need for ocean dumping with potential environmental, social, and economic impacts of land-based disposal options.

[3]Under certain conditions, RCRA precludes land disposal without requiring that alternative disposal methods first be evaluated (box P).

Box P.—The Importance of the 1984 RCRA Amendments

The *Resource Conservation and Recovery Act* (RCRA) of 1976 (42 U.S.C. 6901 et seq.) defines and lists "hazardous" wastes and controls their generation, transport, treatment, storage, and disposal. Some hazardous wastes currently enter marine waters from disposal sites located on land but near estuaries or coastal waters, or as part of "indirect" industrial discharges into municipal treatment plants that subsequently discharge into these waters. The Domestic Sewage Exemption of RCRA, for example, allows legal "indirect" discharges of some hazardous wastes into municipal treatment plants.

The 1984 *Hazardous and Solid Waste Amendments* to RCRA dramatically increased the scope and complexity of the RCRA program and represented an attempt by Congress to discourage most land-based disposal methods for managing hazardous wastes. Section 201 is the most important amendment affecting marine waste disposal. It prohibits land-based disposal of dioxin-containing and spent-solvent wastes by November 1986. Eight months later, all "California list" wastes must be banned from land-based disposal unless EPA determines that land-based disposal is safe for a particular waste. The California wastes include liquid hazardous wastes and sludges containing specified levels of metals, arsenic, halogenated chemicals, PCBs, or highly acidic liquids. Underground injection of dioxins, solvents, and California wastes would stop by 1988, unless EPA finds that they can be safely disposed of in this way. For all other hazardous wastes, EPA is given deadlines of 44, 55, and 66 months to review and set standards for the most hazardous and highest volume wastes. If the first two deadlines are missed, the wastes are automatically banned from land-based disposal if adequate alternative disposal facilities exist.

These "hammer provisions" are intended to force the phasing out of land-based disposal for hazardous wastes. In many cases, EPA probably will find it difficult to meet the deadlines or to determine that land-based disposal is safe, so the provisions may effectively encourage alternatives such as physical and chemical treatment methods (487,684). These provisions also could lead to the consideration of marine disposal of hazardous wastes. One preliminary study estimated that the restrictions might cause annual shortfalls in land-based treatment and disposal capacity of over 50 million gallons of certain solvents, dioxins, and California list wastes; based on the ocean dumping regulations, however, none of these would be legally acceptable for ocean dumping (241). Finding sites on land for the disposal of sewage sludge and dredged material also could become more difficult. In addition, regulations now require that hazardous waste generated by small quantity generators (those businesses generating between 100 and 1,000 kilograms of hazardous waste per month) be disposed of in permitted or interim status facilities. Enforcing this requirement is difficult; illegal discharges to municipal sewers could increase and such discharges could further contaminate municipal effluent and sludge.

"Material" is defined in MPRSA as all wastes except effluent discharged through an outfall, oil, or sewage from vessels, all of which are regulated under CWA. Thus MPRSA governs solid wastes, incinerator residues, sewage sludge, industrial wastes, dredged materials, low- and high-level radioactive waste, and chemical and biological warfare agents. High-level radioactive waste and chemical and biological warfare agents are specifically prohibited from ocean disposal, while other materials are allowed under some circumstances. Fish cleaning wastes are generally not regulated except if disposed of in harbors or other protected or enclosed coastal waters, although seafood processing is regulated under CWA through National Pollution Discharge Elimination System permits.

Under the first two titles of MPRSA, commonly referred to as the Ocean Dumping Act, four Federal agencies have responsibilities: the Environmental Protection Agency (EPA), the Corps of Engineers (COE), the National Oceanic and Atmospheric Administration (NOAA), and the Coast Guard. Title I of the Act authorizes EPA to designate specific ocean disposal sites, establishes a permit system for the use of such sites, and directs EPA to establish ocean dumping criteria based on specified factors. The permit system is administered by EPA for all materials except for dredged material, which is under the jurisdiction of the Corps of Engineers, although EPA does retain review authority.

Title II requires EPA and NOAA to conduct research and monitoring on ocean dumping and to study alternative disposal methods. The Coast Guard is charged with maintaining surveillance of ocean dumping. Section 203 was amended in 1986 (as part of the Title II reauthorization of MPRSA included in the Consolidated Omnibus Budget Reconciliation Act of 1985, Public Law 99-272, Apr. 7, 1986) to direct EPA to cooperate with other appropriate government agencies to assess the feasibility of regional management plans for waste disposal in coastal areas. The plans would integrate all the waste disposal activities in an area into a comprehensive regional disposal strategy.

Title III of MPRSA gives the Secretary of Commerce authority to establish marine sanctuaries. Through the National Marine Sanctuary Program, marine areas as far seaward as the outer edge of the continental shelf, including inland waters, can be designated if this is determined necessary to preserve or restore an area for conservation, recreational, ecological, or esthetic purposes (Sec. 302). The designation of certain sanctuary sites has created controversy when it entailed prohibiting oil and gas development activities or conflicted with other economic interests (e.g., the creation of the Channel Islands National Marine Sanctuary in California). Although this is an important program, it is not directly concerned with the control of disposal activities and thus is not discussed further in this report.

In 1974, MPRSA was amended so that all U.S. criteria covering the dumping of wastes in marine waters would be consistent with and contain all the basic constraints set forth in the London Dumping Convention (LDC) (box Q). In practice, however, a number of administrative and court actions have not always taken full account of the Convention's requirements (12,214,712).

Permitting—Sections 102 and 103

Section 102 of MPRSA authorizes the EPA Administrator to issue permits, following notice and opportunity for public hearings, for the transpor-

tation and dumping of nondredged material in ocean waters provided that it:

> will not unreasonably degrade or endanger human health, welfare, or amenities, or the marine environment, ecological systems, or economic potentialities.

The Administrator is further directed to establish Ocean Dumping Criteria, based on nine factors specified in the statute (box R), and use these to review permit applications for both dredged and nondredged material. The factors include the need for the proposed dumping; its effect on human health, the environment, and economic and recreational values; and alternative disposal options and their potential impacts.

In 1973, EPA issued final regulations that established these Ocean Dumping Criteria (40 CFR 227). The criteria reflected EPA's policy at that time of terminating all ocean dumping, even if the dumping could be shown not to "unreasonably degrade" the marine environment. EPA also established, however, "interim" and "special" permit procedures to allow the dumping of some materials prohibited by MPRSA. Emergency and research permits were also allowed. The criteria were not entirely consistent with LDC constraints when the United States became a signatory in 1974, which led to later revisions of the regulations and the Act. In 1977, EPA again revised the Ocean Dumping Criteria (42 FR 2462, Jan. 11, 1977), in part as a response to a case brought by an environmental group challenging the dumping regulations and permit criteria already promulgated by EPA (*National Wildlife Federation* v. *Costle*, often referred to as Costle I; 14 E.R.C. 1680 (D.C. Cir. 1980)) (420).

Section 103 of MPRSA authorizes the Secretary of the U.S. Army, acting through the Chief of Engineers of COE, to issue permits for the dumping of dredged material. Federal responsibilities under this section are bifurcated. COE applies criteria developed by EPA pursuant to the Section 102 environmental impact criteria. EPA has the authority to review the application before COE issues a permit and also has the authority to approve site designation. EPA initially exempted COE from several of the more stringent criteria and site designation and evaluation procedures generally applied to nondredged material permits. However,

Box Q.—Relevant International Conventions

Several international conventions affect marine waste disposal activities (165). The two most significant are the London Dumping Convention and the Oslo Convention. Some conventions (e.g., the Barcelona and Kuwait Conventions) were developed under the United Nations Regional Seas Programme,* while other conventions and agreements (e.g., the Helsinki Convention, the Bonn Agreement, and MARPOL) were developed under other auspices.

The London Dumping Convention (LDC) of 1972, officially called "The Convention on the Prevention of Marine Pollution by Dumping of Wastes and Other Matter," is the primary international agreement dealing with marine waste disposal and is the only dumping convention to which the United States is a signatory. As a signatory nation, all U.S. criteria covering marine disposal must, at a minimum, be equivalent to and contain the basic constraints in the LDC. The LDC has been ratified by 61 countries, and the International Maritime Organization serves as the administrative mechanism for cooperation among the contracting States. The LDC's jurisdiction includes all waters seaward of the inner boundary of the territorial sea.

The LDC prohibits dumping of "black-list" substances defined in its Annex I (e.g., organohalogens, mercury and mercury compounds, cadmium and cadmium compounds, persistent plastic oils and oily mixtures, radioactive materials, and agents of biological and chemical warfare) and allows dumping of "grey-list" substances defined in its Annex II only by special permit. Substances that are not on either list require a general permit for dumping, from either the flag State or the loading State.

The Oslo Convention of 1974, titled "The Convention for the Prevention of Marine Pollution by Dumping from Ships and Aircraft," was the first international agreement to regulate the dumping and incineration of wastes at sea by most European countries. Discharges from rivers, estuaries, pipelines, and outfalls are not included. The jurisdiction of the Oslo Convention includes a portion of the Arctic Ocean, the northeastern Atlantic Ocean, and the North Sea. The Oslo Convention has black and grey lists for different pollutants, although the lists vary slightly from those of the LDC. The major difference between the two conventions is that the Oslo Convention has stricter limits for incineration at sea.

The Paris Convention was developed in 1978 by the signatory nations of the Oslo Convention to prevent marine pollution from land-based sources. Also, the contracting parties can adopt discharge standards and environmental quality standards regulating the composition and use of waste substances and products.

The Barcelona Convention for the Protection of the Mediterranean Sea Against Pollution (1978) was developed as part of the Regional Seas Programme of the United Nations Environment Programme. It addresses only dumping from aircraft, ships, and platforms, and pollution from land-based sources.

The Kuwait Convention entered into force in 1979 under the title "Kuwait Regional Conference of Plenipotentiaries on the Protection and Development of the Marine Environment and the Coastal Areas." It is part of UNEP's Regional Seas Programme and focuses on oil pollution from tankers, refineries, and petrochemical industries.

The Helsinki Convention, titled "The Convention on the Protection of the Marine Environment of the Baltic Sea Area," was adopted in 1974 by the seven Baltic Sea States and came into force in 1980. It is the first international marine protection convention that encompasses all pollution sources, including nonpoint agricultural runoff, and it has resulted in some reduction in ocean dumping.

The Bonn Agreement, the 1969 "Agreement for Cooperation in Dealing with Pollution of the North Sea by Oil," is the first regional agreement to promote the development of contingency plans for responding to oil spills and other similar types of accidents.

The International Convention for the Prevention of Pollution from Ships (1973 and Protocols of 1978), often referred to as MARPOL 73/78, attempts to reduce pollution by prohibiting discharges from ships; currently, additional annexes to control substances such as nondegradable plastics, noxious liquids in bulk, and sewage are being considered for adoption.

---

*The Regional Seas Programme of the United Nations Environment Programme encourages international cooperation to abate marine pollution and protect living marine resources. More than 120 coastal nations are part of the Programme, grouped into 10 regions. Each region develops "action plans" that delineate areas of cooperation and adopts conventions which provide legal frameworks for activities in the region. Note that the United Nation's Law of the Sea Convention, of which the United States is not a signatory, is potentially relevant to ocean dumping practices but is not yet in force.

Box R.—Comparison of Factors To Be Considered Before Issuing Permits Under MPRSA Section 102 and CWA Section 403

Ocean Dumping Permits—MPRSA Section 102(a)

Effect of dumping on human health and welfare, including economic, esthetic, and recreational values.

Effect of dumping on fisheries resources, plankton, fish, shellfish, wildlife, shorelines, and beaches.

Effect of dumping on marine ecosystems, particularly with respect to the transfer, concentration, and dispersion of such material and its byproducts through biological, physical, and chemical processes; potential changes in marine ecosystem diversity, productivity, and stability; and species and community population dynamics.

Persistence and permanence of the effects of the dumping.

Effect of dumping particular volumes and concentrations of such materials.

Appropriate locations and methods of disposal or recycling, including land-based alternatives and the probable impact of requiring use of such alternate locations or methods upon considerations affecting the public interest.

Effect on alternate uses of the oceans, such as scientific study, fishing, and other living resource exploitation, and non-living resource exploitation.

Need for the proposed dumping.

In designing recommended sites, the Administrator shall use, wherever feasible, locations beyond the edge of the continental shelf.

SOURCES: 40 Code of Federal Regulations, Secs. 125 and 227.

Ocean Discharge Permits—CWA Section 403(c)(1)

Effect of disposal of pollutants on esthetic, recreation, and economic values.

Effect of disposal of pollutants on human health or welfare, including but not limited to plankton, fish, shellfish, wildlife, shorelines, and beaches.

Effect of disposal of pollutants on marine life including the transfer, concentration, and dispersal of pollutants or their byproducts through biological, physical, and chemical processes; changes in marine ecosystem diversity, productivity, and stability; and species and community population changes.

Persistence and permanence of the effects of disposal of pollutants.

Effect of the disposal of varying rates, of particular volumes and concentrations of pollutants.

Other possible locations and methods of disposal or recycling of pollutants including land-based alternatives.

Effect on alternate uses of the oceans, such as mineral exploitation and scientific study.

[No comparable factor to consider under CWA Section 403]

[No comparable factor to consider under CWA Section 403]

---

the different treatment of dredged and nondredged material was successfully challenged in court in *National Wildlife Federation* v. *Costle* (often referred to as Costle II; 629 F.2d 118 (D.C. Cir. 1980)). The court held that EPA must consider all Section 102(a) criteria in developing regulations, but that it is not bound to apply all criteria to every permit decision or to every type of waste material.[4]

---

[4]Prior to 1974, under Sec. 103(d), COE could apply to EPA for a waiver of the environmental impact criteria. Only one waiver was ever applied for and it was not granted; the 1974 amendments to MPRSA prohibited EPA from issuing such waivers.

The Continuance of Ocean Dumping and the *City of New York* Decision

Throughout most of the 1970s, EPA invoked a policy of phasing out all ocean dumping and encouraging municipal and private dumpers to seek land-based alternatives. In 1977, Congress statutorily mandated phasing out all "harmful" sewage dumping by December 1981 and later imposed a similar deadline for terminating the dumping of industrial wastes. These stringent deadlines were set primarily because several severe marine pollution incidents in the mid-1970s had heightened public awareness of actual and potential adverse health and environmental impacts from marine disposal.

The 1981 deadlines for phasing out dumping of harmful sewage sludge and industrial waste initially seemed a way to bring an end to ocean dumping. In fact, since 1973 about 319 permits or permit applications have been withdrawn, phased out, or denied. Some large municipalities (e.g., Philadelphia) ceased dumping sewage sludge in the ocean.

In 1981, New York City brought suit against the EPA to stop implementation of the regulations. In *City of New York* v. *United States Environmental Protection Agency*, the Federal District Court in New York ruled that dumping of municipal sewage sludge in the New York Bight could not be banned without full consideration of the costs and environmental consequences of alternative disposal methods. According to the court, EPA's conclusive presumption that many materials which fail ocean environmental impact criteria will unreasonably degrade the environment was arbitrary and capricious. Many factors, including the environmental and socioeconomic impacts of alternative disposal options, also needed to be considered when analyzing the acceptability of a given disposal alternative. The decision granted New York City and several other sewerage authorities in New York and New Jersey permission to continue dumping sludge on an interim basis, even though MPRSA did not allow interim permits to be granted after December 31, 1981. Thus, the court decision effectively postponed the December 1981 deadline.

The interim permit procedure under Section 102 has been considered by some observers to be a ''substantial loophole'' which allows the dumping of materials that do not meet ocean disposal criteria (12,214), even though the justification for interim permits was to provide time for research and the development of alternative, land-based options.[5] Twenty-two interim permits had been granted by 1980. After 1981, fewer than 10 such permits remained in effect, but the terms of these permits were extended (291). The amount of sewage sludge dumped in marine waters steadily increased during this time, while the disposal of industrial wastes

declined dramatically (531). A number of cities, including Philadelphia, Boston, Washington, D.C., Seattle, and San Francisco have indicated that they would consider ocean dumping as a potential disposal option in the future if it was permitted.

EPA did not appeal the 1981 court decision. In light of the decision and various arguments that a total ban on marine waste disposal was unnecessary and perhaps counterproductive (377), EPA began to focus on developing a more comprehensive management strategy. EPA is still in the process of promulgating new regulations based on the decision. Major questions remain, however, about how much analysis will be required when the economic and technical feasibility and environmental soundness of alternative options are considered. It is also unclear how decisions will be made when alternatives to marine disposal of sludge are not environmentally superior or readily implemented (291,502). The philosophical shift from ocean protection to management is not yet incorporated into MPRSA, but amendments passed by the House of Representatives in 1985 showed some movement in that direction (H.R. 1957).

The Federal Water Pollution Control Act

The Federal Government has played a role in the abatement of water pollution since the turn of the century. Initially, the Federal role was limited to offering assistance to the States in cases that involved interstate waters. This role has gradually increased over the last several decades: today's Federal Water Pollution Control Act, the Clean Water Act (CWA) (33 U.S.C. 1251 et seq.), has jurisdiction over all U.S. waters, establishes standards for industries and municipalities, and contributes billions of dollars to the construction of municipal waste treatment plants.

When enacted in 1972, CWA was the most comprehensive and expensive environmental legislation to date. It set the ambitious goal of eliminating all discharges of water pollutants by 1985, and had an interim objective, where possible, of making the Nation's waters ''fishable and swimmable'' by 1983. Major revisions were made in 1977 and 1981, which among other things modified these deadlines. The Water Quality Act of 1987 further amended the act (see box C in ch. 1).

---

[5]For example, an interim permit was used to phase out the disposal of sewage sludge by Philadelphia, which was able to develop land-based options (578).

The primary purpose of CWA is to restore and maintain the chemical, physical, and biological integrity of U.S. water resources. To accomplish this, Congress established a combined Federal and State system of controls to implement water programs. CWA consists of two major parts: the Federal grant program to help municipalities build sewage treatment plants (Title II); and the pollution control programs, which consist of regulatory requirements that apply to industrial and municipal dischargers. Responsibility for implementing and administering CWA programs is delegated to States that can demonstrate that they have the legal authority and resources to do so.

NPDES and the National Pretreatment Program—Sections 402 and 307

Under Section 402, all facilities—industrial and municipal—discharging directly into the navigable waters of the United States are required to obtain a National Pollutant Discharge Elimination System (NPDES) permit. "Direct" discharges regulated under NPDES must:

1. comply with applicable effluent limitations;
2. not result in violation of applicable water quality standards; and
3. for marine discharges, comply with the Ocean Discharge Criteria (Sec. 403).

Industrial effluent limitations are based on national guidelines developed by EPA for major industrial categories.[6] Municipal effluent limitations are based primarily on requirements to provide "secondary" levels of treatment (ch. 9).

Section 307 established the National Pretreatment Program (40 CFR 403.5), which authorizes and mandates municipalities operating publicly owned treatment works (POTWs) to develop a pretreatment program capable of regulating industrial discharges into municipal sewers ("indirect" discharges). General pretreatment standards prohibit the discharge of pollutants that can create a fire or explosion, or damage or interfere with POTW operations. Categorical pretreatment standards have also been developed for major industrial categories; these are intended to remove pollutants that might otherwise pass through POTWs into U.S. waters.

Implementing the NPDES and pretreatment programs has affected marine waste disposal in at least two major ways.[7] First, it has resulted in the generation of large quantities of treatment sludges, some of which have been considered for marine disposal—particularly municipal sludges. Second, it provides direct control over the discharge of pollutants from point sources to marine environments, or to other bodies of water that eventually reach marine waters.

Types of Pollutants Regulated.—When first adopted, CWA focused primarily on the control of highly visible *conventional* pollutants such as suspended solids and, as added later, oil and grease. There was, however, increasing recognition of the serious impacts associated with the discharge of *non-conventional* and *toxic* pollutants.[8] The development of a list of so-called toxic "priority pollutants," resulting from settlement of a suit (commonly known as the "Flannery Decree") brought against EPA by the Natural Resources Defense Council (NRDC), reflected this growing concern.[9] This list was incorporated into CWA in the 1977 amendments. Pollutants listed were to be the first toxic pollutants for which EPA would develop pollution control standards.

Types of Standards Governing Pollutant Discharges.—Each NPDES permit contains effluent limitations on specific pollutants that are present

---

[6]The "fundamentally different factors" (FDF) variance procedure (40 CFR 125.30-32) allows a discharger to apply to EPA for modification of an effluent limitation when additional information demonstrates that the characteristics of the discharge are "fundamentally different" from those considered when the effluent limitation was set. If a variance is granted, EPA or a delegated State tailors an effluent limitation to the discharge. Some observers have expressed concern that the use of such variances, as encouraged by the courts (e.g., *Chemical Manufacturers Association* v. *NRDC;* 105 S. Ct. 1102, 1985), could lead to less stringent controls on toxic water pollution (176). The Water Quality Act of 1987 authorized EPA to grant FDF variances under strictly limited conditions.

[7]NPDES and the National Pretreatment Program, including problems associated with their implementation, are discussed in detail in ch. 8.

[8]See box A in ch. 1 for definitions of these classes of pollutants.

[9]The current list of 126 regulated toxic pollutants is largely the result of two court settlements, *Natural Resources Defense Council* v. *Train* (Civ. A No. 2153-73 (D.D.C. 1976)) and *Natural Resources Defense Council* v. *Costle* (636 F.2d 1229 (D.C. Cir. 1980)), which require EPA to develop technology-based effluent limitations based on Secs. 301 and 304 for these priority pollutants. The Flannery Decree also included a list of primary industrial categories for which EPA was to develop specific effluent limitations. See ch. 8 for further discussion.

in the discharge. These effluent limitation standards are either technology-based, as set forth in Sections 301 and 304, or water quality-based, as set forth in Section 302. Technology-based standards are derived from estimates of the removal of pollutants that could be achieved through application of best practicable technology (BPT), best available technology (BAT), or best conventional technology (BCT). EPA or a State with an approved NPDES program is responsible for translating the applicable standards into specific effluent limitations on a permit-by-permit basis.

The 1972 CWA (Sec. 307(a)(2)) mandated that EPA establish toxic effluent standards based on health and environmental considerations such as water quality (567). For a variety of reasons, including lack of needed scientific information, only six toxic effluent standards of this type were ever developed (177).[10] The 1977 amendments, largely through the incorporation of the Flannery Decree, further directed EPA's efforts toward the development of technology-based standards. These standards are derived by estimating the extent of pollutant removal accomplished through use of a particular level of control technology.

The legislation required the use of increasingly stringent control technology. For existing sources discharging directly to U.S. waters, BPT primarily designed to control conventional pollutants was to be employed initially; later, BAT was to be introduced for toxic and non-conventional pollutants and BCT for further reduction of conventional pollutants (Secs. 301 and 304). For new sources, compliance with new source performance standards (NSPS) equivalent to BAT/BCT was mandated (Sec. 306). Finally, indirect dischargers using municipal sewers were required to comply with pretreatment standards for existing sources (PSES) and pretreatment standards for new sources (PSNS), which were analogous to BAT and NSPS, respectively.

Since the 1977 CWA Amendments, EPA has promulgated BPT, BAT, BCT, NSPS, and pretreatment standards for most of the primary industries. Industrial sources were originally to have

achieved BPT by July 1, 1977 and BAT/BCT by July 1, 1984. However, final compliance dates for many of these standards have yet to be reached (ch. 8).

CWA retained provisions to allow the development of water quality-based standards. Section 303 requires States to set water quality-based standards for their waters. If a permitted discharge is likely to violate these standards, Section 302 requires that water quality-based effluent limitations be incorporated into the discharge permit to ensure achievement of the standards. Several States with approved NPDES programs have instituted a number of innovative approaches to water quality-based permitting (130).

Dredged Material Disposal—Section 404

The disposal of dredged material in U.S. waters is regulated by several statutes (582). Under the Rivers and Harbors Act of 1899, COE has authority to regulate any activity in rivers and coastal waters which could directly interfere with their navigability. Although much of the law has been superceded by CWA and other laws, COE still uses this authority, for example, to regulate dredge and fill activities beyond the 3-mile limit. As noted above, Section 103 of MPRSA controls the *dumping* of dredged material in coastal waters and the open ocean. The *discharge* of dredged or fill material is regulated under Section 404 of CWA.

The 404 program is complicated and somewhat controversial (572,582). COE evaluates permit applications using guidelines developed jointly with EPA, and in light of review comments by EPA, the Fish and Wildlife Service, the National Marine Fisheries Service, and the States. EPA can veto any proposed sites for dredged or fill material disposal. Where COE's jurisdictions under Section 103 of MPRSA and Section 404 of CWA overlap in the territorial sea, COE typically issues an ocean dumping permit.

Provisions Specific to Marine Waters

The Ocean Discharge Criteria.—Section 403 of CWA requires that all NPDES-permitted discharges from point sources into certain marine environments—the territorial seas, the contiguous zone, or the open oceans—must not "unreasona-

---

[10]This slow rate of progress was one of the primary factors that motivated NRDC to bring suit against EPA in the first place.

bly degrade the marine environment'' (225).[11] Under this delineation, marine waters shoreward of the baseline are excluded, and thus the criteria do not apply to discharges into estuaries and coastal waters such as Chesapeake Bay, New York Harbor, and Puget Sound (45 FR 65944, Oct. 3, 1980). Section 403 only began receiving dedicated funding in fiscal year 1987. EPA, however, is considering applying the criteria to estuaries and other waters *inside* the baseline of the territorial sea. Given that the criteria are considered relatively stringent, this could provide an additional level of protection for these waters.

The ocean dumping and ocean discharge regulations (40 CFR 227 and 40 CFR 125, respectively) rely on similar data and require similar decisions, but for different activities.[12] This has led some observers to argue that the criteria should at least be consistent (377). The main differences between the two sets of criteria is that MPRSA has additional requirements to consider the *need* for the proposed dumping and to use locations, when possible, beyond the edge of the continental shelf (see box R).

Waivers from Secondary Treatment.—Section 301(h) of CWA exempts qualified POTWs that discharge into marine waters from the requirement to achieve secondary treatment; yet, it still requires monitoring, implementation of existing pretreatment requirements, and compliance with existing water quality standards. EPA adopted final amended rules in 1982 (47 FR 53666, Nov. 26, 1982), and a total of 208 applications were received by the administering EPA regional offices. By January 1987, EPA had approved 46 applications; another 125 applications were withdrawn or denied, and no final action had been taken on the remain-

---

[11]The definition (40 CFR 125.121) of ''unreasonable degradation of the marine environment'' is:
  - significant adverse changes in ecosystem diversity, productivity, and stability of the biological community within and surrounding the discharge area;
  - threat to human health through direct exposure to pollutants or through consumption of exposed aquatic organisms; or
  - loss of aesthetic, recreational, scientific, or economic values that is unreasonable in relation to the benefit derived from the discharge.

[12]In a 1977 case, *Pacific Legal Foundation* v. *Quarles* (440 F. Supp. 316), the court found that these criteria could be applied concurrently to ocean discharges or dumping. This combination of discharge and dumping criteria was subsequently challenged and in 1979 the court ordered EPA to issue new guidelines for Sec. 403 ocean discharge permits (*Pacific Legal Foundation* v. *Costle*, Civ. No. 5-79-429-PCW).

ing 37 (see ch. 9, figure 34). The 301(h) program was initially envisioned as appropriate primarily for west coast municipalities discharging effluent into deep, cold waters, and for the most part EPA's decisions reflect this intent.

Provisions Addressing Comprehensive Waste Management in Estuaries and Coastal Waters

A number of statutory provisions potentially bear on estuaries and coastal waters. Currently, 21 programs—under 8 different statutes administered by 11 different Federal agencies—affect these waters in some way (670). Clearly, efficient management of estuaries and coastal waters requires careful integration and coordination of these various programs. Several provisions of CWA address or could address long-term planning and management efforts in estuaries and coastal waters:

- estuarine programs (Sec. 104),
- estuarine management conferences, and
- area-wide planning (Secs. 208 and 303).

National Estuary Program—Section 104.—CWA is the primary statute governing pollution in estuarine and coastal marine environments. Section 104(n) directs EPA—through appropriate coordination of interagency, intergovernmental, and public and private sectors—to conduct comprehensive studies on the effects of pollution on estuaries and estuarine zones. EPA was directed to coordinate interstate pollution abatement and management in the waterbodies and to transfer funds to NOAA to develop a comprehensive water quality sampling program. To carry out these responsibilities, EPA created the Office of Marine and Estuarine Protection (OMEP) and NOAA created the National Estuarine Program.

Appropriations were first made in fiscal year 1985, when $4 million was designated as part of Public Law 99-160 for water quality research, monitoring, and assessments in four waterbodies: Long Island Sound, Narragansett Bay, Buzzards Bay, and Puget Sound. This initiative has since been known as the National Estuary Program (NEP). Additional waterbodies—San Francisco Bay and the Albemarle and Pamlico Sounds in North Carolina—were added to NEP in April 1986. The 1986 budget was $5.6 million. The Water Quality Act of 1987 authorized additional funding and provided more direction for NEP (ch. 1).

A Comprehensive Master Environmental Plan is being developed for each waterbody. Ideally, each plan will address the control of point and nonpoint sources of pollution, implementation of environmentally sound land-use practices, the control of freshwater input and removal, and the protection of living resources and pristine areas. In addition, the plans are supposed to delineate public participation and monitoring programs, and identify personnel and funding needs. The focus of the Federal effort is on planning and management; given statutory limits, implementation of the plans will generally be left to local or State authorities. In most cases, this means that EPA supports the efforts of a particular local or State planning or management agency, rather than serving as the lead agency for an area.

In some areas, however, the coordination between Federal and State efforts has not been entirely smooth. In the Puget Sound region, for example, programs of the State of Washington and of EPA Region 10 currently are separate, but loosely coordinated. One source of contention is that EPA has kept control of the $1.4 million received by the area from the Federal Government and restricted the participation of all other agencies to "review and comment." The Puget Sound Water Quality Authority contends that greater Federal-State coordination in deciding on the priorities for spending these moneys will be needed to avoid having the two programs operate in different directions (K. Skinnarland, Puget Sound Water Quality Authority, pers. comm., 1986).

Drawing on experience with the Chesapeake Bay and Great Lakes Programs, EPA has developed a draft manual that will provide guidance on the development of comprehensive management plans for current and future sites. In a related effort, EPA's Near-Coastal Waters Strategic Planning Initiative is identifying implementation options that EPA could pursue to better control point and nonpoint sources, protect living resources, and manage land use in and around estuaries and coastal waters (670).

Estuarine Management Conferences.—The Water Quality Act of 1987 authorized EPA to convene management conferences to solve pollution problems in estuaries. The conferences would be authorized to:

1. collect data on toxics and other pollutants within an estuary,
2. develop comprehensive conservation and management plans that recommend priority corrective actions and compliance schedules to control point and nonpoint sources of pollution,
3. monitor for program effectiveness, and
4. develop plans for intergovernmental coordination for implementation.

Areawide Planning—Section 208 and 303.—Two existing sections of CWA address regional or areawide planning and can be applied to estuary management. The intent of Section 208 is to link various water pollution control requirements on the basis of watersheds, primarily to control nonpoint source pollution. Section 303(e) provides for a Continuing Planning Process by States and is another regional approach to water quality management. This provides for coordination with Section 208 and emphasizes better implementation of water quality standards.[13]

Under Section 208, an agency of local governments is selected by the governor(s) of the State(s) to coordinate regional planning. The emphasis has frequently been on controlling nonpoint sources and linking their control with controls on wastewater and storm discharges. This involves coordinating State and local efforts, with at least partial guidance and funding by the Federal Government. However, the program encountered numerous problems resulting from disagreements among State and local officials over authority for implementation, discontinuity in funding levels, inadequate technical information on nonpoint pollution, and delays by EPA in issuing rules and guidelines (88, 557,570,699). Section 208 funding was terminated in 1981, although some funds for areawide planning continue to be distributed under Section 203(j) of CWA.

---

[13]The Water Quality Act of 1987 included a provision that would require the inclusion of proposed treatment works in areawide Sec. 208 and Sec. 303(e) plans. The act also included a provision that would establish a program for management of nonpoint sources of pollution. The program would provide $400 million for 4 years to States to develop nonpoint source management programs.

Section 208 plans have been developed for some coastal areas. In San Francisco Bay, for example, a regional body (the Association of Bay Area Governments) received Section 208 funding and produced a comprehensive Environmental Management Plan in the late 1970s. The plan covered air, water, and solid waste management for the Bay, and called for, among other things, establishing a research program to improve monitoring and understanding of pollutant impacts in the Bay. As a result, in 1982 the San Francisco Bay Regional Water Quality Control Board adopted the Aquatic Habitats Program Plan to assess pollutant effects in the San Francisco Bay/Delta estuary (R.H. Whitsel, California Regional Water Quality Control Board, pers. comm., November 1986; also see app. 1). In the Puget Sound area, the Washington Department of Ecology used Section 208 funding to develop a dairy waste management plan. In the Chesapeake Bay watershed, both Sections 208 and 303(e) were used to prepare and adopt a number of river basin plans to help alleviate water quality problems; these programs, however, have achieved only limited success (168).

KEY ISSUES AFFECTING MARINE WASTE DISPOSAL PROGRAMS

The most important findings and policy options discussed in this report relate to the need to improve current water pollution control programs, the need for and desirability of more comprehensive management in estuaries and coastal waters, and the great need for information in these areas (ch. 1). Three sets of issues are critical to understanding these findings and options and to the development of sound marine waste disposal management:

1. issues associated with the management of industrial effluents under current water pollution control programs,
2. issues related to the effectiveness of existing comprehensive waterbody management programs, and
3. issues related to the status and needs of relevant information programs.

Industrial Effluents and Current Water Pollution Control Programs

OTA's analysis of current water pollution control programs discovered several key problems related to the adequacy of the regulatory framework for controlling point source pollution. These issues are briefly summarized here and discussed in detail in chapter 8.

Delays in Program Implementation.—Federal regulations for some significant industrial categories have yet to be promulgated or have compliance dates that have not yet been reached, and enforcement actions cannot be taken until compliance dates have been reached. Incomplete and inconsistent identification and permitting of dischargers is also a widespread problem. Finally, some POTWs have been slow to develop pretreatment programs and have them approved by States or EPA, and thus many indirect industrial dischargers remain essentially unregulated.

Gaps and Deficiencies in Program Coverage. —For a variety of reasons, several significant industrial categories and many toxic pollutants—both priority and nonpriority—remain unregulated under the current framework. Moreover, the incorporation of new or upgraded effluent limits even for regulated pollutants and regulated industries has been sporadic and slow. Finally, only marginal development and use of water quality-based standards for toxic pollutants has occurred.

Inadequacy of Regulatory Compliance and Enforcement.—Problems in three major areas exist:

1. the quality and completeness of data submitted by dischargers;
2. the extent of noncompliance with effluent standards or other permit requirements; and
3. the extent, timeliness, and effectiveness of enforcement actions taken in response to violations.

Additional Issues Facing the Pretreatment Program.—Other issues that must be addressed include the potential for conflict between the need for local control and national consistency; the lack of incentives for full implementation and enforcement of pretreatment programs; and the adequacy of controls over the legal discharge of hazardous waste into sewers.

Waterbody Management Programs

One key finding of this report is that estuaries and coastal waters are in need of further protection if even the current level of water quality is to be maintained. Several recently established programs are attempting to provide more comprehensive and coordinated management of estuaries and coastal waterbodies. This section briefly illustrates the general approaches, capabilities, and deficiencies of several of these "waterbody management" programs, including some non-Federal programs.[14]

A variety of local, State, and national programs exist to manage estuaries and coastal waterbodies (table 12 and app. 1). Some programs address only one waterbody, while others address multiple areas. The Chesapeake Bay Program, for example, focuses on a single estuary, while the National Estuary Program currently is conducting activities in six areas. Programs are also initiated at various levels of government. For example, some programs are initiated primarily by the Federal Government (e.g., the Chesapeake Bay Program), while others are initiated by the States (e.g., Puget Sound Water Quality Authority), or by local authorities (e.g., the Southern California Coastal Waters Research Project).

Regardless of the level at which a program is initiated, a number of agencies from different levels of government are likely to be involved in implemention. The Federal EPA and State environmental protection departments generally are involved in various aspects of a program. In addition, other Federal agencies (e.g., COE, NOAA, Fish and Wildlife Service), their counterparts at the State level, and various municipal and county authorities (e.g., port districts, sewerage authorities) can have specific responsibilities or interests in managing the water quality of estuarine or coastal waters. The Puget Sound Water Quality Authority (PSWQA), for example, involves the coordina-

tion of many State, regional, and local government agencies (464,513).

Waterbody management programs are designed to serve a variety of functions and their structures vary accordingly (table 12). The wide variety of programs is understandable. Some programs are designed primarily to share information about research needs or findings; some are given decision-making authority only for distributing research funds; others have clear goals for improving water quality and have authority for planning and/or coordination.

For example, the Southern California Coastal Waters Research Project (SCCWRP) focuses its research on the environmental effects of marine disposal of municipal wastes; in contrast, the Aquatic Habitat Institute (for San Francisco Bay) directs its research to facilitate coordination of the efforts of other regional, State, and Federal agencies in the area. For the Chesapeake Bay, all these functions—research, planning, and program coordination—are the responsibility of one management body, the Chesapeake Bay Program.

At least two broad needs must be met by any program designed for the management of waterbodies: 1) it must possess sufficient statutory and regulatory authority to carry out its assigned functions, and 2) it must have the ability to coordinate other agencies and programs already involved in some aspect of managing the waterbody. Since numerous Federal, State, and local programs and agencies are typically involved in the management of individual waste types and/or sources for a given waterbody, it is essential that the various programs be informed of each other's actions and that lines of authority and jurisdiction be clearly defined.

Within these two general areas of need, a number of specific functions can be performed by a program, including:

- **Planning**: Includes setting priorities among pollution sources, waste types, or pollutants; setting goals and target dates for their achievement; scheduling research and other programmatic activities; and planning the allocation of resources.
- **Initiating research and establishing data requirements**: Includes identification of research needs, and initiation and coordination of re-

---

[14]It does not attempt to evaluate the successes and failures of the selected programs, to identify all existing programs, or to identify geographic areas in great need of such programs. Box S describes some selected international perspectives on waterbody management.

Box S.—Selected International Perspectives on Marine Waste Disposal

Although this report does not analyze the management of marine waste disposal by other nations in detail,* several innovative water pollution management programs in Europe suggest approaches that could be considered by the United States. This box highlights two river basin management programs in France and West Germany that use fee systems and regional plans to improve the management of waste disposal. In addition, it describes an estuary management program in Cubatao, Brazil, which exemplifies both the severity of marine problems in a developing nation and the strong effort that can be made, despite political and economic constraints, to cope with such problems.

France

In 1964, France became the first country to adopt a fee system on a nationwide basis (224). The system consists of independent river basin agencies that develop plans (under national supervision) and establish fees for both water extraction and waste discharge. The purpose of the system is to provide an integrated approach to water management, by complementing the existing system of permits and regulations. The overall approach to water management thus is a mix of fees and standards, which is typical in Europe (e.g., the Netherlands, West Germany) and Japan (43,56).

Different fees are applied to domestic and industrial dischargers; for the latter, fees are levied on suspended matter, oxidizable matter, dissolved salts, and toxic matter. The fee system makes the river basin agencies self-financing, and any surplus money generated by the charges over the operating costs of the agencies is used for water management projects.

West Germany

In 1976, West Germany enacted the Federal Water Act (which closely parallels the U.S. Clean Water Act) and the Effluent Charge Law to address water pollution. These statutes established a two-part permit system that consists of technology-based effluent limitations for particular discharges and charges on the amount of discharge expected during the permit period. The charges are based on characteristics such as settleable solids, chemical oxygen demand, ''fish toxicity,'' and cadmium and mercury concentrations (56,305).

The charge system allows flexibility in achieving compliance with discharge standards and promoting waste reduction efforts, although its applicability to most **toxic** pollutants may be limited (43,56,305,657). The West German system has not been fully evaluated, but it appears to be: 1) more manageable than first anticipated, and 2) effective in providing incentives to dischargers to achieve and maintain compliance with the Federal Water Act standards (56). Some areas in the United States have experimented with similar user fee approaches (e.g., the Fox River area in Wisconsin; 305).

Brazil

The nature of marine pollution problems facing rapidly industrializing nations and the availability of resources to address such problems often differ dramatically from those of older, more industrialized nations (304). Much of the initial challenge in rapidly industrializing nations has been to understand the nature of their marine environments, define specific problems, and clarify the relationship between these problems and various land-based activities (3).**

Brazil is one example of an industrializing nation attempting to better manage its marine resources. It has initiated efforts to increase ocean research, education, and the use of marine resources, for example by

---

*Information about the policies and practices of other countries regarding the dumping of wastes in marine waters was obtained, in response to a 1985 OTA questionnaire, from 17 countries: Canada, Belgium, Denmark, the Federal Republic of Germany, France, German Democratic Republic, India, Israel, Japan, Korea, Norway, the People's Republic of China, Portugal, South Africa, Sweden, the Netherlands, and the United Kingdom. As might be expected, policies range from prohibition, to varying degrees of regulated usage, to lack of regulation. This information is summarized in FIT (165). Information about the *amounts* of wastes dumped by other countries is summarized in box I, ch. 3 of this report.

**The U.S. Agency for International Development and the University of Rhode Island's Coastal Research Center, instituted the International Coastal Resources Management Project to aid such countries. Under this project, coastal resources management programs have been established in Ecuador, Thailand, and Sri Lanka to promote ecologically and economically sound management of marine resources and coastal regions; data collection and technical assistance are primary features of the programs (L. Zeitlin-Hale, Coastal Research Center, pers. comm., December 1986).

establishing an Interministerial Committee for Marine Resources in 1979 and issuing a "National Policy for Marine Resources" in 1980.

Brazil also has attempted to tackle some specific marine pollution problems, for example in the Santos Estuary, one of the most polluted estuaries in the world. Santos is a major port and tourism area, located downstream from Sao Paulo (a city of 15 million people) and Cubatao, major industrial area. The Santos Estuary receives raw sewage from Sao Paulo and industrial and hydroelectric powerplant discharges from Cubatao. High levels of oxygen-demanding substances, phenols, metals (e.g., copper and zinc), and pesticides have been detected in the water, and metals and pesticides have been found in sediments. Some observers suggest that the chronic pollution of the estuary could cause a total collapse of its ecosystem.

In 1983, the Brazilian environmental agency (CETESB) established the Program for Environmental Pollution Control to survey pollution sources, inventory emissions to the estuary, and develop environmental control plans for each industrial source in Cubatao. Public participation has been encouraged throughout the process; for example, CETESB held quarterly public meetings to discuss progress of the plans. Thus far, measurable emission reductions of different pollutants (as well as improved air quality) have been recorded. The program is particularly noteworthy for its development and use of epidemiological studies, biological methods and criteria for assessing toxicity, and models for evaluating environmental risks (196).

search to support the management program; conducting ambient monitoring and establishing databases.
- **Obtaining and allocating funding**: Includes obtaining and allocating financial resources for research, planning, and other program activities.
- **Implementation**: Includes integrating and coordinating basinwide cleanup efforts, for both water quality and resource management (e.g., through the use of management committees).
- **Establishing public participation**: Includes developing and implementing effective mechanisms for public education and participation in decisionmaking.

Critical Function: Planning

One of the central features of most waterbody programs is planning. It can involve research planning or management planning, both of which may include establishing goals for improving water quality and setting priorities for action. Almost all programs are involved in research planning. For example, programs such as SCCWRP and the Aquatic Habitat Institute are oriented primarily towards research. Some programs are involved in both kinds of planning. For example, the Chesapeake Bay Program (CBP) emphasizes comprehensive management for the Bay estuarine system. The Great Lakes Program, which was the first compre-

hensive waterbody management program, and CBP, which is the oldest estuarine management program, are serving as models for the development of the National Estuary Program (NEP). NEP is attempting to identify conditions and trends in the systems and develop comprehensive plans for selected estuaries and coastal waters. Management committees are established in each selected NEP area to carry out the planning function.

The Gulf Coast Waste Disposal Authority (GCWDA) is a program that falls somewhere between a research-oriented regional program and a comprehensive management plan. Its focus is on planning and developing regional facilities for treatment of industrial and municipal wastewater, hazardous wastes, municipal solid wastes, and sludge. It also provides technical assistance to area industries (ref. 215; L. Goin, GCWDA, pers. comm., 1986).

Despite variation among existing programs, certain planning elements are generally necessary. For example, four factors are associated with the Chesapeake Bay Program's success thus far:

1. **Preliminary research**: The effort began by conducting research which was then synthesized and used along with other Bay studies to understand the conditions and trends of the Bay and the sources of pollutants; this scientific information on the Bay's ecological con-

Table 12.—Selected Waterbody Management Programs

Program features	Chesapeake Bay Program	Puget Sound Water Quality Authority	Great Lakes National Program	Gulf Coast Waste Disposal Authority	Southern California Coastal Water Research Project	San Francisco Bay Regional Water Quality Control Board	Aquatic Habitat Institute	National Estuary Program
Primary purpose	Overall Federal-State efforts to control point and nonpoint source pollution to the Bay	State agency to study and report on impacts of pollution on marine and human life; and to devise plan for management whose recommendations are binding on other State and local agencies	Federal-State effort to control point and nonpoint pollution of Great Lakes and basin area, emphasis on toxic pollution; primary activity is research and monitoring	Multi-county unit to control point sources by constructing regional treatment plants to abate pollution of Houston Ship Channel and Galveston Bay	Primarily local research program to study and monitor impacts of municipal discharges on marine life in coastal waters of southern California	State's regional board for water resources control in S.F. Bay area, oversees programs (e.g., Aquatic Habitat Program[a]) to monitor municipal and other point source discharges	Quasi-public organization, conducts independent research on S.F. Bay resources. Also coordinates other research efforts, conducts public education	Federal program to study water quality and pollution effects in selected estuarine waters;[b] coordinates efforts with other Federal, State, and local agencies
Date of initiation	1976	1983	1977	1969	1969	1970	1983	1985
Participating authorities	• Chesapeake Bay Executive Council:* —EPA* (Federal) —State agencies of MD, VA, PA —District of Columbia —other Federal agencies: SCS, NOAA, FWS, Corps of Engineers, USGS, DOD	• Composed of appointees of diverse interests and geographical areas in the Sound region; ex-officio members are the Director of Ecology and the Commission of Public Lands	• EPA (Federal) Great Lakes Nat'l Program	• Composed of 9 members, 3 each from counties of Chambers, Galveston, and Harris	• Commission Members: —Sanitation Districts of Orange, L.A., and Ventura Counties —Cities of San Diego and L.A.	• San Francisco Bay Regional Water Quality Control Board	• Board of Directors:* —U.C. Berkeley —3 dischargers (municipal, industrial, nonpoint) —3 environmental group members —3 regulators (Cal. F&G, Cal. Reg. Bd., EPA)	• EPA (Federal) • NOAA • appropriate State and/or local authorities (lead authority varies in different areas)
Funding	• EPA • Maryland • Virginia • Pennsylvania • Other Federal agencies	• State of Washington	• EPA (Federal)	• Counties • Pollution Control Board • Fees and service charges from dischargers	• Joint Points Agreement (Commission Members) • NOAA • EPA • other local authorities	• California Water Resources Control Board	• EPA (Federal—NEP) • S.F. Bay Regional Water Quality Control Board • California Water Resources Control Board • Donations from dischargers	• EPA (Federal) • State and/or local authorities

[a]This program evaluates present and future effects of pollutants on Bay resources and encourages integration of all Bay-Delta water-related studies.

[b]Individual programs have been established in Puget Sound, San Francisco Bay, Narragansett Bay, Buzzards Bay, Long Island Sound, and Albemarle-Pamlico Sounds, as well as Chesapeake Bay and the Great Lakes.

*Indicates lead authority if more than one authority involved.

SOURCE: Office of Technology Assessment, 1987.

Photo credit: S. Dollar, University of Hawaii

Monitoring waste disposal impacts is a critical component of any waterbody management strategy. Here a diver is surveying a coral reef community near a sewage outfall from Oahu, Hawaii.

dition provided a relatively objective base from which to generate cooperation and develop control programs.

2. **Adequate funding**: The Federal Government and States provided sufficient funds (e.g., nearly $30 million was spent on research over a 7-year period).

3. **Long-term effort**: The pace of the program was deliberate, allowing adequate time to develop a database, clearly define the major problems, and lay the foundation for the institutional relationships necessary for sustaining later efforts.

4. **Strong public participation**: Strong public support existed and an active public participation program was encouraged (21).

The Ability To Set, Review, and Achieve Specific Goals

Comprehensive planning for an estuarine or coastal waterbody involves many elements. First, conditions in the waterbody must be understood and specific goals set to improve trends. An effec-

tive management structure and an effective public participation program must be established; these can involve, for example, the scientific community, periodic review of progress toward achieving those goals, and a master plan endorsed by the public, scientific community, and managers. The implementation of any plan depends to a large extent on the planning agency's ability to involve other entities in the process.[15]

One of the oldest programs is CBP. After 7 years of study, CBP produced the Chesapeake Bay Restoration and Protection Plan. It identifies the Bay's most important problems, assesses current pollutant control efforts, and sets general goals for achieving pollution abatement. The plan addresses both point and nonpoint sources of pollution and sets a goal of restoring the Chesapeake Bay to its condition of the 1950s, recognizing the need for long-term strategies to achieve this goal.

---

[15]In addition, achieving goals would of course depend on the enforcement of existing regulations under various pollutant control programs.

Currently, the Chesapeake Bay Program is implementing a $100 million cleanup effort directed at nutrient control to improve the dissolved oxygen problem in the central Bay. Additional control efforts underway include: 1) a phosphate ban in Maryland and the District of Columbia; 2) new nonpoint source programs in Maryland, Pennsylvania, and Virginia; 3) major point source reductions from municipal sewage treatment plants basinwide; 4) land use controls in Maryland; 5) a moratorium on harvesting rockfish; and 6) submerged aquatic vegetation restoration efforts (96). These efforts have only recently been initiated, so it is too soon to judge the success of CBP's transition from planning to implementation.

Two important issues regarding the Chesapeake Bay Executive Council are its ability to: 1) define specific goals for program managers in the various State and local governments, and 2) influence actual practices in these jurisdictions. Some observers have suggested that the Council should recommend water quality standards, establish baywide goals for inputs of CWA priority pollutants and nutrients, and identify point and nonpoint control strategies to achieve them. These observers further suggest that the Council adopt some features of the Great Lakes Water Quality Agreement of 1978 (168,590), which focuses on toxic pollutants and establishes specific goals and standards.

A program is likely to be more effective if it has adequate review procedures so it can adjust to changing conditions and priorities. Yet, neither the CBP nor the Great Lakes Water Quality Board, for example, have specified review periods to update their management plans or agreements.[16] On the other hand, determining the appropriate time periods for reviews can be difficult. In fact, CBP officials have expressed resistance to updating the 1985 plan anytime soon, maintaining that in sufficient time the existing plan will lead to more stringent control efforts (590).

One example of a program that has relatively broad authority is the Puget Sound Water Quality Authority (PSWQA). One of its greatest

strengths is its clear statutory authority to be the lead agency for managing and protecting water quality in Puget Sound (K. Skinnarland, PSWQA, pers. comm., September 1986). PSWQA has developed a comprehensive management plan for the Sound and is authorized to produce biennial reports on the state of the Sound. Its recommendations are binding on all other State and local government agencies involved in Puget Sound water quality management. The lines of authority and coordination among the various jurisdictions are specified by PSWQA in the management plan. The authority also can revise its management plan, which should allow for quicker assessments of its success in meeting goals and for changing priorities.

In addition, the Puget Sound Water Quality Management Plan is more comprehensive and detailed than those for other areas. The emphasis is on preventing pollution by effectively implementing programs, having adequate staff and funding, and developing a nonpoint source pollution program to address problems that cross jurisdictional lines. It proposes specific programs for several critical areas: water quality, fish and shellfish, wetlands, and wildlife habitat. While based on up-to-date scientific information, the plan also recognizes the need (given the uncertainty surrounding many issues) for continued support of research and monitoring in the Sound (464). Clear goals are defined, guidelines for priority-setting are established, standards for development and implementation of the program are specified, and a schedule for completing the planning of programs is set (subject to revision). Although the plan appears to be a promising approach to water quality management, it was adopted in late 1986 and it is too soon to judge its effectiveness.

The Importance of Coordination

Adequate cooperation among multiple jurisdictions and among various agencies is likely to be crucial to successful waterbody management. Difficulties arise because of jurisdictional disputes and because land-use management issues are involved. Such problems can be overcome, however. For instance, CBP has achieved a remarkable degree of cooperation between the multiple jurisdictions of the Bay. Maryland and Virginia each are developing some land-use management programs, but

---

[16]Despite the 1978 agreement, high levels of toxic pollutants continue to flow into the Great Lakes (200,389). This may in part be the result of the Great Lakes Commission's lack of enforcement authority and insufficient authority to encourage participating governments to follow its recommendations (389,538).

the amount of resources available varies with a State's level of interest in the Bay; in particular, Pennsylvania's efforts are relatively small. Yet Pennsylvania, which borders the Bay in only one small region, is the Bay's main nonpoint contributor of nutrients (543).

PSWQA has an advantage in achieving coordination because the sound is located within one State. Moreover, most of its funding is from the Washington State legislature. Even so, there has been some difficulty in coordinating the State and Federal efforts (see above). Other areas such as San Francisco Bay are located within one State, but efforts to develop comprehensive water quality management have been frustrated by the lack of a lead agency with clear authority for coordination of various program efforts.[17]

Two additional factors appear crucial to effective interagency or multiple jurisdiction cooperation: 1) the number of agencies already attempting to manage the waterbody, and 2) the degree of environmental degradation in the waterbody. For example, at the time PSWQA was established there were no well-developed, independent government programs working on comprehensive management plans for the Sound.[18] In San Francisco, on the other hand, several agencies—none with any greater lead authority than the others, and each focused only on particular aspects of management of the Bay's resources—exist and compete for funding and greater authority. The San Francisco Bay Regional Water Quality Control Board could conceivably be the lead agency for San Francisco Bay, but it lacks authority as well as necessary resources (R. H. Whitsel, San Francisco Bay Regional Water Quality Control Board, pers. comm., September 1986). In addition, the Sound is generally consid-

ered to have less severe environmental problems than other areas such as San Francisco Bay or Chesapeake Bay (464).

One purpose of the management committees established by EPA's NEP is to encourage cooperation by bringing together the managers of the various organizations involved. Recently, mediation has been used to help resolve conflicts among various jurisdictions involved in development planning for an estuary (i.e., the Columbia River estuary in Oregon and Washington; (216)). This technique could be applied to developing waste management plans for an estuary or coastal area where there are disagreements among the controlling jurisdictions.

Additional Functions—Research, Funding, and Public Participation

Research.—One major objective in most waterbody programs is the study of the existing conditions in a particular waterbody and, in some cases, the coordination of research efforts in the area. For example, EPA's Great Lakes National Program monitors water, sediments, fish tissue, and air deposition to identify critical areas in the lakes that need remedial action. It also prepares plans for phosphorous control and for nonpoint source control of conventional and toxic pollutants (1,663). CBP and PSWQA also play lead roles in coordinating and conducting research for Chesapeake Bay and Puget Sound, respectively, and research is the sole purpose of SCCWRP.

A unique approach is being tested for San Francisco Bay, where numerous agencies are involved in research and management and no one program has the lead authority to coordinate these efforts. Disagreements over the interpretation of research findings led the State and Regional Water Quality Boards to create the Aquatic Habitat Institute. Its purpose is to conduct independent research on the Bay and serve as an unbiased authority on scientific and technical matters related to the Bay (D. Segar, Aquatic Habitat Institute, pers. comm., September 1986).

Funding.—Adequate funding is obviously essential for any waterbody management program. Existing programs obtain funding in several ways, for example, from government agencies and from user fees or other revenue-generating mechanisms.

---

[17]For point sources, the San Francisco Bay Regional Water Quality Control Board has authority to formulate and adopt water quality control plans and in the process must consider recommendations of affected State and local agencies. The basin plan documents for this must be approved by the State Water Resources Control Board. In this way, for *point sources* the Regional Board acts to encourage regional planning and takes any action required within its authority to achieve water quality control; however, no authority in the Bay area coordinates *comprehensive* management activities (i.e., both nonpoint and point source controls) (R. H. Whitsel, California Regional Water Quality Control Board, pers. comm., November 1986).

[18]One exception is EPA's Puget Sound Estuary Program which focuses on problems of contaminated sediments in the urbanized bays of Puget Sound.

Strong governmental funding, for example, has been key to CBP's accomplishments. CBP has obtained more research money from the Federal and State governments than any other program, in part because Chesapeake Bay is the Nation's largest estuary and has high commercial and recreational value. Initial funding for CBP came from the Federal Government, but recently the States have assumed more responsibility. Currently, States (with Maryland as the main contributor) contribute $47 million and EPA contributes approximately $10 million each year for research and monitoring (334, 543).

Other programs have been funded quite differently. For example, PSWQA is funded by the State of Washington. One of the innovative features of the Aquatic Habitat Institute is its funding: the institute was created by the California legislature and receives about one-third of its funding from the State, one-third from EPA's National Estuary Program, and the remaining third from sources such as donations from municipal and industrial dischargers. The institute is required, however, to develop its own funding strategies to eventually support itself as an independently funded, nonprofit organization. It is likely to use discharger taxes or user surcharges, rather than line-item appropriations as most waterbody programs do. The Gulf Coast Waste Disposal Authority (GCWDA) funds its operations with non-public sources of revenue by issuing bonds to build waste treatment facilities that are then repaid by the industries or municipalities involved. GCWDA, although a unit of government, is designed to operate much like a business; any excess funds generated by its pollution control programs are used for other experimental or innovative programs (L. Goin, GCWDA, pers. comm., 1986).

Public Participation.—Public participation is also critical to the success of waterbody management programs. It provides people an opportunity to have a say in decisions that affect them, and it can help ensure that economic and technical issues are not considered in isolation from relevant social and political aspects of environmental problems. In addition, waterbody management is likely to be given higher priority if the public is greatly concerned and well-informed about protecting a particular waterbody. For this reason, public education programs, as well as public participation in citizen advisory panels or through other means, are important aspects of any management program.

Most waterbody management programs make some provision for public participation (e.g., PSWQA, Great Lakes Water Quality Board). CBP has encouraged particularly strong citizen involvement through the Citizen's Program for the Chesapeake Bay, Inc., an alliance of nonprofit organizations formed in 1971. The public is also involved through the Chesapeake Bay Foundation, a nonprofit organization with an endowment of $3 million and an annual budget of $400,000. The foundation has initiated educational and land acquisition projects, as well as activities in legislative, administrative, and judicial proceedings (21,543).

Summary of Waterbody Program Functions

Estuaries and other waterbodies do not recognize political boundaries, so programs for their comprehensive management often require the coordination of many political jurisdictions and agencies. This can greatly complicate the functioning of any such programs. Frequently, the implementation of several statutes is also involved.

It is critical that there be a lead agency to coordinate the efforts of everyone involved and establish clear lines of responsibility and authority in any effort to better manage estuaries and coastal waters. The success of establishing such authority often depends on how well-established existing institutions are, because agencies are generally reluctant to surrender authority to other agencies.

Other factors are also critical to the successful functioning of waterbody programs. These include:

1. adequate study and assessment of the waterbody, including peer review of the findings as part of the development of an adequate scientific basis for decisionmaking;
2. setting specific goals and priorities;
3. the ability to evaluate the program on a continuous basis and shift priorities for action accordingly;
4. sufficient funding and staff to support these efforts; and
5. strong public participation programs.

The role of the Federal Government in waterbody management programs varies greatly. The

Federal Government has been very active in initiating and participating in CBP, and has been referred to as "the glue which binds the Bay Program together" (168). Its role in other programs has been more peripheral. Even in the National Estuary Program, which is administered by EPA, the Federal role is primarily one of guidance. In some areas, for example, the Federal Government uses NEP to channel Federal funding to the lead agency of a waterbody management effort. In other cases, Federal money is sprinkled among regional, State, and local agencies, which can reinforce the tendency toward fragmented efforts or lead to duplication of or competition between efforts. Enforcement may be an area where a strong Federal presence is appropriate; it can also be argued, however, that the States should have greater control of enforcement programs because they are in closer proximity to the problems, and that EPA's role should be one of strong oversight.

Information Programs

An assortment of public and private entities generate and disseminate the information that is needed to develop and implement sound waste disposal policies. Much has been done in recent years to improve the Nation's ability to obtain and use such information, but serious gaps still persist in understanding waste disposal and its impacts. These gaps exist partly because some important types of information are not gathered or analyzed, and partly because existing information often is difficult to access and use.

Cutbacks in the funding of information-related activities can further limit our ability to detect and understand trends. Yet, such cutbacks are particularly likely during periods of economic constraints. If current and future efforts (e.g., monitoring, research, analysis) are not maintained at a sufficient level, then the utility of information collected in the past may be seriously compromised and accurate determinations of past trends and future changes may not be possible.

Types of Information Activities

To develop and implement sound waste disposal policies, information is needed about ecosystem characteristics, the status and value of marine resources such as commercial fisheries, the types and quantities of pollutants entering marine waters, and the ecological and human impacts of these pollutants. Several major Federal programs are designed to generate, analyze, and disseminate such information (app. 2).

Ecosystem Characteristics.—The effects of waste disposal activities on marine waters and resources cannot be evaluated unless the basic characteristics of different marine ecosystems are understood. Among the important characteristics are those of the water (e.g., flow patterns, temperature, turbidity, and chemical parameters); sediment (e.g., composition); and biological relationships (e.g., diversity of organisms, food chains). Many of these characteristics are affected by natural and anthropogenic activities that occur over different periods of time and over varying areas.

Many public and private agencies are engaged in efforts to increase our understanding of ecosystems and their basic characteristics. Numerous studies are supported by Federal agencies, such as EPA, the Fish and Wildlife Service (FWS), and NOAA's National Marine Fisheries Service (NMFS) and National Ocean Service. The National Ocean Service, for example, is developing a National Estuarine Inventory that describes the physical, hydrological, and biological characteristics of many estuaries. When completed, this should provide a sound basis for comparing and assessing conditions in these estuaries. The agency also is generating atlases that include detailed information on the physical and biological characteristics of U.S. coastal regions.

Despite these and other efforts, information about these characteristics often is not sufficient to identify or understand the impacts of waste disposal (226,341). Of necessity, most studies are restricted to small areas, short periods of time, or limited groups of variables. Objectives, methods, and the quality and quantity of results vary considerably among waterbodies or watersheds. More research needs to be conducted on changes over relatively large scales—for long periods of time and for entire ecological communities.

While some marine waters have received adequate attention, other waterbodies have not, including some that—while relatively free from waste disposal activities or impacts in the past—are now threatened with imminent and rapid increases in waste inputs. These areas include many waterbodies that receive wastes from the rapidly growing regions of the Southeastern United States, such as parts of North Carolina, Florida, and Louisiana.

Status and Value of Marine Resources.— Many resources (e.g., commercial fisheries and uncontaminated swimming beaches) are of obvious and substantial value. Because commercial fisheries and shellfisheries are of considerable economic value, important data about them have been collected and analyzed for many years. Information tends to be more sparse and widely scattered, however, about trends in quality, quantity, and value of other resources.

At the national level, several regular analyses provide data on the quantity and quality of commercially important fish and shellfish populations. The National Shellfish Register, for example, provides information on the degree to which shellfish waters are contaminated, although its usefulness in fully characterizing stocks is restricted (app. 2). Data on the quantities of fish and shellfish landed commercially are frequently and regularly collected, analyzed, and disseminated by Federal and State agencies. However, these data often are inadequate to evaluate population conditions because, for example, they may not reflect fishing effort or natural population fluctuations. Information on the economic value of fish and shellfish can be used, but it has limitations because a sizable amount of commercial activity is not reported (38,443,514,705). For example, small commercial fishermen or sport fishermen may not report their catch but still sell it in roadside stands. The Federal budget for analyzing commercial fishery statistics was stable during the 1980s, and no immediate changes are expected (ref. 618;; S.W. McKeen, NOAA, pers. comm., August 1986).

Federal, State, and local goverments and other public and private groups also provide supplemental information on commercial fish resources, varying from studies of particular fishing industries to analyses of pollutants in marketed fish and shellfish. The Food and Drug Administration (FDA), for example, routinely samples a wide assortment of fish and analyzes edible tissues for the presence of specified pollutants. While these sources of information are useful in specific situations, they offer only a fragmented and incomplete picture of the nationwide status and value of commercial stocks.

Considerably less information is available on the value of recreational fishing, although in many areas it may far exceed the importance of commercial fishing. Two major Federal sources provide information: surveys conducted every 5 years by the FWS and Bureau of the Census (628), and annual surveys conducted by NMFS (605,606). Other public and private entities also generate information on recreational fisheries. For example, some States require licenses for marine recreational fishing and thereby generate information on the number of fishermen in those States. In addition, several government studies address the health risks suffered by recreational fishermen who consume contaminated catches.

Little information is available, on a national scale, about the value of recreational resources such as beaches and coastal parks. The primary data come from a few Federal surveys that summarize visits to coastal wildlife refuges and National Parks. The National Ocean Service's *Economic Survey of Outdoor Marine Recreation in the USA* will include a comprehensive inventory of publicly provided outdoor recreation when completed (348, 611). Additional information on these resources (as well as on commercial and recreational fisheries) in more geographically limited areas—at the State level, for example—comes from studies supported by the Federal Sea Grant College Program (604, 617,706). The Sea Grant program received about $40 million in fiscal year 1986, but was slated for elimination in the proposed fiscal year 1987 budget.

Pollutant Inputs and Transport.—Information on the levels of pollutants in discharges, dumped material, or runoff is gathered by multiple sources—from Federal, State, and local governments; from dischargers themselves; or from private research efforts. The levels are measured both directly and indirectly; for example, permits can indirectly indicate the expected level and composition of discharges, and discharges or runoff can sometimes be described on the basis of pollutants found in nearby sediments, water, and organisms.

Information from these sources varies widely in accuracy, completeness, and accessibility.

Related activities can affect the quality and quantity of information on permitted discharges. For example, Federal laws and regulations prescribe what information dischargers must report. Enforcement generally reduces the deviation of discharges from legal levels, which enhances the value of permits as indicators of discharge quality and quantity.

Federal agencies—the primary ones being EPA, the Corps of Engineers (COE), and NOAA—are involved in efforts to generate, analyze, and disseminate information on pollutant inputs. EPA and COE generate a great deal of information on discharges from particular industries or waste disposal activities. Other Federal agencies, such as the U.S. Geological Survey (USGS) and Department of Agriculture, generate information on nonpoint runoff. The National Stream-Quality Accounting Network of USGS provides information on pollutants discharged into freshwaters and ultimately transported into estuaries. Compiling this information accurately and comprehensively on a national scale is difficult.

One notable Federal effort is NOAA's National Coastal Pollutant Discharge Inventory, which pools information from numerous sources on pollutant inputs into U.S. estuaries and coastal waters. When complete, and if updated periodically, it could provide useful overviews of trends in pollutant inputs into these waters. It also could be useful as a tool for evaluating the effects of different pollution control policies (610).

It is not possible to accurately estimate the total Federal expenditures being directed toward information gathering and dissemination because of the number and variety of agencies and programs involved, and the large overlap between program objectives. Given current economic restrictions and past trends in analytical program funding, it is likely that accurate and comprehensive information on discharges will remain difficult to obtain.

Many States also conduct programs to monitor pollutant inputs into marine waters. Since 1977, for example, California's Mussel Watch has used strategically located ''sentinel'' organisms—in this case, mussels—to detect pollutants in coastal waters. The Mussel Watch has generated data on the geographic and temporal variations in the concentrations of many pollutants. By 1985, this effort had identified at least eight areas where metals or organic chemicals had contaminated mussels to alarming levels (71).

Impacts on Resources and Human Well-being.—Large amounts of information about actual impacts on marine ecosystems and humans are generated each year by local, State, and Federal agencies, numerous private organizations, and academic groups. At the Federal level, many of the efforts are concentrated within the National Marine Pollution Program (NMPP). Despite these efforts, most experts agree that much more remains unknown and that the assessment of impacts should be more coordinated and integrated than it has been in the past (161,170,505,508).

The level of effort expended to generate information on marine impacts has varied around the country and from year to year. While general trends are difficult to ascertain because of the diversity of individuals and organizations involved and the variety of efforts, some specific trends are apparent. For example, the number of samples of certain toxic pollutants in marine organisms that were archived each year increased dramatically and peaked in the late 1960s and then declined, with large fluctua-

Photo credit: Woods Hole Laboratory, National Marine Fisheries Service

Laboratory work should be closely linked with field work, and both are time-consuming and expensive. Here a technician is sectioning fish liver tissues to analyze pollutant impacts.

Figure 26.—Number of Archived Samples Per Year of Chlorinated Pesticides and Polychlorinated Biphenyls in Fish and Invertebrates From U.S. Marine Waters, 1945–85

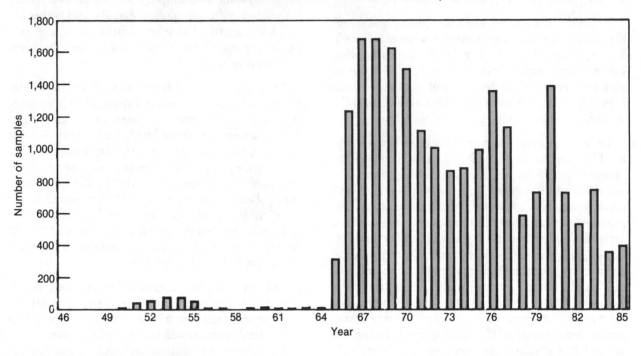

NOTE: Years lacking bars, no data are available.

SOURCE: A.J. Mearns, et al., *The Historical Trend Assessment Program, PCBs and Chlorinated Pesticide Contamination in U.S. Fish and Shellfish: An Assessment Report* (Seattle, WA: National Oceanic and Atmospheric Administration, Ocean Assessments Division, Coastal and Estuarine Assessment Branch, November 1986).

tions during the late 1970s (figure 26). A similar decline has occurred in the number of water quality samples collected in individual States such as Florida (figure 27), reflecting monitoring cutbacks by various Federal and State agencies (658).

Understanding and predicting how humans will be affected by marine waste disposal activities involves additional layers of complexity. Determining the full impact to humans of waste-induced changes in marine environments, for example, presupposes that adequate information is available about pollutant inputs and the ecological processes that affect the fate of pollutants.[19] This is rarely possible.

Figure 27.—Number of Waterbody Segments in Florida for Which Water Quality Monitoring Data Are Available, 1970-85

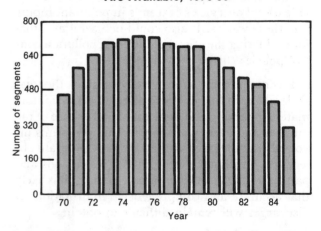

The decline in the number of segments for which data are available reflects cutbacks in monitoring by Federal and State agencies. Only data entered into STORET (the U.S. Environmental Protection Agency's system for storing and retrieving data on water quality) are included. Twenty percent of the 926 segments of water bodies in Florida are estuaries.

SOURCE: J. Hand, et al., *Water Quality Inventory For the State of Florida: Technical Appendix* (Tallahassee, FL: State of Florida Department of Environmental Regulation, June 1986).

---

[19]The complexity and difficulty of establishing impacts on marine resources and humans, specifically economic injury, is reflected in regulations proposed by the Department of the Interior for ''43 CFR Part 11. Natural Resource Damage'' (51 FR 27674-27753, Aug. 1, 1986).

The Need for Integration and Coordination

Information must be accessible and integrated so it can be used by a wide variety of people, including policymakers. Greater integration and coordination of information-related activities, as well as increased financial resources for their implementation, is considered essential by most observers (266,412,413,415,505,506,692,693).

Many suggestions to improve the planning, execution, and usefulness of these activities have been incorporated into the recommendations of NMPP (app. 2). Federal agencies have heeded NMPP recommendations to a degree; for example, NOAA's Ocean Assessments Division has undertaken several projects to develop more comprehensive databases and disseminate increasingly sophisticated analyses.

In addition, programs that are directed toward specific waterbodies and that are capable of cutting across social, institutional, and scientific boundaries may be of special value. Some crosscutting programs have been established for individual estuaries and coastal waters. Notable examples include the Puget Sound Water Quality Authority, the Aquatic Habitat Institute for San Francisco Bay, the Chesapeake Bay Program and the Chesapeake Research Consortium, and the individual programs established under EPA's National Estuary Program. These efforts are promising approaches to the many problems inherent in generating and applying useful information to an issue as large and complex as marine resource management. Similar efforts, however, do not exist for most other waterbodies.

APPENDIX 1: EXISTING WATERBODY MANAGEMENT PROGRAMS

Numerous programs have been initiated at the Federal, State, and local levels to address water quality problems in a particular estuary or coastal waterbody, including:

The Chesapeake Bay Program

The primary purpose of CBP is to develop a comprehensive understanding of the Bay's ecosystem. It is a combined State-Federal effort initiated by EPA in response to legislation passed by Congress in 1976. The intensive study of the Chesapeake Bay's water quality and resources was the result of heightened concern in the early 1970s about the health of the Nation's largest estuary. Specifically, EPA was directed to assess and make recommendations on how to improve water quality management in the Bay, to coordinate all research in the Bay, and to establish a system of data collection and analysis. The study of Chesapeake Bay was authorized for 5 years, but was twice extended by a year and was completed in 1983 at a cost of nearly $30 million.

The study focused on the Bay's 10 most critical water quality problems, three of which were studied intensively: 1) nutrient enrichment, 2) toxic substances, and 3) the decline of submerged aquatic vegetation. The findings documented a historical decline in living resources in the Bay and indicated the need for better management (95). As a result, several State and Federal entities signed the Chesapeake Bay Agreement. The Agreement established the Chesapeake Executive Council to facilitate the implementation of coordinated plans for the improvement and protection of the Chesapeake Bay estuarine system.

The Chesapeake Bay Restoration and Protection Plan, issued in September 1985, was the first planning effort to result from the Agreement. The Plan describes Federal and State strategies and programs designed to coordinate, evaluate, and oversee the Bay's restoration and protection (95). The first annual progress report, published in December 1985, discusses the plan; the coordinated monitoring program which has been developed; and modeling, research, and data management efforts (96). In 1986, the Council began reporting on the Bay's water quality conditions and working with the agricultural community on nonpoint source pollution control programs.

The Puget Sound Water Quality Authority

PSWQA, established by the Washington State legislature in 1983, was authorized by the legislature in 1985 to develop a comprehensive management plan for Puget Sound and its related waterways. The plan will be revised every 2 years and a "State of the Sound" report completed. The first Puget Sound Water Quality Management Plan was adopted in late 1986, with implementation beginning in early 1987 (464). The plan focuses on protecting Puget Sound from toxic pollutants and pathogens, both of which have contaminated sediment and harmed resources such as fish and shellfish, and on the control of nonpoint pollution. It emphasizes a lead role for local governments in identifying and controlling important nonpoint sources.

The Authority will oversee the implementation of the plan, propose funding mechanisms and, if necessary, propose new legislation. Although many Federal, State, and local agencies are involved in the study and regulation of the Sound, the Authority is the only agency specifically responsible for planning, oversight, and coordination of programs related to Puget Sound. It has considerable authority because State agencies and local governments are required to evaluate, incorporate, and implement applicable provisions of the plan.

The Great Lakes Water Quality Board

The Great Lakes together represent an extremely large surface expanse of freshwater, yet the Great Lakes system also functions somewhat like a large-scale estuary. The experience of the Great Lakes National Program thus serves as a model for the development of other waterbody management programs, such as those of the National Estuary Program. The Great Lakes Water Quality Board was established by the United States to implement agreements with Canada, reached under the auspices of the International Joint Commission, regarding the water quality of the Great Lakes. The Great Lakes National Program Office of EPA staffs the Board and ensures that U.S. commitments are met.

Two major agreements have been reached by the Commission: the *Great Lakes Water Quality Agreements* of 1972 and 1978. The 1972 agreement established water quality objectives and focused on pesticide control. The 1978 agreement added an ecosystem management approach and the goal of essentially zero discharge of pollutants; it also calls for the control of all toxic substances.

The agreements thus encourage the protection of the Great Lakes and call for remedial actions against pollution, as well as research and monitoring programs. EPA's Great Lakes National Program Office and Region 5 are most involved in coordinating activities relating to the Agreements. In May 1986, the Great Lakes States issued *The Great Lakes Toxic Substance Control Agreement*. Intended to be consistent with both the Federal Clean Water Act and the Great Lakes Water Quality Agreement, the agreement establishes a framework for coordinating regional action to control toxic pollutants entering the Great Lakes system (200, 389,663).

The Gulf Coast Waste Disposal Authority

GCWDA is a unique example of a within-State waste coordination effort. The authority is a three-county unit

of local government, established by Texas statute in 1969 to abate point source pollution in the heavily industrialized Houston Ship Channel and Galveston Bay area. It has established numerous waste management facilities, primarily sewage treatment plants, but also some industrial treatment facilities and a land-based incineration facility. The system is funded by issuing bonds for construction and these are repaid through user charges (ref 215; L. Goin, GCWDA, pers. comm., 1986).

The authority is active in pollution control financing and itself owns and operates four industrial wastewater treatment facilities. These facilities treat and dispose of liquid wastes from over 40 industrial plants. In addition, its 22 municipal wastewater treatment plants and 7 drinking water treatment plants serve over 40 water districts or cities. The objective is for at least one-third of these facilities to become large, regional waste treatment facilities. The authority is also pursuing regional approaches to municipal sludge disposal and resource recovery for municipal solid wastes.

The Southern California Coastal Water Research Project

SCCWRP is dedicated to researching and monitoring the effects of municipal wastewater discharges on marine life. The project publishes a report on recent research efforts every 2 years. It is sponsored by the sanitation districts of Orange County and Los Angeles County and the cities of Oxnard, San Diego, and Los Angeles. These wastewater dischargers created SCCWRP through a joint powers agreement. Each has representatives on a commission that oversees the operation of the project.

The project is not intended to study specific pollution sources or needed controls. Instead, its focus is a variety of specific environmental problems, such as predicting sediment quality around outfalls, fish reproduction near outfalls, and the influence of chlorinated hydrocarbons on fish. The goal of the research is to develop predictive models that would help determine what levels of wastewater treatment are needed to protect marine life (527,528).

The San Francisco Bay Regional Water Quality Control Board

The San Francisco Bay Regional Water Quality Control Board is one of the major agencies involved in managing the Bay's waters. It operates independently of, but is responsible to, the State of California Water Resources Control Board. The Regional Board, comprised of nine appointed members who are involved in activities to control water quality in the Bay, is primarily an enforcement agency and is limited to activities such as controlling and monitoring sewage outfalls and other point source discharges. It has no authority to control impacts caused by pollutants carried by the Sacramento and San Joaquin Rivers or from other areas; these are under the exclusive jurisdiction of the State Board. It also has no authority to coordinate activities of the other agencies involved in Bay water quality management.

The Regional Board is active in planning, reviewing, and amending the Basin Plan for the Bay area and in reviewing water quality standards. The plan, last amended in 1982, is the basis for distributing both State and Federal grants for water quality programs such as building and upgrading wastewater treatment facilities. The Board is also active in the study of shellfish through the San Francisco Bay Shellfish Program and in the State's Mussel Watch Program. Its Aquatic Habitat Program studies the effects of toxic pollutants on aquatic life in the Bay (67,70).

The Aquatic Habitat Institute

The Aquatic Habitat Institute is a nonprofit, quasi-public corporation independent from, but highly supportive of, the Regional Board's Aquatic Habitat Program. Although established in 1983, funding only began in 1986. Its purpose is to produce independent research acceptable to all agencies and interests concerned with the management of the Bay area. The Institute is planning a number of scientific assessments and education programs, and will attempt to better coordinate research and monitoring in the San Francisco Bay/Delta area.

The Institute's 10-member Board of Directors consists of representatives from a wide range of government and nongovernment interests (see table 12). Currently, the program's largest single source of funding is EPA's National Estuary Program. This funding will continue for 5 years. The California State and Regional Water Quality Control Boards also contribute funds, and donations are accepted from municipal and industrial dischargers. Eventually, however, the Institute is required to rely on its own funding strategy. This is another unique requirement of the program and will most likely involve the use of discharger taxes or sewer user surcharges, rather than direct appropriations (ref. 66; also D. Segar, Aquatic Habitat Institute, pers. comm., 1986).

The National Estuary Program

NEP was created within EPA in 1985 to oversee the implementation efforts in the Great Lakes and Chesapeake Bay, and to initiate comprehensive programs in other estuaries in the United States. Programs are underway in Puget Sound, Long Island Sound, Buzzards Bay, Narragansett Bay, San Francisco Bay, and Albemarle-Pamlico Sounds. NEP uses existing authorities under the Clean Water Act (Sec. 104), other Federal statutes, and State legislative authorities to control sources of pollution. The program emphasizes the need to focus and integrate existing programs at the Federal, State, and local levels to maximize benefits of pollution abatement. The objective of each program in NEP is to characterize the conditions and trends in the system and develop an integrated management program to maintain or restore the estuary.

The Water Quality Act of 1987 expanded the scope of NEP and authorized additional funding for the development, under its auspices, of individual waterbody management programs.

APPENDIX 2: FEDERAL INFORMATION AND MONITORING PROGRAMS

National Marine Pollution Program

The National Marine Pollution Program (NMPP) was established to coordinate the 11 departments and agencies that are engaged in research or monitoring related to marine pollution (including the Great Lakes) (598,599). The program was mandated under the National Ocean Pollution Planning Act of 1978. Figure 28 indicates the overall Federal marine pollution research budget for fiscal year 1984; figure 29 indicates funding of selected activities.

In late 1985, the program issued a Federal plan for fiscal years 1985 to 1989 (601). The plan recommended, among other things, a greater Federal emphasis on:

1. resource-oriented monitoring to provide national assessments of the status and trends in environmental quality,
2. better coordination of monitoring efforts,
3. research and monitoring programs related to municipal and industrial effluents,
4. research and monitoring on nutrients and pathogens (with less emphasis on metals and petroleum),

Figure 28.—Total Federal Funding for the National Marine Pollution Program, Fiscal Years 1981–87

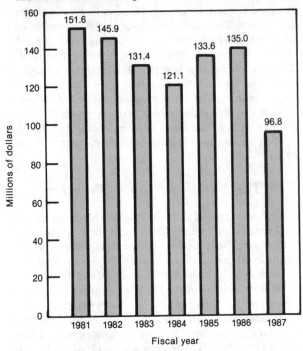

NOTE: 1987 amount is from the proposed budget.

SOURCES: U.S. Department of Commerce, National Oceanic and Atmospheric Administration, National Marine Pollution Program, *Federal Plan for Ocean Pollution Research, Development, and Monitoring, Fiscal Years 1985–1989* (Washington, DC: September 1985); U.S. Department of Commerce, National Oceanic and Atmospheric Administration, National Marine Pollution Program, *Summary of Federal Programs and Projects, FY 1985 Update* (Washington, DC: 1986).

5. data synthesis, interpretation, and information dissemination; and
6. studies conducted in estuaries and coastal waters.

The level of funding of marine pollution research conducted by the Federal agencies has declined during the 1980s, from approximately $152 million in fiscal year 1981 to about $135 million in fiscal year 1986. The presidential budget for 1987 was approximately $97 million (see figure 28). The effects of the decline on the above recommendations are uncertain.

The Northeast Monitoring Program

The Northeast Monitoring Program (NEMP) monitors waters from the Gulf of Maine to North Carolina's Cape Hatteras. Established in 1979 by NOAA, NEMP monitors physical, chemical, and biological variables over long periods. It establishes benchmarks for both the concentration and distribution of pollutants and for their effects. Since 1980, NEMP has issued several reports summarizing the health of these estuaries and coastal waters (592,597). The reports condense a large

body of monitoring information, present it to a wide audience, and provide extensive references for those seeking further information. The information facilitates efforts to assess the effects of pollutants on ecosystems and resources, and to detect and respond to important environmental changes.

NOAA's Ocean Assessments Division

NOAA's activities involving marine pollution assessment, monitoring, and research are conducted primarily by the Ocean Assessments Division (OAD), housed within the National Ocean Service (610). Two branches of the division are especially active in matters pertaining to waste disposal and its effects: the Strategic Assessment Branch and the Coastal and Estuarine Assessment Branch.

Strategic Assessment Branch

The Strategic Assessment Branch evaluates and inventories coastal resources and their exploitation, and also assesses national policies and strategies with regard to these resources and uses (611). Its activities, which accounted for 19 percent of OAD's fiscal year 1985 budget, include:

* assembling Strategic Assessment Data Atlases that summarize key ecological, economic, and political characteristics of each major marine region of the United States (607);
* producing a series of maps on the health and use of U.S. coastal waters;
* surveying Federal, State, and local government expenditures on outdoor marine recreation (348);
* assessing levels of pollutants entering marine waters (the National Coastal Pollutant Discharge Inventory (600));
* inventorying estuaries around the Nation (National Estuarine Inventory), which will allow comparisons of their use and health (362,602,619); and
* periodically inventorying the status of shellfish areas (National Shellfish Register of Classified Estuarine Waters).

Coastal and Estuarine Assessment Branch

The primary function of this branch is to assess the consequences of human activities on marine environments; its activities accounted for 35 percent of OAD's fiscal year 1985 budget. The branch has two relevant programs: National Status and Trends, and Consequences of Contaminants. The bulk of the branch's budget goes into the Status and Trends Program.[20]

---

[20]Up-to-date information on the Status and Trends Program and on another program, the Quality Assurance Program, is available in the biannual Newsletter issued by these programs.

Figure 29.—Federal Funding of Selected Activities Within the National Marine Pollution Program in Fiscal Year 1984

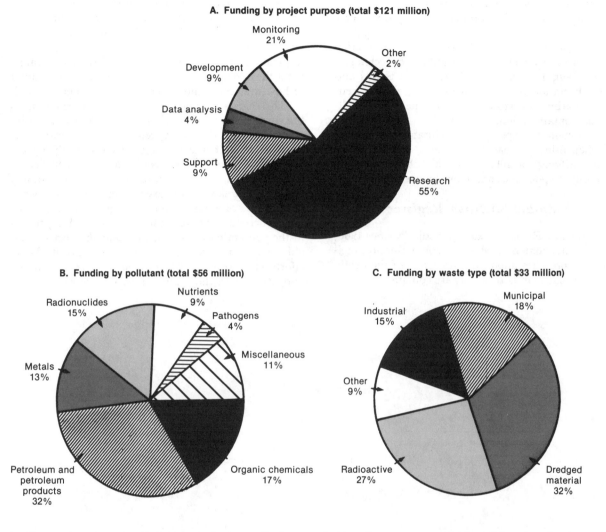

A. Funding by project purpose (total $121 million)

Monitoring 21%
Other 2%
Development 9%
Data analysis 4%
Support 9%
Research 55%

B. Funding by pollutant (total $56 million)

Radionuclides 15%
Nutrients 9%
Pathogens 4%
Miscellaneous 11%
Metals 13%
Petroleum and petroleum products 32%
Organic chemicals 17%

C. Funding by waste type (total $33 million)

Municipal 18%
Industrial 15%
Other 9%
Radioactive 27%
Dredged material 32%

Diagrams for B and C only include funding that can be specifically categorized by pollutant or waste type; hence totals are less than the total for A.

SOURCE: Office of Technology Assessment, 1987; after U.S. Department of Commerce, National Oceanic and Atmospheric Administration, National Marine Pollution Program, *Federal Plan for Ocean Pollution Research, Development, and Monitoring, Fiscal Years 1985–1989* (Washington, DC: September 1985).

The Status and Trends Program.—The objective of the National Status and Trends Program is to document the current status and long-term trends in the quality of estuaries and coastal waters (615,616). The program consists of four components which perform three major tasks:

1. providing data on concentrations of pollutants in finfish, shellfish, and sediments;
2. measuring biological parameters that reflect stress associated with human-induced perturbations; and

3. assessing marine environmental quality and recommending Federal responses.

The fiscal year 1985 budget for the program was $3.3 million, the fiscal year 1986 budget was $2.7 million (a decline of 18 percent), and the proposed budget for fiscal year 1987 was $4.3 million (J. Calder, pers. comm., August 1986). A major component of the program is the Benthic Surveillance Program, which collects samples of sediment, bottom-dwelling mollusks, and bottom-feeding fish from numerous sites throughout the country. The samples are analyzed for substances such as toxic metals, polynuclear aromatic hydrocarbons, and chlorinated organic chemicals (610,621).

Consequences of Contaminants Program.—This program develops techniques to determine how pollutants in marine waters have affected or can affect marine fish and shellfish and human health (620). The techniques then can augment the capabilities of the Status and Trends Program. Recent activities have emphasized:

- evaluating indicators that signal the risk of shellfish contamination,
- documenting exposure to pollutants that results when fishermen eat their catches, and
- quantifying the relationship in fish between exposure to pollutants and reproductive impairment.

National Shellfish Register

The Shellfish Register, issued periodically since 1966, contains information on shellfish contamination incidents and provides important indicators of the extent to which shellfish in U.S. waters are contaminated; the latest register was published in 1985 (603). It uses a classification system based on concentrations of coliform bacteria and natural marine biotoxins, although it also includes information on substances that might be considered hazardous in the shellfish. Productive shellfish waters can be classified as approved, prohibited, conditionally approved, or restricted. Most States implement the system voluntarily, although they may differ in how they meet the general requirements.

The register provides only limited information on the current status of shellfishing areas and still less information regarding past trends, in part because the classification scheme is not used consistently by the States. For example, the classification of a shellfish area could be changed from approved to restricted simply because an area was not surveyed in a particular year, not because of actual contamination. Thus, the register is currently not well-suited for establishing trends in the contamination of shellfish and shellfish waters, although efforts are being made to improve it. These efforts are, however, constrained by severely limited budgets at both the State and Federal levels.

Chapter 8
Managing Industrial Effluents

CONTENTS

Tables

Figures

Box

Managing Industrial Effluents

INTRODUCTION

The direct and indirect discharge of industrial effluents[1] is a widespread practice that contributes substantial quantities of pollutants to marine as well as non-marine waters. The location, composition, and magnitude of these discharges are all important in assessing their ability to cause or contribute to environmental and human health problems.

Industrial effluents are only a subset of the range of materials disposed of in marine waters, yet they provide a good illustration of the complexity of the issues involved in marine waste disposal. The information presented here generally applies to all industrial dischargers, whether their wastewaters enter marine or freshwater environments, because the statutory and regulatory framework governing industrial discharges usually does not distinguish between discharges into freshwater and marine environments.[2]

Estuaries and coastal waters have been used for disposal more frequently and have been more severely affected than have open ocean waters. Essentially all direct and indirect industrial discharges to marine environments occur in estuaries and coastal waters, the majority of discharges occur-

ring in estuaries.[3] In addition, direct and indirect industrial discharges to rivers that subsequently drain into estuaries and coastal waters are another significant source of pollutants to marine waters, though this source is difficult to quantify.

Large quantities of toxic pollutants are entering marine environments, particularly estuaries and coastal waters. Legal discharges of industrial effluents (contributed either directly to receiving waters or indirectly to publicly owned treatment works, or POTWs) often contain substantial amounts of toxic pollutants; indeed, **in the aggregate, industrial discharges represent the largest source of toxic pollutants entering the marine environment.**

The large quantities of waste entering estuaries and coastal waters through industrial discharges reflect several factors:

1. the concentration of population and industrial activity in coastal regions,
2. the cost savings to waste generators that use marine disposal as opposed to other alternatives, and
3. a statutory and regulatory approach that authorizes discharge levels based more on technological capabilities than on resulting water quality.

The net effect of these factors is a considerable degree of ''acceptance'' of the routine (but environmentally very significant) discharge of effluents into estuaries and coastal waters, especially when contrasted with the public and government attention focused on dumping of industrial and municipal wastes in the open ocean. This dichotomy is one reflection of the very different philosophical approaches embodied in the two major statutes governing marine waste disposal: the managerial perspective of the Clean Water Act and the more restrictive perspective of the Marine Protection, Research, and Sanctuaries Act (see ch. 7).

---

[1]Industrial effluents are wastewaters that are discharged through pipelines. They can be either legally discharged *directly* into receiving waters or *indirectly* via sewerage systems operated by publicly owned treatment works (POTWs). Both practices are regulated through programs established under the Clean Water Act (CWA). Direct dischargers are regulated through the National Pollutant Discharge Elimination System (NPDES) and indirect dischargers fall under the National Pretreatment Program. Both programs have State or local, as well as Federal, components.

[2]Industrial discharges into coastal waters (as defined in this report) are distinguished by the CWA and thus are a partial exception to this generalization. CWA Sec. 403 requires that discharges into the ''territorial sea, contiguous zone, or open ocean'' be in compliance with the Ocean Discharge Criteria (40 CFR 125, Subpart M). Under this provision, most industrial discharges into estuaries would not be required to meet the Ocean Discharge Criteria; however, industrial discharges into coastal waters would be subject to the criteria. When promulgating the criteria in 1980, EPA estimated that about 230 land-based point source discharges, as well as fixed offshore facilities required to obtain NPDES permits (e.g., the approximately 3,000 offshore oil and gas platforms), would be affected (45 FR 65944, Oct. 3, 1980). See ch. 7 for further discussion of the Ocean Discharge Criteria.

[3]Ch. 3 presents data on the extend of this activity that specifically affects marine environments. This chapter presents data for all U.S. waters.

Photo credit: Northeast Technical Services Administration, U.S. Food and Drug Administration

More than 1,300 major industrial facilities discharge wastes directly into U.S. marine waters, mostly into estuaries. Thousands of others discharge into rivers that carry pollutants into these waters, or into POTWs that subsequently discharge into marine waters. These various industrial discharges are the largest sources of the organic chemicals and many metals found in many estuaries and coastal waters.

EXISTING REGULATORY FRAMEWORK FOR POINT SOURCE POLLUTION CONTROL

The NPDES Program

The Clean Water Act (CWA) regulates the discharge of wastes into the navigable waters of the United States, including estuaries and coastal waters. This act, passed in 1972, established the National Pollution Discharge Elimination System (NPDES, 40 CFR 122) to regulate these point source discharges. Under NPDES, all point sources—both industrial and municipal—that *directly* discharge into waterways are required to obtain permits that regulate their discharges.[4]

[4]In addition to complying with general treatment requirements specified in NPDES permits, municipal sources (i.e., POTWs) and industries discharging to POTWs must also comply with applicable requirements under the National Pretreatment Program.

General Structure

The NPDES program contains four essential operational elements:

1. The *discharge permit* is the basic "currency" of the NPDES program. Each permit specifies effluent limitations for particular pollutants, monitoring requirements (including a schedule), and reporting requirements for information characterizing and quantifying a facility's actual discharge. If compliance with final effluent limitations has not yet been achieved, a permit (or more typically, a separate order) is issued that contains interim limitations and a compliance schedule.

2. A system based on *self-reporting* by permittees is the basic approach used to provide the information necessary to determine whether effluent limits specified in a permit are being met. The entire system therefore heavily relies on permittee integrity in reporting results of self-monitoring.

3. *Compliance monitoring* encompasses those activities intended to determine whether facilities are achieving the requirements contained in their discharge permits. Such activities include conducting inspections, reviewing reports submitted by permittees, monitoring to verify industry-reported data, and compiling statistical information to assess compliance.

4. *Enforcement* includes all actions taken in response to an identified instance of noncompliance, including determination of the appropriate response based on the severity of a violation and other factors, and initiating and escalating the response until compliance is achieved.

Permit issuance, receipt of data submitted by permittees, compliance monitoring, and enforcement are the primary responsibility of States (if they have been approved to administer their State's NPDES program under CWA Sec. 402) or Environmental Protection Agency (EPA) Regions (under CWA Sections 309 and 504), although EPA Headquarters and the Department of Justice can initiate or intervene in enforcement actions.[5] Thirty-seven States and one Territory (the Virgin Islands) have approved NPDES programs (503).

Regulated Pollutants

The 1972 CWA focused on controlling *conventional* pollutants. It soon became apparent, however, that *toxic* and *non-conventional* pollutants in wastes such as industrial effluents, sewage sludge, and dredged material also were causing adverse impacts.[6] In 1977, Congress amended CWA to provide additional regulation of toxic and non-conventional pollutants from specific industrial categories

within the framework of the NPDES permitting and compliance process.[7]

A list of 65 classes of toxic "priority pollutants" and 21 primary industrial categories to be regulated by EPA were included in the 1977 CWA Amendments. This list arose out of a Settlement Agreement of a suit brought against EPA by the Natural Resources Defense Council.[8] EPA subsequently divided the 65 classes into 129 priority pollutants, an amount later reduced to 126 pollutants (40 CFR 122, app. D). The 21 industrial categories were subdivided into 34 categories (40 CFR 122, app. A); 9 of these were specifically exempted from regulation by categorical standards, leaving about 25 primary industrial categories for which categorical effluent guidelines for priority pollutants were to be promulgated (table 13).[9] An additional 35 industrial categories not included in the amendments were designated as secondary; development of effluent guidelines for priority pollutants in these industries was deferred until an unspecified later date (503).

Compliance Monitoring and Data Management

Several tools have been developed to monitor compliance, identify cases of noncompliance, and manage data (box T).

Enforcement

EPA has revised its national Enforcement Management System (EMS) (673) to provide better guidance on enforcement to EPA Regions and delegated States, and to provide a greater degree of nationwide consistency in administrative responses to in-

---

[5]Sec. 305 of CWA also authorizes private citizens or their representatives to bring enforcement suits against dischargers. While this approach represents an increasingly effective means of strengthening enforcement, it is beyond the scope of this report. For further information on this topic, see refs. 45, 453, and 501.

[6]See box A in ch. 1 for definitions of these classes of pollutants.

[7]Municipal dischargers (i.e., POTWs) are required to meet standards different from those for direct industrial dischargers. Regulations for most POTWs specify a minimum level of treatment (termed "secondary") that is measured in terms of reduction of conventional pollutant parameters. As an alternative to requiring POTWs to employ technological means for controlling toxic and non-conventional pollutants introduced through indirect industrial discharges, CWA provides for such control through industrial pretreatment programs. Discharges from both POTWs and industrial facilities must also meet applicable State water quality control standards where they have been developed.

[8]*NRDC* v. *Train* 8 ERC 2120, June 8, 1976; *NRDC* v. *Costle*, 12 ERC 1833, Mar. 9, 1979; modified by additional orders of Oct. 26, 1982, Aug. 2, 1983, and Jan. 6, 1984.

[9]The actual number of categories has changed as categories have been exempted, combined, or separated.

Table 13.—Industrial Categories Subject to Regulation Under the NPDES and Pretreatment Programs as Significant Sources of Toxic Pollutants[a]

	Number of indirect dischargers	Number of direct dischargers
Aluminum forming	64	42
Battery manufacturing	134	15
Coal mining	0	10,375
Coil coating I	39	29
Coil coating II	80	3
Copper forming	45	37
Electrical and electronic components I	244	83
Electrical and electronic components II	21	1
Foundries	499	301
Inorganic chemicals I	21	114
Inorganic chemicals II	17	35
Iron and steel	160	738
Leather tanning	141	17
Metal finishing and electroplating	10,200	2,800
Nonferrous metals forming	151	51
Nonferrous metals manufacturing I	85	79
Nonferrous metals manufacturing II	40	33
Ore mining	0	515
Organic chemicals and plastics and synthetic fibers	535	1,082
Pesticides	39	42
Petroleum refining	47	164
Pharmaceuticals	392	80
Plastics molding and forming	1,145	810
Porcelain enameling	50	28
Pulp and paper	261	355
Steam electric	93	?
Textile mills	1,047	229
Timber processing	47	?
Total	15,597	>18,058

[a]The number and names of categories listed here do not correspond exactly to those indicated in the text or the Code of Federal Regulations, due to subsequent joining or dividing of categories by EPA. The numbers include only those dischargers that are regulated under categorical standards.

SOURCES: Office of Technology Assessment, 1987; based on U.S. Environmental Protection Agency, Office of Regulations and Standards, Monitoring and Data Support Division, *Summary of Effluent Characteristics and Guidelines for Selected Industrial Point Source Categories: Industry Status Sheets* (Washington, DC: Feb. 28, 1986); except for data from the proposed or final rules for the Steam Electric and Timber Processing categories, from Science Applications International Corp., *Overview of Sewage Sludge and Effluent Management*, contract report prepared for U.S. Congress, Office of Technology Assessment (McLean, VA: March 1986).

stances of noncompliance. First developed in 1977, the original EMS was used by few States and EPA Regions; current policy, however, mandates development of a formal enforcement system consistent with the revised EMS. The revised version specifies time frames within which enforcement actions against "significant noncompliers" must be initiated and also specifies procedures for identifying, initiating, and following through with appropriate enforcement actions. In addition to maintaining an inventory of permits and processing submitted data, States or Regions are expected to:

• conduct "enforcement evaluations" to determine an appropriate level of enforcement action and an associated time frame, based on guidelines and procedures developed for the various predetermined categories of violation. Factors to be considered include: the magnitude and duration of the violation; the compliance record of the permittee and past enforcement actions taken; the expected deterrent effect of the response based on experience from comparable situations; and consideration of fairness, equity, and national consistency.

• institute formal enforcement actions and follow-through wherever necessary, usually triggered by a failure to comply through less formal means within a specified period of time.

• initiate field investigations (i.e., inspections) in support of enforcement actions according to a systematic "annual compliance inspection plan."

EPA expected approved States and EPA Regions to revise and formalize their enforcement policies to meet the new requirements and be consistent with the new EMS by the end of fiscal year 1986. However, the Federal Government has only limited ability to ensure that States do so.

Enforcement Tools Available to Administering Agencies.—Once an NPDES permit violation is identified, two primary levels of enforcement responses are used:

1. **Informal enforcement responses** include inspections, phone calls, violation letters, and Federal Notices of Violation. The latter, which is sent to the permittee and the administering State agency, can require certain steps to be taken according to a specified schedule.

2. **Formal enforcement responses** require written notification that specifies: 1) actions to be used to achieve compliance, 2) a timetable, 3) the consequences of noncompliance that are enforceable without having to prove the original violation, and 4) the legal consequences

Box T.—Tools for Compliance Monitoring and Data Management Under NPDES*

Tracking Compliance and Identifying Noncompliance

- **Discharge Monitoring Reports** (DMRs) are the basic vehicle used by permittees to report the results of their self-monitoring activities. DMRs are to be completed periodically (typically monthly) by all permittees (major and minor), are to include the results of all monitoring required by the permit, and are to be submitted to EPA on a monthly or quarterly basis. Certain serious permit violations are to be reported at the time they occur, usually by phone.

- **Inspections** can in principle be used to collect samples to detect violations, but because of limited resources they are primarily used to verify reported noncompliance and to support formal enforcement actions.

- **Quarterly Noncompliance Reports** (QNCRs) are required to be prepared by EPA Regions or approved States. QNCRs document certain instances of noncompliance by major permittees that occurred during the previous quarter. The most recent rules distinguish between Category I noncompliance (quantifiable, and therefore consistent nationwide) and Category II noncompliance (dependent on professional judgment, and therefore subject to variability). All instances of noncompliance that exceed certain criteria** for magnitude, duration, and frequency—and any enforcement actions taken—must be reported each quarter until resolved; permittees under enforcement orders must be listed until compliance with both the orders and permits is achieved.

- **Semi-Annual Statistical Summary Reports** (SASSRs), prepared by EPA Regions or delegated States, are more comprehensive than QNCRs in that they must list all instances in which a major permittee had two or more violations of the same permit limitation in any 6-month period, regardless of whether other criteria are met. Both QNCRs and SASSRs are intended to help EPA provide oversight of Regional and State compliance programs.

Measuring and Ranking Instances of Noncompliance

EPA has evolved a complex system for ranking instances of noncompliance so the limited resources allotted to enforcement activities are directed toward the most serious violations. A multi-step process is used to identify and evaluate noncompliance. First, an initial screening of DMR data is performed to identify all instances of noncompliance with permit terms or enforcement orders. Second, violations that exceed specified criteria require priority review by a professional to determine if an enforcement action should be initiated. Finally, violations are further ranked to determine whether the enforcement response should be informal or formal. Two sets of criteria have been developed for evaluating violations of effluent limits, compliance schedules, or reporting requirements contained in permits or enforcement orders:

- **Violation Review Action Criteria** (VRAC) specify minimum thresholds for the duration and frequency of violations which require professional review, but not necessarily formal enforcement responses. The VRAC have been developed nationally, and can only be modified by States to be more stringent.***

- **Significant Noncompliance** (SNC): EPA has developed a further set of criteria to define instances of **significant** noncompliance. All instances of SNC must be reported on QNCRs. Moreover, EPA specifies that all SNC violations must be corrected (generally within one quarter) or a *formal* enforcement response be initiated prior to its appearance on a second QNCR (generally within 60 days of the first QNCR), unless an acceptable justification for other (or no) action is provided. If States or Regions fail to act on this schedule, EPA Headquarters has the authority and responsibility to do so.

Data Management

Permit Compliance System (PCS): PCS, an automated data system, is intended to serve as EPA's primary management tool for tracking permit issu-

---

*The basic tools for carrying out these activities for indirect dischargers are discussed in the section describing the National Pretreatment Program.

**QNCR criteria are specified in the QNCR Regulation, 40 CFR 123.45.

***For violations of effluent limits specified in permits, VRAC are intended to identify a broader range of violations than the criteria for determining which violations must be reported on QNCRs; e.g., only the number of violations—not their magnitude—is a factor. However, for effluent limits specified in enforcement orders, and for schedule and reporting violations, the VRAC and QNCR criteria are equivalent.

ance, compliance, and enforcement actions. Under current policy, DMR data for **major** permittees must be entered into PCS within 30 days of receipt. Use of the PCS is far from complete, however (327):

- only recently have all 10 EPA Regions converted to PCS;
- only nine States enter all of their data directly into PCS;
- another eight States enter some of their data directly into PCS;

- another eight States use an automated system that interfaces with PCS;
- the remaining States submit data to EPA Regions for manual entry into PCS. Increased funding of PCS and strong encouragement of States to become direct users of the system will be needed to increase its effectiveness.

of noncompliance. Two basic types of formal enforcement actions are available:

—An **administrative order** (AO) generally specifies actions to be taken by a permittee to return to compliance, and a schedule for doing so; AOs issued by EPA cannot be used to assess penalties, although noncompliance is itself an enforceable violation. AOs are generally the first course of formal action because of their expedience and low cost. More than 1,600 AOs were issued by EPA Headquarters or Regional Offices for CWA violations in 1984 (327).

—**Judicial referrals** are civil actions filed by the State attorney general in an approved State, or otherwise by the Department of Justice. Such referrals are much more lengthy and costly procedures, and fewer than 100 such actions were initiated at the Federal level in 1984 (327).

Currently, court action is the only option available to the Federal Government that can result in the imposition of a financial penalty. Because of the slowness of this procedure and the low penalties that often result, however, it is generally considered an insufficient enforcement mechanism (see discussion of the enforcement issue later in this chapter). Moreover, insufficient resources for enforcement in EPA Regions and Headquarters effectively limit the number of judicial actions that can be undertaken. Some States have the legal authority to impose administrative penalties, and EPA recently was granted comparable authority through amendments to the CWA adopted in the Water Quality Act of 1987. Such authority should greatly enhance the capability of EPA to mount appropriate and timely enforcement actions against violators.

Evaluation of Program Performance.—Under the EMS, two levels of review are mandated to evaluate the performance of administering agencies (i.e., approved States or EPA Regions).[10] First, EPA Headquarters is to perform midyear evaluations of the progress of EPA Regions in implementing the EMS. Second, Regional offices are to conduct reviews of approved State programs, including file audits. The new requirements of the EMS will be the benchmark for measuring system performance.

The National Pretreatment Program

In addition to discharging pollutants directly into receiving waters, industrial facilities also discharge into sewerage systems operated by POTWs; these discharges are designated as "indirect" to distinguish them from "direct" discharges into rivers and other waterbodies. In 1977, Congress broadened the effective scope of CWA by mandating additional regulation of pollutants in indirect industrial discharges (CWA Sec. 307(b)). To meet this mandate, in 1981 EPA developed the National Pretreatment Program (40 CFR 403).

General Structure

The National Pretreatment Program is designed to protect POTWs and the environment by preventing the introduction of industrial wastes that might upset or interfere with POTW operations, pass through the POTW untreated, or contaminate sewage sludge. Under this program, all POTWs are responsible for enforcing General Pretreatment Regulations, and some POTWs are also required to enforce National Categorical Standards.

---

[10]These requirements are spelled out in the National Guidance for Oversight of NPDES Programs, issued June 28, 1985.

The General Pretreatment Regulations establish industrial, local, State, and Federal responsibilities for implementing the program. They also contain standards that prohibit the discharge into a POTW of pollutants which could cause fire or explosion, obstruction of flow, corrosion, interference or upset, or excess heating of wastewater entering the POTW. These general regulations apply to all industrial and commercial establishments that discharge into POTWs.

The National Categorical Standards contain specific pretreatment standards for the "categorical" industries that are subject to regulation as significant sources of toxic pollutants (see table 13). A subset of the Nation's POTWs, chosen on the basis of high wastewater flow and/or significant industrial inputs, must develop individual pretreatment programs that meet national specifications and must enforce these categorical standards against their indirect dischargers.

The General Pretreatment Regulations also include a provision for regulating industrial discharges of pollutants not covered by categorical standards. If a POTW experiences operational or pass-through problems that are related to industrial discharges of pollutants, POTWs are mandated to develop their own limitations on these discharges (termed "local limits"). Local limits can be more stringent than categorical regulations if this is needed to prevent pass-through or interference; they can also be developed to cover pollutants or industries not regulated by categorical standards.

Federal, State, POTW, and Industrial User Responsibilities

Responsibilities under the National Pretreatment Program are closely linked to the administration of NPDES program. Two general areas of responsibility are established. The *Approval Authority* (typically a State or EPA Region) is responsible for overseeing the development and implementation of individual POTW pretreatment programs. The *Control Authority* is responsible for ensuring that indirect industrial dischargers achieve and maintain compliance with pretreatment standards and requirements. Once an individual POTW program is approved, the POTW becomes the Control Authority.[11]

States that have received NPDES program authority can also be approved to administer the pretreatment program (i.e., become the Approval Authority). To be approved, the State must demonstrate that it has the authority, resources, and procedures required to approve, oversee, and ensure enforcement of individual POTW pretreatment programs. EPA Regions serve as Approval Authority for nonapproved States, and provide general oversight, guidance, and enforcement assistance for approved States. Under certain circumstances, States with Approval Authority may design a program under which the State is also the Control Authority.

Of the 37 States and 1 Territory with EPA-approved NPDES programs, 24 also have approved pretreatment programs and hence are Approval Authorities (327). Six of these States have opted to assume Control Authority as well, so that POTWs in these States are not required to develop comprehensive individual pretreatment programs.[12]

In general, all POTWs with a total daily flow of more than 5 million gallons, and smaller POTWs with significant industrial inputs, are required to develop and implement individual POTW pretreatment programs. Currently, only about 1,500 of the more than 15,000 POTWs in the United States meet these criteria (327). Of these about 40 percent have an average flow of more than 5 million gallons per day (mgd). These 1,500 POTWs receive an estimated 82 percent of the total industrial wastewater flow entering the Nation's POTWs and over 90 percent of the wastewater flow originating from industries subject to categorical pretreatment standards (666). These POTWs also generate more than 75 percent of the sludge in the United States (503).

As of October 1986, EPA had approved more than 95 percent of the required individual POTW pretreatment programs (data from Strategic Planning and Management System, Office of Water Enforcement and Permits, U.S. EPA). Many of the approved programs, however, are only in the early stages of being implemented (e.g., ref. 497).

Industrial facilities that discharge wastewater to POTWs (termed "industrial users") but which are

---

[11]In large municipalities with several POTWs, the municipality is often designated as the Control Authority.

[12]Texas operates its own permitting program for both direct and indirect dischargers, alongside the NPDES program administered by EPA Region VI.

not subject to categorical standards must still comply with the General Pretreatment Regulations (see above), as well as additional permit, monitoring, or reporting requirements developed by States or POTWs. For categorical industrial users, responsibilities specified in categorical regulations include complying with technology-based effluent limitations on pollutants of concern, monitoring discharges on a periodic basis, and reporting monitoring data and compliance status to control authorities.[13]

Types of Standards for Direct and Indirect Dischargers

Several different types of standards have been developed to regulate industrial discharges of wastewater and pollutants to POTWs or receiving waters. Figure 30 schematically illustrates the relationships between these standards and indicates where they apply within the overall regulatory framework for water pollution control.

Technology-Based Standards

The effluent limits set in NPDES permits to control *direct industrial dischargers* are primarily technology-based standards applied to individual pollutants. Existing direct industrial dischargers were initially required to meet interim standards based on the "best practicable control technology currently available" (BPT). The next levels of limitations imposed on industrial effluents are termed "best available technology economically achievable" (BAT), designed primarily to control toxic and non-conventional pollutants, and "best conventional pollutant control technology" (BCT), designed to control conventional pollutants. Finally, new industrial facilities are required to comply with "new source performance standards" (NSPS), which are generally comparable to BAT and BCT.

[13]Three types of reports are required of categorical industrial users (40 CFR 403.12), each containing information on the composition of a facility's discharge with respect to those pollutants regulated by categorical standards. Baseline Monitoring Reports (BMRs) and Compliance Reports comprise the initial reporting of conditions after pretreatment standards are effective and after the final compliance date is reached, respectively. Semi-annual reports are required so that continued compliance status can be periodically verified. EPA recently has proposed revisions to the General Pretreatment Regulations (51 FR 21454, June 12, 1986) which are intended to clarify and expand requirements applicable to reporting and monitoring for industrial users. Among other things, these revisions would provide the authority for POTWs to extend some of these requirements now applicable only to categorical industries to noncategorical industrial users as well.

Figure 30.—Regulatory Framework and Standards for Industrial Discharges

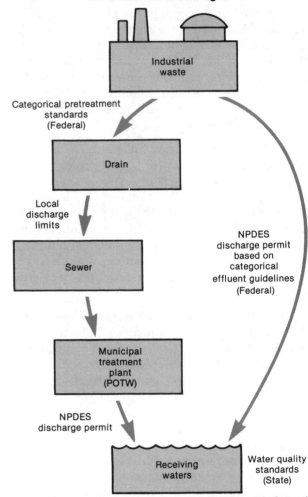

SOURCE: Save The Bay, Inc., *Down the Drain: Toxic Pollution and the Status of Pretreatment in Rhode Island* (Providence, RI: September 1986).

As described previously, *municipal dischargers* (POTWs) are required to meet standards different from those for direct industrial dischargers. Technology-based regulation of POTW discharges focuses almost exclusively on conventional pollutant control through the requirement that POTWs achieve "secondary" levels of treatment (ch. 9).

Indirect industrial dischargers must comply with technology-based pretreatment standards for existing sources (PSES) or pretreatment standards for new sources (PSNS). These standards, in combination with the fact that POTWs incidentally remove some pollutants prior to discharging wastewater (ch. 9), are intended to establish pollutant control which is roughly equivalent to that achieved by the BAT and BCT standards for direct dischargers. In addition, indirect dischargers must comply with local limits where they exist; such limits can be developed by POTWs or States to prevent toxic pollutants from disrupting POTW treatment processes or from passing through a POTW into receiving waters.

Water Quality-Based Standards

Section 303 of CWA mandates that additional standards based on the health and desired use of individual waterbodies be developed for all waters of the United States. Such standards are to be used to supplement technology-based controls on point source dischargers and other sources where the technology-based standards are not capable of meeting water quality objectives. Water quality-based standards have been the subject of much debate and they are analyzed in detail later in this chapter.

QUANTITIES OF PRIORITY POLLUTANTS IN INDUSTRIAL DISCHARGES

Types and Numbers of Industries Regulated Under the Clean Water Act

Almost 60,000 industrial facilities[14] and 15,000 POTWs are regulated as *direct* dischargers under NPDES and are required to comply with technology-based standards, as well as State water quality-based standards where they have been developed. Over 130,000 industries and commercial establishments have been identified as *indirect* industrial dischargers into POTWs.[15] Of these, 14,000 to 16,000 are in industries covered by categorical standards[16] (see table 13). While all indirect dischargers must comply with the General Pretreatment Regulations (as well as any local limits imposed by individual POTWs), only this subset must comply with the National Categorical Standards. (Table 13 lists only those dischargers in selected industries for which

categorical standards have been promulgated or proposed.) Several other industrial categories have been granted exemptions from categorical standards (table 14).

Thus, approximately 75,000 direct and indirect industrial dischargers, and about 15,000 municipal facilities, are subject to federally derived standards. An additional 85,000 or more indirect dischargers must comply only with the General Pretreatment

Table 14.—Industrial Categories Granted Exemptions From Categorical Standards[a]

Category	Number of facilities[b]
Adhesives and sealants	307
Auto and other laundries	90,800
Carbon black	12
Explosives	28
Gum and wood chemicals	21
Paint and ink formulation	1,217
Paving and roofing materials	NA
Photographic equipment and supplies	112
Rubber manufacturing[c]	1,576
Printing and publishing	38,763
Soaps and detergents[c]	NA

NA = data not available.
[a]The number and names of categories listed here do not correspond exactly to those indicated in the text or the Code of Federal Regulations due to subsequent joining or dividing of categories by EPA.
[b]Total includes direct and indirect dischargers but excludes zero dischargers.
[c]Only portions of these industrial categories have been exempted.

SOURCES: Science Applications International Corp., *Overview of Sewage Sludge and Effluent Management*, contract report prepared for U.S. Congress, Office of Technology Assessment (McLean, VA: March 1986); and U.S. Environmental Protection Agency, Office of Water Regulations and Standards, *Report to Congress on the Discharge of Wastes to Publicly Owned Treatment Works (The Domestic Sewage Study)* (Washington, DC: February 1986).

[14]This total includes about 50,000 individually permitted facilities and about 10,000 additional facilities in industrial categories which have been or will be granted general permits (e.g., offshore oil and gas operations). See the discussion below about permit backlogs for further detail on general permits.

[15]This total includes the following number of facilities from exempted or noncategorical industries: 69,000 industrial and commercial laundries; 39,000 printing and publishing facilities; 7,000 timber products processing facilities; 1,100 plastics molding and forming facilities; 970 textile mills; and 750 paint manufacturers or formulators. These data are from EPA's Industrial Technology Division's database (cited in ref. 666).

[16]This range, the accuracy of which is unknown, is derived by adding estimates for the number of individual facilities in each of the categorical industries taken from EPA's development documents (503,666). Approximately two-thirds (10,600) of these facilities are electroplaters or metal finishers.

Regulations and any additional local limits developed by POTWs or States.

Amounts of Wastewater and Priority Pollutants Generated

Total industrial wastewater flow is roughly estimated to be 18 billion gallons daily, or 6.4 trillion gallons per year, with about three-fourths originating from direct dischargers and one-fourth from indirect dischargers (255,256,666).

Prior to treatment, raw wastestreams from direct and indirect dischargers contain large quantities of toxic metals and organic chemicals. EPA's Monitoring and Data Support Division (MDSD) maintains a database of flow and composition estimates for wastewaters of selected categorical industries. These data provide a way to estimate the quantity of metals and organic chemicals present in industrial wastewaters; however, this is a conservative estimate for several reasons:

- Only some of those industries for which categorical standards have been promulgated or proposed are included. In addition, pollutants from noncategorical industries are excluded.
- Only a fraction of all priority pollutants are included.[17]
- Nonpriority pollutants are not included in the estimates.

The MDSD data (tables 15 and 16) indicate that a minimum of about 400 million pounds of priority metals and about 170 million pounds of priority organic chemicals are present in the raw wastewaters generated by *direct* dischargers in categorical industries each year. Raw wastestreams from *indirect* categorical dischargers are estimated to contain a

---

[17]Only those priority pollutants identified in the development document for a particular industrial category are included; only a subset of these are specifically regulated under BAT or PSES standards, based on consideration of the amount present in the wastewater and engineering and economic feasibility. For example, 8 priority metals and 32 priority organic chemicals are listed for the leather tanning industry. However, BAT and PSES specify a limit on only one metal (chromium) and no organic chemicals, even though a number of the other priority pollutants are present in significant amounts. Reductions in total suspended solids and other conventional pollutants—achieved through BPT for direct dischargers and POTW treatment for indirect dischargers—are relied on to achieve incidental removal of most of the unregulated substances; in addition, compliance with a BAT or PSES limit on one priority pollutant might also achieve incidental removal of others.

Table 15.—Expected Reductions in Discharges of Priority Metals and Organic Chemicals by Selected Categorical Industries, Assuming Full Compliance With and Implementation of BAT and PSES Controls (summary of table 16)

Type of discharge	Raw Amount[a]	Full controls Amount[a]	Percent reduction from raw[b]
Priority metals:			
Indirect dischargers	198	6.8	96.6%
Direct dischargers	403	6.4	98.4
Total	601	13.2	97.9%
Priority organic chemicals:			
Indirect dischargers	56	8.3	85.3%
Direct dischargers	172	1.7	99.0
Total	228	10.0	95.6%

[a]All amounts are millions of pounds per year.
[b]Percent reduction in amount of toxics as compared to amount in raw waste stream.

SOURCE: Office of Technology Assessment, 1987; based (except as noted in table 16) on U.S. Environmental Protection Agency, Office of Regulations and Standards, Monitoring and Data Support Division, *Summary of Effluent Characteristics and Guidelines for Selected Industrial Point Source Categories: Industry Status Sheets* (Washington, DC: Feb. 28, 1986).

minimum of 200 million pounds of priority metals and almost 60 million pounds of priority organic chemicals annually (665). Together, well over 800 million pounds of priority pollutants are present in raw wastewaters annually generated by categorical industries discharging to POTWs or the navigable waters of the United States.

Projected Removal of Priority Pollutants Through Full Treatment

Full implementation of, and compliance with, categorical standards would achieve major reductions in the amounts of priority pollutants discharged by these industries. Tables 15 and 16 summarize data on the estimated reductions that could be achieved *under full implementation* of BAT (for direct dischargers) and PSES (for indirect dischargers) categorical standards. These data are drawn from performance models of control technologies mandated under standards already in place or proposed (665). The total quantity of priority pollutants present in discharges of fully regulated wastestreams is projected to be only about 3 percent of the levels in the raw wastewaters. Even at this level of removal, however, more than 13 million pounds of priority metals and 10 million pounds of priority organic chemicals will be discharged an-

Table 16.—Expected Reductions[a] in Discharges of Priority Metals and Organic Chemicals by Selected Categorical Industries, Assuming Full Compliance With and Implementation of BAT and PSES Controls

| | Priority metals | | | | | | Priority organic chemicals | | | | | |
| | Indirect dischargers | | | Direct dischargers | | | Indirect dischargers | | | Direct dischargers | | |
	Raw	PSES[b]	Percent removal[c]	Raw	BAT[b]	Percent removal	Raw	PSES	Percent removal	Raw	BAT	Percent removal
Aluminum forming	1,192	8	99	736	16	98	1	0	100	736	16	98
Battery manufacturing	3,495	2	100	566	1	100	0	0	0	0	0	0
Coil coating (I)	363	2	99	251	1	99	3	0	100	2	0	100
Coil coating (II)	76	18	77	3	1	78	85	2	98	3	0	100
Copper forming[d]	8,780	10	100	7,219	9	100	37	1	97	31	3	90
Electrical and electronic components (I)	147	135	8	63	38	39	281	42	85	102	19	82
Foundries	12,317	13	100	12,827	16	100	686	5	99	562	5	99
Inorganic chemicals (I)	2,300	24	99	6,067	208	97	0	0	0	0	0	0
Inorganic chemicals (II)	195	1	99	394	17	96	0	0	0	0	0	0
Iron and steel	18,402	361	98	269,606	750	100	18,635	847	95	30,957	77	100
Leather tanning	5,321	475	91	899	17	98	517	142	73	76	2	98
Metal finishing/electroplating	136,684	3,754	97	62,446	1,704	97	8,815	92	99	2,429	42	98
Nonferrous metals forming	215	1	100	631	1	100	0	0	0	2	0	100
Nonferrous metals (I)	131	3	98	331	45	86	0	0	0	171	1	99
Nonferrous metals (II)	249	0	100	750	3	100	0	0	0	9	0	100
OCPSF[e]	5,519	23	100	34,999	102	100	18,698	17	100	118,420	123	100
Petroleum refining	291	291	0	1,488	291	80	1,222	1,222	0	6,248	38	99
Pharmaceuticals	125	125	0	26	26	0	212	112	47	43	23	47
Plastics molding and forming	26	26	0	41	29	30	29	29	0	42	37	12
Porcelain enameling	527	15	97	277	8	97	0	0	0	0	0	0
Pulp and paper	864	864	0	2,865	2,752	4	5,645	5,582	1	11,052	1,131	90
Textiles	725	603	17	590	360	39	1,329	214	84	914	173	81
Total	197,944	6,754	97	403,075	6,395	98	56,195	8,307	85	171,799	1,690	99

aAll quantities are in thousands of pounds per year.

bAmounts expected to be discharged under full implementation of, and compliance with, PSES or BAT categorical standards. See text for definition of PSES and BAT; note that indirect removals are mandated through PSES standards; direct removals are mandated through BAT standards.

cPercent reduction in amount of priority metals or organic chemicals relative to amount in raw wastewater.

dData for the Copper Forming Category are derived from U.S. Environmental Protection Agency, Final Development Document for Effluent Limitations and Standards for the Copper Forming Point Source Category, EPA 440/1-84/074, table X-19, p. 467 (Washington, DC: March, 1984); during a review of a draft of this OTA report, EPA found that the primary source of information for this table (see below) contained incorrect data for this category.

eData for the Organic Chemicals and Plastics and Synthetic Fibers (OCPSF) category are taken from a "Correction Notice/Notice of Availability" published by EPA (50 Federal Register 41528, Oct. 11, 1985). These data are the most recent available estimates for the OCPSF category and have been revised downward from previous estimates. They should not be regarded as final, however, as they may be subject to further revision.

SOURCE: Office of Technology Assessment, 1987; based (except as noted) on U.S. Environmental Protection Agency, Office of Regulations and Standards, Monitoring and Data Support Division, Summary of Effluent Characteristics and Guidelines for Selected Industrial Point Source Categories: Industry Status Sheets (Washington, DC: Feb. 28, 1986).

nually in wastewaters originating from categorical industries (665).[18]

Direct dischargers account for more than twice as many of the priority pollutants present in raw wastewaters than do indirect dischargers (69 percent v. 31 percent); in contrast, most (about 65 percent) of the priority pollutants expected to be *discharged* upon full implementation of BAT and PSES controls will originate from indirect sources. However, pollutants in indirect discharges are subject to additional "incidental" removal as a result of treatment at POTWs, so that further reductions are achieved prior to their discharge to waterbodies.[19]

About 80 percent of all indirectly discharged industrial wastewater enters POTWs that are required to have individual pretreatment programs, and a similar percentage receives secondary or higher levels of treatment at POTWs (255,503; also see ch. 9).[20] As discussed previously, the intent of the national categorical standards is to achieve a total reduction—through a combination of pretreatment by industries and incidental removal by POTWs—in discharges of priority pollutants by regulated indirect dischargers that is roughly equivalent to the levels resulting from implementation of BAT and BCT standards for direct dischargers.

Removal of Priority Pollutants Achieved to Date

The partial implementation achieved to date of standards developed under the NPDES and National Pretreatment programs has significantly reduced the amounts of priority pollutants in direct and indirect discharges from some categorical industries. However, **the Nation is far from achieving full implementation of, and full compliance with, categorical standards. Moreover, these standards only apply to some industries and pollutants, so even if fully effective, significant quantities of toxic pollutants will remain unregulated by categorical standards**.

Unfortunately, existing data do not allow an accurate assessment of how close these programs are to achieving full removal of regulated pollutants. The MDSD database does contain estimates of "current" discharges for industries with categorical standards in place. These estimates are highly questionable, however, because the current discharge is assigned the same value as the discharge expected under *full* implementation of BAT or PSES, even though implementation of and/or compliance with the standard is far from complete in many cases (e.g., metal finishing and electroplating). These and other shortcomings cast sufficient doubt on estimates of current discharges so as to preclude their use in estimating the extent of removal of priority pollutants achieved to date.

For particular industries, more reliable data are available in some cases and reveal considerable variation in progress toward achieving full BAT or PSES reductions. For example, EPA's most recently published estimates for the Organic Chemicals, Plastics, and Synthetic Fibers (OCPSF) industry (50 FR 41528, Oct. 11, 1985) suggest that technology already in place is achieving pollutant removals of more than 99 percent for *direct* dischargers, resulting in discharges that are only about fivefold higher than that expected under the most stringent BAT standards proposed.[21] For *indirect* dischargers, however, these same data indicate only a 4 percent removal, vastly less than the removal expected under full implementation of PSES.

---

[18]It is important to remember that these removal estimates are only for a subset of priority pollutants and do not include any nonpriority substances. Nor do they include any pollutants from noncategorical industries.

[19]Many pollutants removed from industrial wastewaters treated at POTWs will become contaminants of municipal sludge, generating a different set of disposal problems. This issue is discussed in detail in ch. 9.

[20]In marine waters areas, a smaller percentage of total wastewater flow into POTWs—and presumably indirectly discharged industrial wastewater as well—receives secondary or higher treatment. About 39 percent of the wastewater entering POTWs discharging to coastal waters, and 60 percent of that entering POTWs discharging to estuaries, receives secondary or higher treatment (503). This is probably due in large part to the fact that a number of large coastal POTWs have been granted or have applied for waivers from achieving secondary treatment under CWA Sec. 301(h). The fraction of total POTW flow contributed by industrial dischargers is also slightly lower in marine waters: 14.5 percent v. 17 percent nationally.

[21]OCPSF industry representatives argue that this level of removal has been accomplished—even in the absence of final regulations—by installing the appropriate pollutant controls either in anticipation of the regulations or in response to permit limits that have been developed using the best professional judgment (BPJ) of the permit writers (R. Schwer, E.I. DuPont de Nemours & Co., pers. comm., November 1986).

Reducing Priority Pollutants in Specific Industries

Table 15 presents data on selected primary industrial categories, indicating a) the quantities of priority pollutants in their raw wastewaters, and b) the quantities expected in their discharges assuming full compliance with categorical standards. Direct and indirect dischargers, and priority metals and organic chemicals, are considered separately. Several conclusions can be drawn from these data:

• Whether for raw or fully regulated (PSES/BAT) wastestreams, a few industries tend to dominate the picture, for both direct and indirect discharges. In each case, the top 3, 4, or 5 industrial categories comprise 90 percent or more of the total amount of priority metals or organic chemicals present in wastewaters.

• The effectiveness of BAT or PSES levels of control varies greatly among different industries, ranging from very low removal (pulp and paper) to essentially full removal (organic chemicals, plastics, and synthetic fibers[22]). Under full implementation of standards, however, most industries are predicted to achieve more than 90-percent removal.

---

[22]Estimates for this industrial category are based on the controls specified in the *proposed* categorical regulation.

ISSUES IN THE MANAGEMENT OF INDUSTRIAL EFFLUENTS

A number of issues remain regarding the adequacy of the current framework that regulates industrial discharges. These issues primarily concern major constraints on the development and implementation of the NPDES and pretreatment programs. Four key areas of deficiency need to be addressed:

1. delays in program implementation, including:
   —delays in promulgation of Federal regulations, and
   —unpermitted sources and permit renewal backlogs;
2. gaps in program coverage, including:
   —nonregulated industries,
   —nonregulated toxic pollutants, and
   —permit deficiencies;
3. inadequacy of regulatory compliance and enforcement, including:
   —self-reporting: quality and completeness of discharge data,
   —extent of noncompliance with effluent discharge limits, and
   —effectiveness of enforcement; and
4. additional issues facing the pretreatment program, including:
   —lack of incentives for POTW program implementation and enforcement, and
   —hazardous waste in sewers.

In addition to these four areas, several other problems that adversely affect the management of industrial effluents are often identified. These include: 1) the inadequacy of resources available for permitting, compliance monitoring, and enforcement activities; 2) a need for better management of the data collected from dischargers to allow centralized access for assessing program performance and progress; and 3) inconsistent policy, methodology, and performance among EPA Regions, States, and local authorities with respect to implementing and enforcing water pollution control programs.

Delays in Program Implementation

Delays in Promulgation of Federal Categorical Regulations

In the absence of final categorical regulations, it is difficult for POTWs and other regulatory bodies to carry out enforcement against dischargers. Regulations for some significant industrial categories (e.g., organic chemicals and plastics) have been proposed, but not yet promulgated (table 17). Uncertainties about the final form of these regulations have caused delays in the implementation of treatment technologies in these industries. Even where final regulations exist, scheduled pretreatment compliance dates have in many cases not yet been

Table 17.—Final and Proposed Regulations for Categorical Industries

	Date of promulgation	Effective date[a]	Date of PSES compliance
Timber processing	1-26-81	3-30-81	1-26-84
Electroplating:[b] Integrated	1-28-81	3-30-81	4-27-84
Nonintegrated	1-28-81	3-30-81	6-30-84
Iron and steel	5-27-82	7-10-82	7-10-85
Inorganic chemicals I	6-29-82	8-12-82	8-12-85
Textile mills	9-02-82	10-18-82	N/A[c]
Coal mining	10-13-82	11-26-82	N/A[d]
Petroleum refining................	10-18-82	12-01-82	12-01-85
Steam electric	11-19-82	1-02-83	7-01-84
Pulp and paper..................	11-18-82	1-03-83	7-01-84
Leather tanning	11-23-82	1-06-83	11-25-85
Porcelain enameling	11-24-82	1-07-83	11-25-85
Coil coating I	12-01-82	1-17-83	12-01-85
Ore mining	12-03-82	1-17-83	N/A[d]
Electrical and electronic components I	4-08-83	5-19-83	7-01-84
(arsenic subcategory)	4-08-83	5-19-83	11-08-85
Metal finishing[b]...................	7-15-83	8-29-83	2-15-86
Copper forming	8-15-83	9-26-83	8-15-86
Aluminum forming	10-24-83	12-07-83	10-24-86
Pharmaceuticals	10-27-83	12-12-83	10-27-86
Coil coating II	11-17-83	1-02-84	11-17-86
Electrical and electronic components II	12-14-83	1-27-84	7-14-87
Battery manufacturing	3-09-84	4-18-84	3-09-87
Nonferrous metals I	3-08-84	4-23-84	3-09-87
Inorganic chemicals II	8-22-84	10-05-84	8-22-87
Plastics molding and forming	12-17-84	1-30-85	N/A[c]
(phthalates subcategory: action due)	6-??-87[e]	?	N/A
Nonferrous metals forming	8-23-85	10-07-85	8-23-88
Nonferrous metals II..............	9-20-85	11-04-85	9-20-88
Foundries	10-30-85	12-13-85	10-31-88
Organic Chemicals and Plastics and Synthetic Fibers ...proposed	3-21-83[f]	?	?
Pesticides ....issued	10-04-85	?	?
withdrawn	12-15-86[g]		

[a]BAT standards take effect as specified by the compliance schedule written into individual NPDES permits issued after the effective date of the regulation.
[b]Existing job shop electroplaters and independent circuit board manufacturers must comply only with the electroplating regulations. All other electroplating subcategories are also covered by metal finishing regulations.
[c]No pretreatment standards were promulgated for these categories because they were exempted under Paragraph 8 of the NRDC consent decree.
[d]No pretreatment standards were promulgated for these categories because they contain no indirect dischargers.
[e]This subcategory is under study to establish treatability data for possible future regulation. Final action is expected in June 1987 (51 *Federal Register* 4526, Apr. 21, 1986). EPA expects to exempt this subcategory from pretreatment standards under Paragraph 8 of the NRDC consent decree (as was previously done for the rest of the industry).
[f]It is unclear when final regulations will be issued for this category (51 *Federal Register* 44082, Dec. 8, 1986).
[g]Final regulations (BAT, PSES) for the pesticides industry were issued in 1985, but subsequently withdrawn by EPA (51 *Federal Register* 44911, Dec. 15, 1986). No date for reissuance of the regulation was provided.

SOURCE: Office of Technology Assessment, 1987; based on *Federal Register* notices cited in the appropriate sections of 40 Code of Federal Regulations, Sections 401 to 460, except as noted above.

reached. As of the end of 1985, final compliance dates had not yet passed for half of the industrial categories for which pretreatment regulations have been issued; only five more compliance dates were reached by the end of 1986, leaving nine categories still without final regulations in effect.[23]

Final regulations for all but two of the primary categorical industries, OCPSF and Pesticides, have now been promulgated. Regulations for pesticides have been promulgated, but were challenged in court and are being raised.[24] Moreover, the one remaining industry for which standards have never been issued (OCPSF) is the category contributing the most priority organic chemicals in its raw wastewaters (table 16).[25] In addition, compliance dates for the seven latest regulations to be promulgated will not be reached until well into 1987 and 1988 (table 17).

The effect of delays in issuing final regulations can be very different for direct and indirect dischargers. In the absence of final regulations, technology-based limits based on the best professional judgment (BPJ) of the permit writer are often written into NPDES permits for *direct* dischargers. In the same situation, however, *indirect* dischargers are only subject to local limits (if they exist), which are based largely on the ability of the POTW to meet its own NPDES permit limits or sludge disposal requirements. This factor may in part account for the major differences in the levels of pollutant removal achieved by direct and indirect dischargers in the OCPSF industry in the absence of final categorical regulations.

---

[23]Final compliance dates with pretreatment standards are specified for indirect dischargers, and are required under CWA Sec. 307 to be no more than 3 years after the date of promulgation. For direct dischargers, compliance dates are written into permits, typically in the context of a compliance schedule. Sec. 307 specifies that compliance should generally be required within 1 year, but in no case more than 3 years, after promulgation of the BAT standards. However, delays in renewing permits (discussed later in this chapter) may further lengthen the period before compliance must be achieved.

[24]Final regulations for the Pesticides Industry actually were promulgated on Oct. 4, 1985 (50 FR 40672). Various aspects of the rule were challenged in the Court of Appeals, however, and EPA subsequently discovered significant errors as well. Under Court order, EPA remanded the regulation on Dec. 15, 1986 (51 FR 44911).

[25]It is unclear when final regulations will be issued for the OCPSF category. EPA recently announced its intention to file for an extension of its deadline (set by the Court at Dec. 31, 1986) for promulgation of final regulations and has asked interested parties to comment on several new proposals (51 FR 44082, Dec. 8, 1986).

Unpermitted Sources and Permit Renewal Backlogs

Lack of Ability To Identify Facilities That Have Not Applied for Permits.—Many (if not most) States and Regions lack a systematic method for identifying nonfilers, instead relying on informal approaches such as citizen complaints. EPA and State permit officials generally believe that all major dischargers have been identified and have applied for permits (576), although data supporting this claim are not available.

Several studies, however, suggest that unpermitted facilities may be significant sources of pollutants in at least some areas (462,576). In Puget Sound, for example, many nonpermitted commercial and industrial facilities were recently identified and as much as 20 percent of the toxic pollutants entering Puget Sound are estimated to originate from such nonpermitted discharges (462).

Backlog in Processing Submitted Applications for Initial Permits.—The backlog in issuing new NPDES permits was a major problem in the late 1970s and early 1980s. The extent of this backlog varies greatly: 1) among EPA Regions and approved States, 2) between municipal and industrial dischargers, and 3) between major and minor dischargers. EPA national statistics for 1982 indicated more than 16,000 unprocessed permit applications, only about 200 (1.3 percent) of which were from major dischargers (576).[26] Studies of individual States documented a similar situation (576). The State of Washington has eliminated its backlog of unissued initial permits for major dischargers, and is now concentrating efforts on "significant minor" dischargers (462).

Resource limitations are routinely cited as the major factor that forces permitting efforts to focus primarily on renewals, new sources, and major dischargers. This factor appears to be the primary reason for the backlog in general and for the much lower rate of permitting for minor dischargers.

Backlog in Renewing Expired Permits.—EPA and delegated States also face the continual, and

---

[26]EPA argued that the backlog was probably overstated because it included an unknown number of facilities that did not require a permit. This argument would apply almost exclusively to minor dischargers.

in many cases increasing, task of renewing permits that have expired. While an expired permit is still enforceable, the opportunity is lost to review and upgrade permits in a timely manner. Moreover, each expired permit that is not reissued represents a de facto permit length longer than the 5-year term intended under CWA.

Most initial NPDES permits were issued in 1973 and 1974, with expiration dates set for 1978 and 1979. As of the end of 1982, EPA reported 34,000 expired (and not reissued) permits. About 13 percent (4,400) of these were for major dischargers, the remainder for minor dischargers. Over half of these permits had been expired for more than 22 months (576). A similar picture existed in the five States that GAO examined in detail.

Several factors have been cited as causes for this backlog:

- the lack of BAT guidelines for use in upgrading effluent limits,
- heavy reliance on BPJ as a substitute for BAT and BCT guidelines,
- a shortage of resources devoted to permit issuance,
- the need to develop general permits for certain categories of minor dischargers which do not require individual permits, and
- low management priority placed on renewing permits for minor dischargers.

EPA recently has increased the resources devoted to permit issuance and renewal, and has promulgated most of the BAT regulations it was required to develop. These efforts were taken in part to meet a national goal of eliminating the permit backlog for major dischargers by the end of fiscal year 1985, one that was largely met by EPA Regional offices, but not by many approved States.

Recent EPA data suggests a considerable reduction in the national backlog for major dischargers, although a substantial fraction of major permits and an even larger fraction of minor permits remain expired (327). Thirty-four percent of all major industrial and 13 percent of all major municipal permits are expired. This is a total of 1,810 expired permits, compared to the 4,400 reported in 1982. Many approved States continue to have even larger backlogs, however. For example, in Washington half of all permits (and one-quarter of those for major dischargers) are currently expired (462).

EPA is addressing the minor discharger backlog by developing general permits to cover an estimated 10,000 minor dischargers.[27] A second, more controversial approach has been EPA's legislative proposal to extend the term of NPDES permits from the current 5 years to 10 years (242). EPA argues that this change would reduce the annual permitting workload, and presumably the backlog; modification of permits would still be required to incorporate "significant" new standards. Opponents view this proposal as a "paperwork" solution that would further reduce the opportunity to review and upgrade permits at the frequency originally intended by CWA.

Gaps and Deficiencies in Coverage of Toxic Pollutants

Nonregulated Industries

Some entire categories of industrial dischargers, such as car washes and other commercial laundries or paint and ink formulators, are exempted from BAT effluent guidelines and pretreatment standards (see table 14).[28] In addition, certain subcategories of other industrial categories, for example, adhesives and sealants, are exempted. Finally, pretreatment standards for some industrial categories, such as textile mills and plastics molding and forming, were proposed but never promulgated.

These and other categories can contribute significant amounts of toxic pollutants to surface waters or POTWs. The laundries and textile mills categories together account for approximately 22

---

[27]Such permits have been issued or proposed for activities such as offshore and coastal oil and gas facilities, coal mining, animal feedlots, construction sites, noncontact cooling water, petroleum storage and transfer, deep seabed mining, and seafood processing. Currently, general permits cover about 7,700 facilities (E. Ovsenik, Office of Water Enforcement and Permits, EPA, pers. comm., January 1987).

[28]Reasons given for exempting these industries from effluent limitations for toxic pollutants include: presence of insignificant levels of pollutants, no or few direct or indirect dischargers, economic constraints, no new plants expected, presence of pollutants for which removal technology does not exist, etc. These facilities must still obtain and meet effluent limitations specified in their NPDES permits; in the absence of categorical standards, however, limitations are likely to be specified only for conventional pollutants or to rely on the best professional judgment (BPJ) of the permit writers for toxic pollutants.

percent of the total industrial flow into POTWs. Approximately 91,000 laundries, including 22,000 car washes, have a total wastewater flow of 526 million gallons per day; this wastewater contains at least 13 priority pollutants. About 1,000 textile mills discharge 312 million gallons per day to POTWs; pollutants identified in these wastewaters include several priority metals and organic chemicals (503).

Other industries not included by EPA on its original list of industrial categories may also be significant sources of toxic pollutants. These industries include treatment, storage, and disposal facilities for hazardous wastes; drum and barrel reconditioners; solvent reclaimers; battery salvagers; septage haulers; and automotive radiator shops.[29]

Many small dischargers within industries that have categorical regulations are exempted from the regulations because of the potentially heavy economic burden of meeting effluent standards. For example, electroplating job-shops that discharge less than 10,000 gallons per day of wastewaters containing chromium, copper, nickel, or zinc are specifically exempted from pretreatment regulations. In some cases, however, such low-volume discharges can contain high enough concentrations of toxic pollutants to upset POTW operations. Although local limits (which are authorized and enforceable under Federal law) could be used to regulate such discharges, POTWs face many obstacles in developing such limits, particularly in the absence of Federal standards or guidance (503,653,666).

Nonregulated Toxic Pollutants

Many toxic pollutants in industrial wastewaters are not regulated by national standards for a variety of reasons—for example, lack of data on the presence of certain pollutants in a wastestream; lack of analytical means for measuring certain pollutants; lack of technological means to control certain pollutants; low regulatory priority; or the diversity and complexity of individual plants or processes within an industrial category.

Categorical regulations generally contain standards for only a fraction of all priority pollutants. Development of a standard only occurs if three con-

ditions are met: 1) the pollutant is present in high amounts, 2) the technology for its control is available, and 3) the implementation of that technology is economically feasible. Thus, many priority pollutants—in particular, priority organic chemicals—may be present in high concentrations but remain unregulated for technological or economic reasons. For example:

- EPA has not developed limits for phthalates generated by the plastics molding and forming industry because it does not have sufficient data on technologies for controlling these chemicals (51 FR 14526, Apr. 26, 1986).
- While pretreatment standards have been developed for the petroleum refining industry, they contain no limits on priority organic chemicals, even though this category is a very significant source of such pollutants, accounting for almost 15 percent of the expected discharge of priority organic chemicals to POTWs under full PSES implementation.[30]
- As a result of the much greater focus on metals than on organic chemicals, full implementation of EPA's categorical pretreatment standards is predicted to greatly reduce total inputs of priority metals to POTWs, but to reduce priority organic chemicals by only 47 percent (666).[31]

Other facilities that discharge priority pollutants but are not regulated by national standards include those that: 1) contribute wastes to POTWs that are either not required to or have not yet developed an individual pretreatment program, and 2) are in *noncategorical* industries. **About 30 percent of the priority pollutants that enter POTWs originate from noncategorical sources** (503). Available data suggest that roughly equal amounts originate from domestic households and from noncategorical industries or commercial establishments.

---

[29]Estimates of the amount of hazardous chemicals introduced by these industries into POTWs are presented in ref. 666.

[30]These data and comparable information for numerous other industrial categories are discussed in detail in ref. 666.

[31]This low percentage results from two factors: the relative lack of standards specified for organic chemicals, and the significant contribution of priority organic chemicals by noncategorical (and relatively unregulated) industries. The latter factor largely accounts for the difference between this removal estimate and the higher estimate (85 percent) indicated in table 15.

Finally, both NPDES and the pretreatment programs have limited ability to address discharges of nonpriority pollutants. Categorical standards have focused almost exclusively on the 126 priority pollutants. However, numerous additional toxic pollutants are known to be present in significant quantities in both direct and indirect discharges. **Data collected by EPA for the OCPSF industrial category indicates that this industry's raw wastewaters contain 2.5 pounds of nonpriority organic chemicals—including such toxic compounds as formaldehyde and methanol—for every pound of priority organic chemicals.** Other categories, such as the pesticides and pharmaceuticals industries, are also significant sources of toxic nonpriority pollutants (666).

In principle, categorical standards, local limits, or water quality-based standards can be used to control nonpriority pollutants, and in some cases these approaches have been developed. However, these types of controls do not currently provide a systematic means for addressing pollutants that fall outside the primary focus of the CWA pollutant control programs.

At least two legal mechanisms for regulating these additional toxic pollutants are available, but they have only been used to a limited extent. Section 307(a) of CWA gives EPA the authority to add substances to the priority pollutants list, but this authority has not been used to date. Under paragraph 4(c) of the Consent Decree reached between EPA and the Natural Resources Defense Council, EPA is required to identify, and possibly regulate, toxic pollutants that might violate the objectives of the NPDES and National Pretreatment programs but that are not listed as priority pollutants. In one survey of industrial wastewaters, EPA detected the presence of over 1,500 compounds; of the more than 400 that were specifically identified, 6 compounds were chosen as candidates for future regulation due to their presence in significant amounts and their human or aquatic toxicity (666). No standards have yet been developed, however, due to lack of information on the ability of in-place or other available control technologies to remove these pollutants.

Permit Deficiencies

Lack of Limits on Toxic Pollutants.—Typically, discharge permits specify numeric limits for most or all conventional pollutants, but far fewer limits are specified for priority metals or organic chemicals. While the development and introduction of BAT guidelines will help alleviate this deficiency, several aspects of this problem remain:

* Monitoring for priority pollutants other than those *known* to be present (and therefore specified in the discharge permit) is rarely required, so that the presence of additional priority pollutants in a discharge often is not documented.
* Even where the presence of priority pollutants has been reported by permittees or discovered through sampling, limits for such pollutants often have not been included in permits. For example, an audit of 44 permittees discharging into Puget Sound that reported the presence of priority pollutants revealed only 14 that had any limitations on the reported substances (462).
* While technically a violation of CWA, the discharge of a pollutant for which no standard exists in the permit is unlikely to be identified or treated as a violation.
* Water quality-based standards have not been developed by most States for most priority pollutants, hampering the introduction of water quality-based discharge limits on these substances into individual permits. Even where developed, the standards do not address sediment contamination, which is probably the more important "sink" for most metals and organic chemicals of concern (see section later in this chapter on water quality standards).

Heavy Reliance on Best Professional Judgment (BPJ) in the Absence of Standards.—Whenever national standards for a particular industry or pollutant do not exist or have not yet been developed, individual permit writers must rely on their BPJ. At least two levels of discretion are involved: determining which, if any, pollutants should be limited, and setting the actual level. Both elements can be significant sources of inconsistency in setting discharge limits.[32]

"Backsliding" on Permit Limits During Renewal.—The CWA's goal of achieving zero discharge of pollutants by 1985 was based on a "ratchet" approach in which discharge limits would be made increasingly stringent in successive permits. Indeed, current Federal regulations require that a reissued permit "be at least as stringent" as the previous permit, unless certain exceptional conditions are met (40 CFR 122.44(l)).[33]

However, some "second round" permits incorporating BAT limits on toxic pollutants have been found to contain weaker limits than those imposed in the "first round" permits. For example, in a 1985 study of 16 major industrial dischargers in the Puget Sound basin, 14 of the renewed permits had been weakened for at least 1 pollutant, and 8 were weakened for at least 3 pollutants (458). Many of the weakened limits were for conventional pollutants based on BPT standards promulgated a decade ago. Overall, for those pollutants specified in both sets of permits, standards for 40 percent of the pollutants had been weakened, 27 percent had been strengthened, and 33 percent remained unchanged.

Justification for such changes may often exist, particularly given the extensive use of BPJ in setting initial limits. However, there appears to be a disturbing lack of appropriate means for communicating or explaining such changes to the public during the renewal process (458,462).

Inadequacy of Regulatory Compliance and Enforcement

Various elements of the NPDES and pretreatment programs determine their effectiveness in ensuring compliance with permits or other means of controlling the discharge of pollutants from point sources. Significant problems occur in three areas: quality and completeness of data submitted by regulated facilities, extent of noncompliance, and effectiveness of enforcement.

Information available on these issues is often far from complete or is regional and selective in nature, and thus may not always be representative of the national situation. Moreover, data may not be entirely current, an inherent problem in light of frequent changes in program design, permit status, available resources, agency priorities, and regulations. Finally, much of the information needed for a thorough national analysis is often unavailable, or is inaccessible due to the relatively primitive state of development of national databases.[34]

Self-Reporting: Quality and Completeness of Discharge Data

The efficacy of the entire NPDES program rests on the ability of agencies to obtain reliable data characterizing the discharges of permitted facilities. Given the immense number of such facilities, NPDES relies on a self-reporting system in which facilities are required to monitor their discharges and regularly report the results to the appropriate EPA Region or delegated State. Administrative review of reported data is the only *systematic* means of identifying instances of noncompliance, although inspections are occasionally used to supplement industry reporting.

Several problems with this self-reporting system have been identified. The foremost and most obvious of these is that such a system relies to a large extent on the integrity of permittees to submit accurate information. In order to generate reliable data, a self-reporting system must include effective deterrents to counter the obvious incentive to fal-

---

[32]BPJ has been identified as the "least consistent link in the current discharge permit system" (462), and causing a "movement from the issuance of consistent conditions to those tailored to the needs or pressures from each individual permittee" (392).

[33]For example, a permit limit may be loosened if proper operation of the required control technology still does not achieve the required limit, or if lower levels are based on newly issued national standards and the initial levels were set using BPJ. In any case, the new standards cannot be set below water quality-based standards or technology-based effluent guidelines.

[34]This discussion of necessity relies heavily on particular sources of information, for example the thorough and up-to-date information compiled by the Puget Sound Water Quality Authority (461,462) and the less current "random surveys" conducted by GAO (574,576,689). While such information may not fully identify or accurately represent problems in all States or EPA Regions, these data are generally consistent with other available information and provide a reasonable picture of problems facing water pollution control programs in all parts of the Nation. Wherever possible, the most recent data available from EPA are included; in many cases, these data indicate renewed attention to or substantial improvements in existing problems.

sify information rather than comply with discharge limitations. Such deterrents require that the system be able to detect and penalize those who violate reporting requirements.

These deficiencies call into serious question the adequacy of the mechanisms available to EPA and the States to verify information received through the NPDES self-monitoring system. In the face of declining resources available for such activities, this crucial link in the current compliance and enforcement program is unlikely to be strengthened.

Violations of Reporting Requirements.—Failure to submit a discharge monitoring report (DMR), or submission of incomplete data, are obvious means of concealing serious noncompliance with discharge limitations. One study of major industrial and municipal NPDES permittees in six States found that 8 percent failed to submit a DMR at least once during an 18-month period, and that 37 percent had submitted one or more incomplete DMRs (576).[35] However, this rate of compliance is an apparent improvement in the DMR submission rate for industrial dischargers compared with that found in an earlier study (574).

EPA or State response to reporting violations varies, but in general such violations appear to have received little or no attention.[36] As an extreme example, a recently identified major discharger into Puget Sound had not submitted a DMR for 30 consecutive months, yet no action had been taken by the permitting authority (462).

Quality of Reported Effluent Sample Analyses.—In several surveys of the quality of chemical analyses of effluent samples submitted by major NPDES permittees, EPA found that significantly more than half of all permittees reported unacceptable data[37] for one or more effluent parameters, and

that 20 to 25 percent of all analyses were of unacceptable quality (576).

In many cases, poor performance was due to a high rate of *reporting* errors, rather than analytical errors. While in principle easier (and less expensive) to correct, such errors call into question the integrity of permittees as well as the reliability of submitted data.[38]

Adequacy of EPA's and States' Abilities To Verify Information.—Compliance sampling inspections (CSIs)—during which effluent samples are collected for the purpose of verification—are the primary tool available to EPA and States to verify the data submitted by dischargers.[39] CSIs are employed primarily in cases where noncompliance has been reported or is suspected (327,576). In 1982, for example, only 7 percent of New Jersey's and 5 percent of New York's major dischargers received CSIs. Nationwide, EPA statistics showed a large decrease in CSIs performed during the period 1979 to 1981; there was a 20 percent reduction for municipal sources and an almost 50 percent decrease for industrial dischargers.[40] Inadequate resources were cited as the primary reason for the decline, a problem which may have been partially alleviated in subsequent years (468).

As further illustration of the infrequent use of CSIs, guidelines in the State of Washington strongly encourage annual CSIs for all major dischargers, including collection and analysis of effluent samples. A recent analysis found, however, that only one-sixth of dischargers into Puget Sound had ever received a CSI (over a 10-year period);[41] analysis for priority pollutants had been conducted for only five dischargers (462). Moreover, it is standard practice in the State to announce inspections and field sampling in advance, raising concerns over how representative of typical effluent such samples actually are.

---

[35]DMRs are required to be submitted quarterly or monthly. Two percent of GAO's sample did not submit DMRs for 3 or more quarters or 5 or more months during the survey period; 11 percent submitted incomplete DMRs for 3 or more quarters or 5 or more months during this period.

[36]This type of violation is now an instance of Category I noncompliance, and therefore must be reported in Quarterly Noncompliance Reports (QNCRs; see 40 CFR 123.45(a)(2)(ii)(D)).

[37]"Unacceptable" results were those that were either higher or lower than the acceptable limits established by EPA's Quality Assurance Program, determined on a case-by-case basis for pollutants actually specified in a discharger's permit.

[38]The data cited by GAO (576) did not indicate what fraction of errors were *below* acceptable limits; consistently low errors would be expected if deliberate falsification were the cause.

[39]Several other types of inspections of NPDES permittees are also conducted, but do not involve collection of effluent samples.

[40]The number of other types of inspections—compliance evaluations and performance audits—actually increased significantly during fiscal years 1979 to 1981. These are considerably less expensive and time-consuming than CSIs.

[41]This fraction included only about half of Puget Sound's major industrial dischargers.

Extent of Noncompliance With Effluent Discharge Limits

This discussion separately considers compliance for direct dischargers (municipal and industrial facilities regulated under the NPDES program) and for indirect dischargers (regulated under the National Pretreatment Program).

Direct Dischargers.—Estimates of the extent of permittee noncompliance with NPDES permits are quite disparate (238,392,462,574,576,689). A major source of variability—and controversy—surrounding these estimates is the criteria used to define noncompliance, particularly *significant noncompliance* (SNC).[42]

In particular, different criteria have been used by EPA and Congress' General Accounting Office (GAO) to determine SNC (table 18). (EPA has since partially revised its criteria, as discussed in

table 18.) As a result, GAO and EPA have reported considerably different SNC estimates, even using the same raw data. For example, based on its review of dischargers in six States, GAO found that about 80 percent exceeded one or more permit limits at least once during an 18-month period; almost half of the permittees who exceeded their permit limits did so for more than 6 of the 18 months, and about 20 percent did so for more than 12 of the 18 months. Based on GAO's criteria, one-quarter of all dischargers were in SNC at least once (576).

EPA took issue with several of GAO's findings, in particular those regarding SNC. EPA's data for the same six States indicated a SNC rate 7 to 12 percent lower than that found by GAO over the same time period. Nationally, EPA reported SNC rates of 18 percent for municipal and 16 percent for industrial dischargers.

Regardless of the criteria used, it is clear that noncompliance is a major and continuing problem. At the same time, some progress has been made: *industrial* compliance has improved considerably over the last several years, and it has consistently exceeded the degree of *municipal* compliance. EPA data for the first quarter of 1985 indicated that only 5 percent of major industrial facilities were in SNC. Similarly, as of the third quarter of 1985, only 6

---

[42]Another source of variability is in the interpretation of how significant a particular rate of permit violations really is. Essentially all studies express noncompliance in terms of the number of facilities with at least one permit limit violation in a given time period. These facilities, however, may be in full compliance with many other permit limits during the same period. EPA argues that it is also important to examine the number of limits that are exceeded relative to the total number of possible exceedances (i.e., the number of limited parameters in each permit times the number of permits) to fully appreciate both the magnitude of the compliance problem and the success or failure rate (651).

Table 18.—Criteria Used To Identify "Significant Noncompliance"

Environmental Protection Agency	General Accounting Office
Based on magnitude and frequency:	***Based on magnitude and frequency:***
• 2 exceedances of a monthly average limit in any 6-month period that meet the following criteria: —40% over limit for conventional pollutants and "nontoxic" metals —20% over limit for toxic pollutants —discretionary for fecal coliforms or pathogens	• 4 consecutive exceedances of an average limit by at least 50% during its 18-month survey period
Based on frequency only:	
• 4 exceedances of a monthly average limit (by any amount) in any 6-month period for any pollutant	
Other criteria:	***Other criteria:***
• Excludes permittees on interim limits and/or construction schedules[a] • Excludes permittees returned to compliance by end of quarter[b]	• Includes permittees on interim limits and/or construction schedules (about 25% of GAO's sample)

[a]This criterion was used by EPA at the time of the GAO analysis; EPA has since revised its definition so as to include any violation which is of concern to the agency, including those by facilities on interim limits or construction schedules.
[b]Using this criterion, if a permittee were in significant noncompliance during the first two months of a quarter, but returned to compliance in the third month, it would not be reported. This criterion was used by EPA at the time of the GAO analysis; EPA has since revised its definition so as to eliminate this possibility.
SOURCES: Office of Technology Assessment, 1987; based on 50 *Federal Register* 34648, Aug. 26, 1985 (for EPA's definitions); and U.S. Congress, General Accounting Office, *Wastewater Dischargers Are Not Complying With EPA Pollution Control Permits,* Report to the Administrator, Environmental Protection Agency (Washington, DC: Dec. 2, 1983) (for GAO's definitions).

percent of *completed* major POTWs were in SNC (table 19). However, over one-third of all major POTWs had not yet completed the construction needed to meet treatment requirements, and 15 percent of these were in SNC with their interim limits.

In marked contrast to EPA's latest national statistics, a considerably less optimistic picture was recently reported by the State of Washington (462). Using EPA's definition of SNC, 41 percent of the State's major municipal and industrial facilities were in SNC during the second half of 1985; moreover, the SNC rate was considerably worse for industrial facilities than for municipal facilities (50 percent v. 31 percent).

Whether the Washington survey is representative of other dischargers in the State or in other parts of the country is not known. Nevertheless, these results clearly indicate that, even if national average compliance rates are as high as reported by EPA, certain regions of the country are experiencing compliance rates far below average.

Two related factors are responsible for the differences seen in SNC rates for industrial and municipal dischargers. First, far fewer municipal facilities (65 percent) have completed the construction needed to achieve compliance than have industrial dischargers (94 percent) (327). Second, EPA initially adopted a less aggressive enforcement policy toward municipal facilities, in part because of the uncertainties or delays associated with Federal funding for construction (576). EPA has subsequently issued a new National Municipal Policy (49 FR 3832,

Jan. 30, 1984), which adopts a more aggressive stance.[43]

Data concerning compliance rates for *minor* dischargers generally are not available, although the reduced attention paid to them in permitting and enforcement strongly argues that their extent of compliance is likely to be considerably lower than for major dischargers. Of an estimated 12,000 minor POTWs nationwide, almost 3,400 (28 percent) had not met the statutory deadlines of CWA or were not in compliance with their NPDES permits as of October 31, 1985 (327).

Indirect Dischargers.—Relative to direct discharges, less information is available on the compliance of indirect dischargers with pretreatment regulations. The National Pretreatment program is newer and less developed than the NPDES program for direct dischargers. Moreover, the specified deadlines for final compliance with pretreatment standards in seven industrial categories will not occur until 1987 and 1988, and final regulations for two additional categories (OCPSF and Pesticides) are yet to be issued.

Another major obstacle is the fact that the primary authority and responsibility for regulating such facilities and determining compliance is far more decentralized than is the case in the NPDES program. The primary authority can be a POTW, a State, or an EPA Region. Of the more than 15,000 POTWs in the United States, the EPA required about 1,500 to develop pretreatment programs.[44] To date, 24 States have received authority to approve and oversee individual POTW pretreat-

Table 19.—Municipal Treatment Plants (POTWs) in Significant Noncompliance as of Sept. 30, 1985

	Number of permits	Number in SNC[a]	Percent of permits
Completed major POTWs	2,506	158	6%
Major POTWs on compliance schedules and/or interim limits ...	1,219	180	15%
Total major POTWs ...	3,725	338	9%

[a]SNC = significant noncompliance.

SOURCE: Management Advisory Group to the EPA Construction Grants Program, *Report to EPA: Municipal Compliance With the National Pollutant Discharge Elimination System* (Washington, DC: June 1986).

[43]The compliance deadline for municipal facilities was originally 1977 but is currently July 1, 1988. EPA's National Municipal Policy now indicates that compliance with the 1988 deadline is mandatory for all municipal facilities, regardless of whether they received Federal funding.

[44]These POTWs were to have developed programs by Sept. 30, 1985, or be referred to the Department of Justice. As of June 1985, only 1,100 programs had been approved, and 9 civil actions had been initiated against POTWs lacking approved programs (503). As of October 1986, however, all but 30 had approved programs; 18 of these remaining POTWs had been referred for judicial action. An additional 60 POTWs were identified as needing to develop pretreatment programs because of the presence of new industrial users or environmental problems; these POTWs are currently on schedules to develop programs. (These most recent data were obtained through personal communication from the Strategic Planning and Management System (SPMS), Office of Water Enforcement and Permits, U.S. EPA, January 1987).

ment programs; EPA Regions bear these responsibilities for the remaining States.

Several additional factors complicate the measurement, as well as achievement, of compliance for indirect dischargers. No national tracking system (comparable to the PCS for direct dischargers) currently exists for compiling and analyzing the self-reported data required of industrial users. Moreover, the size of the regulated universe of the pretreatment program is considerably larger than that of the NPDES program. An estimated 100,000 to 140,000 indirect dischargers are subject to General Pretreatment Regulations; approximately 15,600 of these fall into industries which are also subject to Categorical Pretreatment Standards (see table 13).

Despite these limitations, some attempts have been made to examine compliance rates for indirect industrial dischargers, particularly in the electroplating industry. In a 1984 survey of selected major national electroplating firms, baseline monitoring reports (BMRs)[45] for 22 percent of the facilities were either lost or never submitted (258). For those facilities for which some compliance information could be located, only 54 percent were in compliance with categorical standards; 28 percent were clearly not in compliance; and the status of the remaining 18 percent could not be determined.[46]

A 1984 study of electroplaters in California also documented widespread noncompliance: 40 percent of the facilities in the San Francisco Bay area were in violation, and 70 percent in the Los Angeles area were classified either as "compliance unknown" (61 percent) or "out of compliance" (9 percent) (97). In response, EPA (which administers the pretreatment program in California) initiated a number of enforcement actions against major violators.

A 1985 study examined compliance for 1,600 major facilities in a broader range of industrial cat-

egories (98).[47] One-fourth of the major facilities studied were in noncompliance with Federal standards during 1985. Noncompliance was three times higher in southern California than in the San Francisco Bay area, ranging from 18 percent (L.A. County) to 50 percent (Orange County). Virtually all reported violations in southern California were from electroplaters and metal finishers, which accounted for 93 percent of all industrial users in the region.

For the electroplating and metal finishing industries, the noncompliance rate improved considerably in the Bay area, decreasing from 40 percent to 11 percent between 1984 and 1985. In southern California, progress during this period was made primarily in determining the compliance status of industrial users. However, 32 percent of these facilities were still reported to be out of compliance (98).

Several other problems associated with determining or measuring pretreatment compliance also have been identified: the absence of a consistent definition of, or means of quantifying, noncompliance; the use of different and inconsistent data sources for determining noncompliance; and the use of varying methods for monitoring (98,502).

Effectiveness of Enforcement

The final, essential link in the NPDES and National Pretreatment programs is enforcement. EPA's philosophy and policy toward enforcement has undergone major fluctuations over the last decade. The number of enforcement actions initiated by EPA steadily declined from more than 1,500 in 1977 to about 400 in 1982 (576). Part of this decline was caused by explicit changes in EPA's enforcement policy, which placed greater emphasis on pursuing "voluntary compliance" and negotiated settlements (238,462,576).

Since that time, EPA and some States have taken several steps to strengthen and codify their enforcement policies. The revision of EPA's Enforcement

[45]BMRs must be submitted by indirect dischargers to the control authority within 6 months of the effective date of a pretreatment standard. They contain information on the composition of the facilities' discharges, with respect to those pollutants regulated by categorical standards.

[46]The electroplating facilities in this survey were all associated with major corporations, and hence may reflect a greater degree of compliance than the industry as a whole.

[47]Six industries were included: electroplating; metal finishing; semiconductor manufacturing; pulp, paper, and paperboard; steam electric power generation; and textile mills. The 1,600 facilities examined were all industrial users of POTWs required to develop pretreatment programs.

Management System, the commitment of new resources to the Agency's enforcement programs, and the increase in the number of formal enforcement actions all point to such changes (462,468). In general, these efforts have been based on the realistic assumption that the resources available to identify and effectively respond to all violations will remain limited, so the development of a consistent way to rank violations is essential. It remains to be seen how effective such actions will be in improving enforcement and compliance under CWA.

Several enforcement problems have been identified during recent years: the nature and timeliness of the response to a violation once it is identified; the effectiveness of the response in eliminating the violation in a timely manner; the ability to impose meaningful penalties; and the adequacy of resources for enforcement activities. Problems in these areas can in turn lead to delays in achieving compliance on the part of industrial firms or municipal plants, or to unfair economic advantages to violators.

Unfortunately, few data systematically evaluate these factors. Available information indicates two phases: an initially very poor record of enforcement, followed by a general trend toward gradual improvement in recent years. These phases illustrate the nature and extent of obstacles (some of which are being overcome) that face enforcement of water pollution control programs.

Extent and Timeliness of Response.—Several studies have examined EPA and State responses to reported violations, with special attention to whether and how quickly action is taken. In one study of enforcement in EPA Region II, over 4,000 violations by 158 major industrial dischargers between 1975 and 1980 were identified (391). Overall, only 13 percent of these violations received any response after they were reported; in contrast, about twice as many violations discovered during onsite inspections received a response. The vast majority of responses taken were informal: phone calls or warning letters. Moreover, almost a year elapsed, on average, before authorities first responded, during which time an average of three additional violations occurred.[48]

A continuing low level of response to permit violations, many of which are instances of SNC, has also been documented in the State of Washington (462). During the second half of 1985, less than half of the reported SNC violations received a formal enforcement response, despite the issuance of a new enforcement policy that required all instances of SNC to be subject to a formal action. During this same 6-month period, no civil penalties were assessed against municipal dischargers, while a substantial number of industrial facilities were fined.

A study of State and POTW enforcement of pretreatment standards in Rhode Island also documented widespread noncompliance by indirect industrial dischargers over the last several years; during this period, only one judicial enforcement action was initiated (497).

Recent national data show some improvement in the extent of response to permit violations, at least at the Federal level. Since 1980, both the number of administrative orders (AOs) issued and the number of judicial actions undertaken (i.e., number of cases referred to the Department of Justice) have been on the rise (figure 31).[49]

More limited data are available on the level of State activity during this period. For 1985 and 1986, the number of AOs issued by NPDES States was almost three times higher than the number issued at the Federal level; the number of civil actions initiated by these States was 50 percent higher in 1985, and 100 percent higher in 1986, than the number of cases initiated at the Federal level (672).

Effectiveness of Response.—The effectiveness of EPA's response to violations at 33 Louisiana facilities, many of which had "frequently and extensively" violated their permit limits during a 2-year period has been examined (576,689). While EPA initiated numerous informal and formal actions against these facilities, GAO concluded that they were generally ineffective as judged by the continuation of noncompliance. Formal enforcement actions appeared to be no more effective than informal actions: of the 17 facilities that received one

---

[48]No data were presented on the ultimate effectiveness of these actions in restoring compliance.

[49]Data for AOs was further broken down into actions against major and minor, and municipal and industrial, permittees. Each year, the majority of the AOs were issued to major dischargers; in addition, the largest number of AOs were consistently issued to major municipal permittees.

Figure 31.—Federal Enforcement Activity, Fiscal Years 1980–86

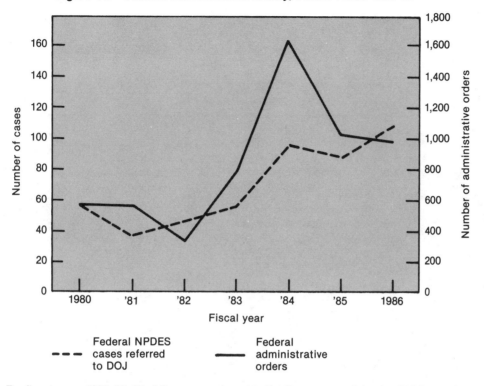

For fiscal years 1980–84, 25 of the cases referred to the Department of Justice (DOJ) were for violations of the Safe Drinking Water Act, not for violations of NPDES permits; the distribution of the 25 cases during these years is not known. Data for 1985 and 1986 are for NPDES violations only (D. Drelich, U.S. Environmental Protection Agency, personal communication, January 1987).

SOURCES: U.S. Environmental Protection Agency, "Summary of EPA Enforcement Activity for 1980–1986," draft of press release issued Dec. 16, 1986; U.S. Environmental Protection Agency, Office of Water, "Water Enforcement Actions Report (Summary)," received by personal communication from EPA on Jan. 15, 1987 (data current through Jan. 12, 1987).

or more AOs, only one came into compliance in the following months. In one extreme case, a discharger was in violation of its permit for 35 consecutive months, despite receiving six AOs. In another, a discharger that had violated its permit limits for 21 consecutive months, often by more than 100 percent, was never subjected to a formal enforcement action by EPA. Out of all 33 facilities, only one case was referred to Federal prosecutors.

EPA has contested a number of GAO's findings, including GAO's conclusions about the effectiveness of AOs (651). During the period examined by GAO, EPA issued nine AOs, each of which specified a compliance schedule. EPA reports that six facilities returned to compliance within the time frame established in the AO, and that in the other

three cases the AOs had a net positive effect, albeit delayed, on ultimately resolving the situations.

Ability to Impose Meaningful Penalties.— Several studies have documented the infrequent imposition, infrequent collection, and low level of penalties for violations of permit limits in various areas of the country (238,462,576,689). These studies emphasize the often time-consuming and frustrating nature of civil actions, especially when measured against the lack of effectiveness of the resulting penalties as deterrents to the original violator or to other permittees.

One common recommendation in such studies is the need for EPA to be given statutory authority to impose *administrative penalties* against vio-

Figure 32.—Amount of Average CWA Civil Penalty,ᵃ 1975–86

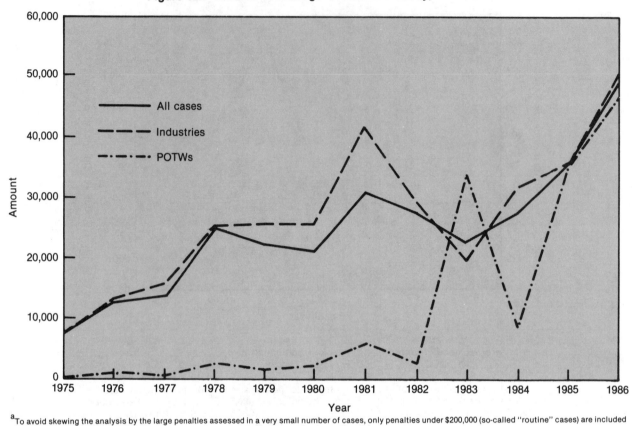

ᵃTo avoid skewing the analysis by the large penalties assessed in a very small number of cases, only penalties under $200,000 (so-called "routine" cases) are included in calculating the average penalty. Only 5 percent of all cases involved penalties larger than this amount.

SOURCE: Office of Technology Assessment, 1987; based on U.S. Environmental Protection Agency, Office of Enforcement and Compliance Monitoring, *1986 Update to Clean Water Act Civil Penalties Analysis* (Washington, DC: Dec. 12, 1986).

lators without having to resort to judicial action (327).[50] Such authority would greatly enhance EPA's capability to mount appropriate and timely enforcement actions against violators.

The penalty situation has improved considerably in recent years (671). Figure 32 illustrates the trend in the size of the average CWA civil penalty collected in cases brought by the Federal Government between 1975 and 1986. These data show an accelerating increase in the average CWA penalty, from $7,500 in 1975 to $48,400 in 1986. Total penalties collected in 1986 were the highest ever, totaling over $5 million. Figure 32 also shows the individual trends for POTWs and industrial facilities. Whereas the average penalty assessed in cases brought against industries showed a gradual increase during this period, the average penalty

assessed against POTWs remained very low (less than $5,000) until 1983, but has since risen substantially. In addition, the fraction of cases settled without a penalty has declined over the last several years, especially for cases brought against municipal violators.

Adequacy of Resources for Enforcement Activities.—A consistent theme encountered in this analysis of enforcement, as well as other activities associated with the implementation and administration of water pollution control programs, is the underfunding of such efforts. Resources available to EPA and approved States for administering the NPDES and National Pretreatment programs and ensuring their enforcement are clearly inadequate. For example:

- Although personnel for the National Pretreatment program doubled from 1984 to 1985 to a total of approximately 65 people, EPA's

---

[50]Language authorizing EPA to impose administrative penalties under certain conditions was included in the Water Quality Act of 1987.

Pretreatment Implementation Review Task Force (PIRT) estimates that an additional 150 positions in Regional offices would be needed to properly implement the program (653). PIRT also recommended increased Federal funding for State programs.

* The State of Washington estimated that it had resources to conduct only 24 percent of the activities needed to effectively administer its NPDES program. Efforts to supplement its budget through increased use of permit fees have been unsuccessful (462). Five staff positions are currently devoted to the State's pretreatment program, despite a 1985 study indicating an actual need for 14 positions (461).

* Under the proposed fiscal 1987 budget, funding for water quality enforcement and permitting will decrease $4.1 million, and funding for implementing the National Municipal Policy will drop $3.1 million. Overall, the proposed budget for water quality programs represents a 15 percent decrease over the fiscal 1986 budget (149).

Additional Issues Facing the Pretreatment Program

Balancing Needs for National Consistency and Local Flexibility

During the last decade, many indirect industrial dischargers and POTWs argued that POTWs with strong locally developed pretreatment programs should be allowed to retain these programs. These locally controlled programs would not have to meet the programmatic and bureaucratic requirements of the National Pretreatment Program as long as they provide control of toxic pollutants that is as stringent as the national program. Proponents of this perspective argue that imposing National Categorical Standards on these POTWs results in overregulation, and unnecessarily increases costs and administrative burdens for POTWs and industries.

Opponents of this approach maintain that a strong national program is necessary to ensure equitable regulation throughout the Nation and to hold POTWs and industries to a set of minimum standards. From an administrative viewpoint, allowing local control would create an atmosphere of uncertainty and might encourage some POTWs or in-

dustries to delay complying with national regulations because of concern about program changes. In addition, EPA would have to devote some of its scarce resources to evaluate local program adequacy and performance.

Most POTWs interested in local control have abandoned the issue for political reasons, and it now appears that the Nation is committed to continued development of a strong national program. At the same time, however, the development of more stringent local limits, where needed, is being strongly encouraged, as is the expansion of water quality-based controls tailored to local needs.

Despite attempts to achieve nationwide consistency, large differences exist in the implementation and oversight of various regulations by different EPA Regions (502,503). Some Regions have allocated resources for both program approvals and oversight and enforcement, while others have only allocated resources for program approvals. In addition, the arrangements among municipalities, States, and EPA Regions are often complex and variable. With more than 95 percent of the required POTW programs approved, the pretreatment program appears to be nearing the end of its development stage. As efforts shift toward implementing these programs, the importance of local limits and water quality-based permitting will grow considerably.

Obstacles to POTW Program Implementation and Enforcement

Local Limits.—Despite the importance of local limits, several obstacles hinder their development and enforcement. POTWs have an incentive to develop local limits on indirect discharges of certain toxic metals or organic chemicals that have the potential to disrupt POTW treatment operations. However, because many pollutants that are harmful to aquatic resources do not disrupt POTW operations, POTWs have had little incentive to develop and impose local limits on these pollutants.

A POTW also might develop local limits to help it meet the specific limitations on its own discharge that are contained in its NPDES permit or to clean up its sludge. However, EPA estimates that only 1 percent of all POTW NPDES permits contain any numerical limits on the discharge of toxic metals or organic chemicals (653). For example, none of Rhode Island's POTWs have any such limits specified in their discharge permits (497). Where they do exist, the limits are typically based on State water quality standards, but those standards have been developed to a much greater extent for metals than for organic chemicals (see last section of this chapter). Moreover, where nationally developed standards for toxic pollutants—in effluent, receiving waters, or sewage sludge—have not been developed, few incentives (and many obstacles) exist for POTWs to develop their own local limits (503,653). The net result of this lack of incentive is that POTWs only rarely impose local limits on their industrial users, especially for priority pollutants.

Monitoring and Enforcement.—These important elements of local pretreatment programs are often inadequate. Even for POTWs with approved pretreatment programs, mechanisms to ensure a program's effectiveness may not exist. Two-thirds of the POTWs examined in one study did not monitor influent for priority metals or organic chemicals, although most large POTWs have generally identified the major sources of toxic pollutants entering their facilities (503). In another study of POTWs with approved programs, only 25 percent had all of the following mechanisms considered essential for controlling industrial wastes: 1) a sewer use ordinance with specific effluent limits, 2) a permitting mechanism, and 3) a monitoring and enforcement program (503). Gradual improvements in these areas have occurred, although for many programs the ability to identify and effectively respond to noncompliance has yet to be demonstrated.

The burden of enforcing pretreatment regulations initially falls on POTWs.[51] Even in the best of situations, it is difficult to determine the extent to which industries have complied with applicable regulations; in part, this stems from heavy reliance on self-monitoring by industries, which commonly is not independently verified by POTWs. Because partial financing for POTW operations comes from the taxes and user fees paid by industrial dischargers, there may be little motivation for a POTW to enforce limits, except where POTW operations or sludge management is impaired (497,503).

Hazardous Waste in Sewers

The current and increasing discharge of hazardous waste to sewers poses a major challenge to the pretreatment program. Hazardous wastes can be legally discharged into POTWs under the Domestic Sewage Exemption of the Resource Conservation and Recovery Act (RCRA). As recent amendments to RCRA come into effect, increasing amounts of hazardous wastes are expected to be discharged into POTWs. New provisions extending RCRA authority to all facilities that generate more than 100 kilograms per month[52] of hazardous waste have increased the total number of generators now subject to RCRA regulations from 14,000 to over 175,000 (50 FR 31285, Aug. 1, 1985). Many of these small quantity generators already discharge into local POTWs and thus could fall under the Domestic Sewage Exemption; it is estimated that as many as 25 percent of all small hazardous waste generators already use this disposal option for their hazardous wastes (568).

EPA recently submitted a major report, known as the Domestic Sewage Study (666), to Congress on the issue of hazardous waste in sewers, and more recently issued an Advanced Notice of Proposed Rulemaking (51 FR 30166, Aug. 22, 1986) that discussed preliminary approaches to implementing the Study's recommendations. For more discussion of this issue and the role of POTWs in treating domestic and industrial wastes, the reader is referred to these sources and to chapter 9.

---

[51]If appropriate action is not initiated by the POTW, higher authorities can step in. For example, in the last 2 years EPA has initiated more than 50 judicial actions against indirect industrial dischargers (primarily electroplaters) for violations of general and categorical pretreatment regulations (J. Moran, Office of Enforcement and Compliance Monitoring, EPA, pers. comm., Dec. 23, 1986).

---

[52]Previously, only generators producing more than 1,000 kilograms per month were subject to RCRA hazardous waste regulation.

USING WATER QUALITY-BASED STANDARDS FOR FURTHER CONTROL OF TOXICS POLLUTANTS

In 1984, EPA released a national policy statement that described a strategy for achieving additional and more efficient control of toxic pollutants beyond that resulting from BAT and other CWA technology-based requirements (49 FR 9016, Mar. 9, 1984). This strategy focuses on **water quality-based permitting** of toxic pollutant discharges to be implemented through the NPDES program. Because water quality-based standards are difficult to set, EPA is adopting an integrated strategy that uses both chemical and biological methods to determine appropriate standards (659). States will be expected to devote more effort to develop water quality-based effluent limits for inclusion in NPDES permits.

This new EPA policy reflects a reevaluation of the regulatory efforts to control toxic pollutants during the last 15 years. The 1972 Clean Water Act shifted the focus of pollutant control from the use of water quality-based standards to the use of technology-based standards. Subsequent implementation of technology-based standards has resulted in significant improvements in the control of toxic discharges and in the quality of some receiving waters, and full implementation would achieve even more control.

However, the level of control provided has not always satisfied all the interested parties. For example, the technology-based standards usually do not consider site-specific circumstances such as the quality of receiving waters. In addition, BAT standards are industry-specific; some industries are required to achieve greater removal of a specific pollutant than are other industries, leading to claims of under- or over-regulation.

As a result, some industries have argued for waivers from complying with technology-based standards, in situations where the quality of the receiving water would not be impaired.[53] In contrast, environmentalists have argued that even the achievement of compliance with technology-based standards has not resulted in sufficient improvement in

the quality of some receiving waters, and that supplemental controls are needed.

The 1972 CWA did not eliminate the use of water quality-based standards. Section 303 required States to adopt water quality standards to protect inter- and intra-state waters through establishment of water quality goals and designation of water uses; specific standards based on Federal water quality criteria are then to be applied to protect these uses.[54] Section 301(b)(1)(C) of CWA requires that all discharges meet water quality-based standards where they have been developed. These standards can in principle be translated, using wasteload allocation techniques, into effluent limits for the various dischargers to a particular receiving waterbody.

EPA maintains and periodically updates a summary of Federal water quality criteria and State standards. According to the most recent summary, Federal water quality criteria have now been developed for most conventional, non-conventional, and toxic priority pollutants (668).[55] The Federal criteria are for guidance only and are not enforceable. Using these criteria, all States have adopted enforceable water quality standards for fecal coliform bacteria, oil and grease, dissolved oxygen, pH, dissolved solids and salinity, and temperature, and almost all have a standard for suspended solids and turbidity (668).[56] EPA estimates that 40 percent of major municipal permits—and perhaps as high a fraction of major industrial permits—are based in some manner on these water quality standards (J. Hoornbeek, Office of Water Enforcement and Permits, EPA, pers. comm., Nov. 13, 1986).

---

[53]These waivers are termed "fundamentally different factor" (FDF) waivers. The conditions (including consideration of water quality) under which EPA can grant an FDF variance were clarified by Congress in the Water Quality Act of 1987.

[54]EPA is authorized to review State standards and may also promulgate standards where State standards have not been developed, although they have not done so.

[55]Federal water quality criteria consist of four criteria based on consideration of aquatic life (acute and chronic criteria specific for fresh or marine waters) and two additional criteria based on consideration of human health (one for both water and fish ingestion, and the other for fish ingestion only). Typically, only a subset of these six parameters is specified for a given pollutant.

[56]Under current policy, the adequacy of State water quality standards is now a consideration in the decision of whether to grant approval to a State to administer the NPDES program. However, the standards developed by States whose programs were approved prior to the development of this policy (or, indeed, prior to the development of most Federal water quality criteria) have not always been subject to a comparable degree of scrutiny by the Federal Government.

In contrast, for most other pollutants—in particular, priority pollutants—none or only a few States have developed standards (668). Specifically:

- Of 85 priority organic chemicals for which Federal water quality criteria exist, no States have developed standards for 37, and only one State has developed standards for another 32. For each of the remaining 16 priority organic pollutants, standards have been developed by an average of 12 States (with a range of 2 to 23 States).
- At least 1 State has developed a standard for each of the 14 priority metals and cyanide for which Federal water quality criteria exist; for each of these substances, an average of 15 States (with a range of 1 to 24 States) have developed standards.

Thus, for no priority pollutants have even half of the States developed a water quality standard. Fourteen States have no water quality standards for priority pollutants whatsoever.

OTA also reviewed water quality standards for priority pollutants that have been promulgated by the 24 coastal States to determine the number of standards that have been specifically developed for or applied to marine waters. This survey revealed the following:

- **Nine of the 24 coastal States have no marine water quality standards for priority pollutants whatsoever, and 16 States have no such standards for priority metals or cyanide.**
- For the 8 coastal States that have any marine standards for priority metals or cyanide, standards have been developed for an average of 4.5 of the 14 priority metals and cyanide.
- For the 15 coastal States that have any marine standards for priority organic chemicals, standards have been developed for an average of 6.8 of the 85 priority organic chemicals.

The development of water quality-based standards poses several problems. First, it is questionable whether EPA has sufficient resources to continue to develop and update the Federal water quality criteria, or to evaluate water quality standards that are developed by States. Moreover, a large increase in compliance monitoring and enforcement burdens would also be anticipated.

Even if resources were sufficient, a number of major technical obstacles would need to be overcome. Only limited data are available on ambient pollutant concentrations in receiving waters, variability in these concentrations, and the fate of these pollutants and their impacts on indigenous organisms. In addition, our ability to monitor water quality in relation to potential environmental or human impacts is relatively primitive.

Nevertheless, further promotion of EPA's policy on water quality-based permit limitations for toxic pollutants could help provide an additional level of control beyond technology-based standards. Several approaches to increase the use of water quality-based standards may be useful for Congress and EPA to consider:

- improve technical assistance to States and local management agencies to aid in the development of State water quality-based standards;
- provide state-of-the-art guidelines to States by updating existing national water quality criteria and accelerating the development of new national water and sediment criteria;
- incorporate water quality standards into POTW NPDES permits as a means of providing incentives for POTWs to develop local limits; and
- promote wider application of whole-effluent toxicity testing, for example, through the expanded use of toxicity-based limitations in NPDES permits.

Chapter 9
Managing Municipal Effluent and Sludge

CONTENTS

Managing Municipal Effluent and Sludge

OVERVIEW

The treatment of municipal wastes generates two products: sewage sludge (the mostly solid material separated from the original waste) and effluent (the liquid remainder). Large quantities of these products are disposed of in marine waters. The treatment of municipal wastes and the management of these products raise many concerns, for example:

- the dumping of sewage sludge in coastal and open ocean waters;
- the impacts of toxic pollutants (in particular, metals and organic chemicals from industrial discharges into sewers) in sludge and effluent on marine resources;
- the constraints imposed by the presence of toxic pollutants on the beneficial use of sludge and effluent;
- the impacts of conventional pollutants (including solids and fecal bacteria), other microorganisms (e.g., viruses), and nonconventional pollutants (e.g., nutrients) in sludge and effluent on marine resources; and
- whether current levels of municipal treatment will be maintained as Federal funding for the construction of treatment plants declines.

Municipal waste management in the United States has been shaped by events that occurred during the past 150 years. In the 19th century, the increased use of water delivery systems and flush toilets dramatically increased the amount of rinsewater and raw sewage flowing from households (175,551). The rinsewater and sewage was usually diverted into cesspools or existing stormwater drains, but these often were unable to handle the increased flow and health problems arose from the contamination of soil and wellwater.

In response, cities began channeling wastewater into newly built sewers that discharged into surface waters, including marine waters (175,551). Initially, these discharges received no treatment because people assumed that the receiving water would dilute the waste and prevent health prob-

lems. In 1909, almost 90 percent of wastewater carried in sewers received no treatment. However, it was soon discovered that discharges into rivers contaminated drinking water supplies in downstream communities, causing major public health problems such as epidemics of typhoid fever.

Cities then began to develop processes to filter wastewater and treat bacteriological contamination prior to discharge (175,551). Many processes developed between 1900 and 1935 are still important components of current municipal treatment (318). One problem arises, however, regarding the nature of the wastes being treated. The original processes were not designed to treat metals and organic chemicals, yet industries discharge wastewater containing these pollutants into municipal sewers. Thus the sludge and effluent products left after treatment are often contaminated with these substances.

The initial responsibility for developing large and efficient disposal systems was usually carried at the municipal level; suburban areas often were annexed and special district agencies (e.g., the Boston Metropolitan Sewerage Commission) were created to facilitate such development. However, the institutional structure to regulate sewage treatment and disposal has grown rapidly and has gradually passed to the State and Federal levels. The Federal Government, for example, spent over $40 billion in the last 15 years to help local sewerage authorities build or upgrade municipal treatment plants (569). The current legal framework for managing municipal effluent and sludge is described in box U.

The generation of both sludge and effluent is expected to increase in coastal areas as populations increase and as more communities are serviced by municipal treatment plants. Effluent discharges into marine and fresh waters probably will increase accordingly because this is the only means currently available for large-scale disposal. In some situations, water conservation or the re-use of effluent (e.g., via water reclamation for irrigation or groundwater

Box U.—Relevant Statutes, Programs, and Policies

Some of the general regulatory programs and policies that apply to sewage effluent and sludge management are:

Clean Water Act

As it applies to sewage management, the Act's two major parts concern construction grants for treatment plants and regulatory requirements that apply to industrial and municipal dischargers (571).

The *Construction Grants Program* helps finance the design and construction of new municipal treatment plants (publicly owned treatment works, or POTWs), and the upgrading of existing POTWs. Using a complex State allocation formula, approximately $44 billion in Federal money has been granted since 1972 (569). The program will be phased out by 1994.

The *National Pollutant Discharge Elimination System* (NPDES) regulates the direct pipeline discharge of municipal wastestreams into fresh and marine waters. Each POTW is issued a permit specifying the maximum amounts of pollutants that are allowed in its effluent; discharges into coastal and open ocean waters also must meet ocean discharge criteria under Section 403(c). In addition, pretreatment of industrial wastes discharged into sewers is often required.

The *National Municipal Policy*, issued by EPA in 1984, applies to POTWs that have not yet met "secondary" treatment requirements or are not in compliance with their NPDES permits. It states that POTWs must meet the compliance deadline of July 1, 1988, regardless of the availability of Construction Grant funds (655). The CWA originally required POTWs to achieve secondary treatment by mid-1977, but in 1981 the deadline was extended to mid-1988.

Section 301(h), added in 1977, allows waivers from secondary treatment under certain conditions, particularly when effluent discharges from coastal POTWs with less than secondary treatment do not degrade marine waters.

Sludge management at the national level consists of a patchwork of regulations and programs that only partially cover sludge disposal options and pollutants. Under the joint authority of CWA Section 405 and the Resource Recovery and Conservation Act, limits have been set on the amounts of cadmium and PCBs allowed in sludge applied to croplands, but few limits have been set on other metals or organic chemicals.

Marine Protection, Research, and Sanctuaries Act

The Marine Protection, Research, and Sanctuaries Act (MPRSA) required EPA to designate disposal sites and regulate the dumping of sewage sludge and other materials in the ocean. Amendments in 1977 prohibited, by the end of 1981, sewage sludge dumping that unreasonably degraded the marine environment.

In 1981, however, a Federal court ruled that ocean dumping of sewage sludge could not be summarily banned; other factors, including the feasibility of alternative options, also had to be considered. The specific conditions under which sludge can be dumped in the ocean are defined in the Ocean Dumping regulations (ch. 7), which currently are being reformulated.

Other Federal Statutes

Under the Resource Conservation and Recovery Act, the Domestic Sewage Exclusion (Sec. 1004) specifies that hazardous wastes discharged into sewers and subsequently mixed with human wastes are not to be regulated as hazardous (666). Under this provision, POTWs receiving hazardous wastes are exempt from regulation under RCRA and instead continue to be regulated under the CWA. The permit-by-rule provision of RCRA allows POTWs to receive hazardous wastes by truck, rail, or dedicated pipe (as opposed to discharges in sewers) without obtaining a RCRA permit. Other pertinent regulations under RCRA include definitions of "hazardous" sludges and controls on the disposal of both hazardous sludges and incinerator ash.

The Clean Air Act regulates the incineration of sewage sludge by imposing emissions standards for a variety of compounds, including mercury. The Toxic Substances Control Act contains provisions concerning PCB-contaminated sludges.

State Programs for Sewage Management

The NPDES, which regulates effluent discharges, is administered by 37 States and by EPA Regions for other States. The situation is more complex for sludge, for which States have instituted many different approaches. In general, though, they tend to focus on the common land-based disposal methods and to follow EPA guidance or regulations, where available, in developing their own regulations (445,638). EPA recently proposed new regulations for State sludge management programs.

recharge) could reduce the need for discharges. Several management options are feasible for sludge, including, for example, dumping in marine waters and beneficial use on cropland. The demand to dump sludge in marine waters could increase, be-

cause the use of other options may be constrained by regulatory and social factors and because dumping sometimes may be economically and even environmentally preferable when compared with other options.

TREATMENT PROCESSES AND PRODUCTS

General Wastewater Treatment Processes

About 70 percent of domestic wastewater in the United States is channeled into publicly owned treatment works (POTWs)[1] for treatment (159). The remainder is discharged into private septic systems or, in some cases, discharged without treatment into various waterbodies. These wastestreams are complex mixtures, generally composed of water, suspended solids, organic material, oil and grease, dissolved nutrients, microorganisms, and metals and organic chemicals.

The exact composition of wastewater entering POTWs is highly site-specific and complex because its components can come from a variety of sources: household chemicals, human wastes, industrial and commercial discharges into sewers (box V), and rainwater and street runoff from combined sewer systems (figure 33). In addition, about 60 percent of the material periodically cleaned from septic tanks is transported to and treated in POTWs (638). Composition also varies with time, particularly in systems that receive large inputs from combined sewers or seasonal industrial discharges.

POTWs treat raw wastewater by removing or degrading organic materials or, in the case of some bacteria, by destroying them. Most of the remaining solid organic and inorganic material is removed, forming a sludge. The remaining liquid effluent typically contains much less than 1 percent solids, while sludge contains from 1 to 7 percent solids (prior to further dewatering).

Treatment levels are defined primarily on the basis of the percentages of two conventional pollut-

ants—biochemical oxygen demand (BOD) and total suspended solids (TSS)—that are removed from the wastewater. At the first or "primary" treatment level, debris is physically screened, and some suspended solids settle in sedimentation tanks. Typically, up to about 60 percent of the suspended solids are removed during primary treatment (633).

"Secondary" or biological treatment uses microorganisms to destroy or remove additional amounts of BOD and TSS. Any additional suspended solids that are removed are added to the sludge, so secondary treatment produces more sludge than primary treatment. More advanced, or "tertiary," treatment generally is used to remove additional suspended solids or nutrients (table 20). It often entails the use of chemicals (e.g., aluminum sulfate, ferric chloride, polyelectrolytes) to precipitate the target pollutants (633,634).

Technologies developed in the early 20th century to treat municipal wastewater still are used, though they have been modified and improved, as the basis for treatment at most municipal plants. One example is anaerobic digestion. Many new technologies have been developed in the last few decades, particularly for sludge treatment (e.g., physical-chemical treatment, pure oxygen-activated sludge systems, and ammonia stripping), but few are widely used in municipal treatment plants (318).

The costs for building and operating municipal treatment plants vary with the particular combination of processes used to achieve a specified treatment level. Costs can escalate rapidly as the required level of treatment is increased, in part because of additional costs for sludge treatment processes (not to be confused with sludge disposal techniques), but economies of scale can counter this to a degree (650,661).

---

[1]The terms "municipal treatment plants" and "publicly owned treatment works" are used interchangeably here.

Box V.—Inputs of Toxic Pollutants and Hazardous Wastes Into POTWs

Many metals and organic chemicals may be present in municipal wastestreams. Some of these substances are classified as Clean Water Act (CWA) toxic or "priority" pollutants, others as RCRA hazardous wastes, and still others are not classified under any statute. These substances are discharged into municipal sewers primarily by industrial facilities.

About 15 percent of the influent received at POTWs that discharge into marine waters comes from "indirect" industrial discharges into sewers (503). The approximately 1,500 POTWs that have pretreatment programs receive about 80 percent of all such discharges, and probably a similar portion of CWA priority pollutants discharged into sewers (666).

At least 102 of the 126 CWA priority pollutants have been found in POTW influents (637). According to EPA, frequently detected organic chemicals include both priority pollutants (e.g., chlorinated solvents, aromatic hydrocarbons, phthalate esters) and nonpriority* but hazardous pollutants (e.g., xylene, methyl ethyl ketone, methanol) (666).

Over 1 trillion gallons of wastewater containing RCRA hazardous wastes are discharged annually into municipal sewers by some 160,000 industrial facilities. Without any treatment at industrial facilities, these discharges would contain at least 160,000 metric tons of hazardous components—including 62,000 metric tons of priority metals, roughly 40,000 metric tons of priority organic chemicals, and at least 64,000 metric tons of nonpriority organic chemicals (666). Several State studies support these conclusions—e.g., an estimated 15 percent of hazardous waste in Minnesota is discharged into sewers (358). Hazardous waste discharges could increase if small-quantity generators choose to discharge more hazardous wastes into sewers under the Domestic Sewage Exemption and if land-based disposal is increasingly restricted.**

Discharges of CWA pollutants or of hazardous wastes into sewers are subject to pretreatment requirements. Under full implementation of the pretreatment program, the levels of CWA priority pollutants might be reduced to 2,540 metric tons of metals and 3,100 metric tons of organic chemicals (666); current levels, however, are much higher.

Under the RCRA permit-by-rule provision, 300 to 700 POTWs also accept hazardous waste shipped by truck, train, rail, or dedicated pipe (152). An unknown number of POTWs also have accepted leachate from Superfund sites; these wastes are not subject to pretreatment regulations, and the POTWs could be liable for impacts from sludge disposal (243).***

---

*Pollutants classified as hazardous under RCRA, but not as CWA priority pollutants, were termed "nonpriority" pollutants in the study.

**The 1984 RCRA amendments expanded coverage to small-quantity generators, increasing the number of regulated generators from roughly 15,000 to about 100,000. About one-fourth of these generators may already discharge hazardous wastes into sewers (568).

***In addition, new toxicity tests proposed by EPA may constrain disposal of sludges from such POTWs (243).

Fate of Microorganisms in Effluent and Sludge

Conventional treatment processes can destroy some of the microorganisms present in municipal wastewater (205,503,662). The level of reduction depends on the techniques used, operating conditions, and the microorganisms in question.

In effluent, for example, the concentration of viruses can be reduced by 90 percent or more, depending on the extent of secondary and tertiary treatment.[2] Bacterial densities can be reduced about fifteenfold through primary and secondary treatment, and additional dilution of about fiftyfold can be achieved when the effluent is discharged into receiving waters. The degree of these reductions is highly variable, however, and these processes are

---

[2]Due to the diversity of viruses found in sewage and the low sensitivity of current methods for their detection, reported densities probably represent a significant underestimation of total virus number (205).

Figure 33.—Generation, Treatment, and Disposal of Municipal Effluent and Sludge

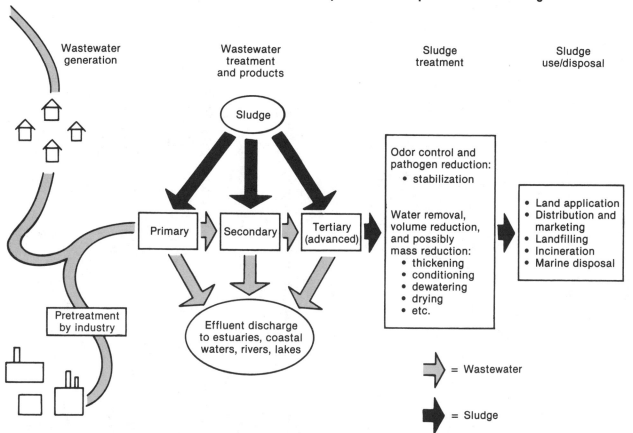

SOURCE: Adapted from U.S. Environmental Protection Agency, Intra-Agency Sludge Task Force, *Use and Disposal of Municipal Sludge*, EPA 625/10-84-003 (Washington, DC: September 1984).

Table 20.—Definitions of Municipal Treatment Levels

Treatment level	Treatment requirements[a]
Primary	Approximately 30% removal of BOD and 60% removal of TSS
Secondary	Removal of both BOD and TSS to levels of 25-30 mg/l, but not less than 85% removal; pH between 6.0 and 9.0
Tertiary	Removal of both BOD and TSS to levels less than 9 mg/l, or removal of over 95% of BOD and TSS; additional requirements for removal of nutrients (e.g., nitrates, phosphates) on site-specific basis

ABBREVIATIONS: BOD = biological oxygen demand
TSS = total suspended solids.
[a]Treatment levels required by the Clean Water Act and codified at 40 CFR Part 133.
SOURCES: Office of Technology Assessment, 1987; based on 40 CFR Part 133; and Science Applications International Corp., *Overview of Sewage Sludge and Effluent Management*, contract prepared for U.S. Congress, Office of Technology Assessment (McLean, VA: 1986).

not very effective against parasites (e.g., protozoan cysts).

Furthermore, the densities of bacteria in effluent (even including the fiftyfold dilution factor) are usually still too high to achieve compliance with water quality standards for recreational and shellfish-growing waters. In addition, both viruses and bacteria tend to become concentrated in sludge because of their tendency to associate with solid material (although large numbers remain in the effluent, as well). For these reasons, "disinfection" techniques (table 21) are often used to further reduce microorganism levels in effluent prior to disposal.

Chlorination is the most commonly used effluent disinfection technique in the United States. It has

Table 21.—Advantages and Disadvantages of Selected Effluent Disinfection and Sludge Treatment Processes

Technique	Used for[a]	Advantages	Disadvantages
Effluent disinfection:			
Chlorination	DI	Commonly used; 98 to 99% destruction for bacteria; high removals of viruses and cysts	Low removal of bacterial spores; formation of chlorine residuals and chlorinated hydrocarbons; residual test not correlated with concentration of microorganisms
Ozonation	DI	99% removal of fecal coliform; high removal of viruses; destroys phenols, cyanides, trihalomethanes; no chlorinated byproducts	Toxic gas; requires onsite generation; expensive; removal of spores and cysts unknown
UV radiation	DI	Initially effective on all microorganisms	Only penetrates a few centimeters; microorganisms sometimes reactivated; potential microorganism mutagenesis; unpredictable
Gamma radiation	DI	Penetrates deeper than UV radiation	High costs; worker safety
Heat	DI	Destroys most pathogens	High energy costs
Chlorine dioxide	DI	98 to 99% bacteria removal; high virus removal; no chlorinated compounds; small measurable residual	Three times cost of chlorination; unstable, sometimes explosive; requires onsite generation
Sludge treatment:			
Aerobic digestion	ST,DI	Removes up to 85% of microorganisms and 40% of volatile solids; more rapid, simpler, less subject to metal upset than anaerobic digestion, PSRP[b]	Costly (requires oxygen); susceptible to upset by pH, organic chemicals, metals; slower at cold temperatures; not always easy to dewater; no commercial gas byproducts
Anaerobic digestion	ST,DI	Removes >85% of microorganisms and 40% of volatile solids; easier to dewater; preferred at larger plants; commercial gas byproducts; PSRP	More susceptible to upset than aerobic digestion; gas explosions; higher capital costs
Thermophilic aerobic digestion	ST,DI	Faster than aerobic; near complete destruction of bacteria and viruses; heat self-generated; PFRP[b]	Requires solids content >1.5% and heat retention equipment; only in emerging/development status
Air drying	DW,DI	Reduces some solids and microorganisms; PSRP	Requires long time (>3 months)
Heat drying	DW,DI	Significant reduction in volume; destroys most bacteria; useful for distribution and marketing products; PFRP	Subject to putrefaction; requires associated digestion; needs prior costly dewatering
Heat treatment (under pressure) ..	DW,DI	Sterilization; readily dewatered; PFRP	Expensive
Liming	ST,DI	Destroys bacteria; binds metals, so less leaching; PSRP	Sludge solids not destroyed, so microorganisms can regrow if pH falls prior to total drying
Chlorination	DI	Free of odors; dewaterable	No significant solids reduction; requires lime to neutralize pH; chlorinated byproducts
Composting	DI	PSRP and PFRP	Requires large space

[a]DI = disinfection; DW = dewatering, ST = stabilization.
[b]PSRP = Process to Significantly Reduce Pathogens; PFRP = Process to Further Reduce Pathogens (see text for details).

SOURCE: Office of Technology Assessment, 1987; after Science Applications International Corp., *Overview of Sewage Sludge and Effluent Management*, prepared for U.S. Congress, Office of Technology Assessment (McLean, VA: 1986).

been considered economical and effective and it usually destroys virtually all fecal coliform bacteria in effluent. Several concerns, however, have been raised about chlorination. First, it may only temporarily inactivate, rather than destroy, microorganisms present in effluent. Second, it is not as effective against pathogenic viruses as it is against fecal coliform bacteria.[3] Finally, chlorinated hydrocarbons can be formed as byproducts of the process and pose significant risks to organisms in the vicinity of municipal discharges, although information on chlorine byproducts in municipal wastewater is limited (503,636).

Alternative effluent disinfection methods such as ozonation and ultraviolet light treatment may be more effective than chlorine against viruses. These methods have rarely been used in the United States for drinking water or municipal wastewater treatment, although ozonation has been widely used in Europe to purify drinking water (142,503). Both methods are hard to apply and expensive, so chlorination may be the only practical means of disinfection in most locations. To combat chlorine's disadvantages, long outfalls that discharge into deep and dispersive ocean waters have been used to dilute pathogen concentrations sufficiently to meet water quality standards.

Sewage sludge also can undergo additional treatment or "conditioning." In particular, sludge that is to be disposed of on land is required to undergo certain technology-based processes to reduce microorganisms; these processes involve varying types of disinfection, stabilization, or dewatering (table 21) (650).[4] Some processes, such as anaerobic digestion, have been in use for over 60 years (318). Others have been developed recently; for example, ionizing radiation has been used on a pilot scale in Boston and on a commercial scale in Miami (59,503).

These sludge treatment processes are grouped into two categories, Processes to Further Reduce Pathogens (PFRP) and Processes to Significantly

Reduce Pathogens (PSRP). If food crops are to be grown within 18 months of land application, the Environmental Protection Agency (EPA) regulations require that a PFRP be used. PFRPs destroy most bacteria and viruses by subjecting sludge to elevated temperatures over a specified time, but again the actual reduction is variable; in addition, some parasites such as protozoan cysts are not readily destroyed (205). If no food crops are involved, sludge treated by a PSRP can be applied to land, subject to certain restrictions (e.g., on public access, grazing, pH, and metals and polychlorinated biphenyls (PCB) content).

Other sludge treatment techniques include chemical fixation and encapsulation for sludges containing high levels of contaminants, earthworm conversion, and emerging processes such as anaerobic fixed-film biological treatment.

Incidental Removal of Metals and Organic Chemicals

POTWs are designed specifically to remove conventional pollutants from wastewater, but not to remove metals and organic chemicals. As some of these pollutants pass through POTWs into receiving waters, they sometimes upset the efficiency of POTW treatment systems.[5] Not all metals and organic chemicals pass through POTWs, however, because treatment processes do result in the unintentional or "incidental" removal or degradation of some of these wastewater pollutants.

Incidental removal can take several forms—volatilization, removal to sludge, or biodegradation. The metals that are incidentally removed tend to be incorporated into sludge (637), and some organic chemicals also can be incorporated into sludge, from which they often are volatilized during subsequent sludge treatment.[6] Municipal treatment

---

[3]The use of fecal coliform bacteria as an indicator species is discussed in ch. 6.

[4]Disinfection reduces odor, as well as bacterial densities; stabilization reduces organic material and, correspondingly, the level of microbial activity, thus reducing levels of odor and microorganisms; dewatering and some types of stabilization can also reduce volume of sludge.

[5]"Upsets" refer to large or sudden changes in the concentrations of metals or organic chemicals that kill the microorganisms used in treatment processes; as a result, municipal effluent can be discharged without adequate treatment (503). For example, an upset at a POTW in Rhode Island, and subsequent discharge of effluent with excess BOD levels, was attributed to the dumping of cyanide into the municipal sewers (497).

[6]For biodegradation of chemicals to occur, however, microorganisms in the treatment plant generally must be acclimated to a small but constant input of the chemicals. However, variable inputs are probably more common and hinder acclimation; thus, many chemicals that could in principle be incidentally removed through biodegradation will instead pass through to receiving waters (503).

processes thus destroy some pollutants and redistribute the remainder.

The degree of incidental removal varies greatly among POTWs, particular treatment techniques, and individual pollutants (637,666). According to one study of 50 POTWs, different treatments showed a range of efficiencies for incidental removal for some selected priority pollutants (637). For example, POTWs with primary treatment removed 10 to 57 percent of metals and 0 to 62 percent of organic chemicals, while POTWs with secondary treatment removed 34 to 85 percent of metals and 40 to 94 percent of organic chemicals.

These results suggest that 15 to 66 percent of the metals and 6 to 60 percent of the organic chemicals could pass through POTWs with secondary treatment. This study has been criticized, however (94). In particular, critics noted that if the pollutant in question was not detected (i.e., its concentration was below the minimum detection level), the concentration was recorded as being at the minimum detection level. This procedure, in combination with other factors, could underestimate actual removal efficiencies.

Effluent and Sludge Composition

Numerous studies have documented the composition of different effluents and sludges (101,173, 524,637,638,639,650,666). The composition of both products is highly site-specific and variable over time, depending on the nature of sources discharging into municipal sewers and on the destruction and redistribution (i.e., incidental removal) of pollutants during treatment processes.

Whatever the degree of incidental removal, however, both effluent and sludge will almost always be contaminated to some degree with metals and organic chemicals (637). In one study of POTWs in New Jersey (including some that dump sludge in marine waters), for example, an average of 27 percent of Clean Water Act (CWA) priority pollutants were present in effluent and sludge (160).

Some waste treatment processes can further alter the composition of sludge and effluent. In particular, the use of chlorination to disinfect effluent prior to discharge can create and increase chlorinated hydrocarbons in the effluent; in one study, the concentration of chloroform in effluent increased by 70 percent (704). In general, though, the majority of chlorine apparently ends up as chloride ions rather than in chlorinated organic compounds (636). In addition, metals tend to associate with solid material, so sludge tends to have higher concentrations of metals than does effluent from the same plant (503).

Although treatment processes destroy high levels of some bacteria, remaining bacteria and other microorganisms can be distributed in both effluent and sludge. Under the right conditions, these organisms can proliferate in effluent and sludge and constrain subsequent management. Furthermore, some bacteria cannot be detected with traditional techniques but apparently can remain viable in marine waters for extended periods of time (in some cases, years) (ch. 6). The apparent absence of human pathogens at or near sludge disposal sites in the open ocean, for example, may actually reflect our inability to detect such organisms rather than their actual absence.

Given the numerous pollutants that have been or could be detected in sludge and effluent, EPA and others have attempted to determine which components are "most important" or of "greatest concern." For example, as part of an effort to develop new regulations for sludge management, EPA identified a list of about 50 metals and organic chemicals that could cause environmental or human health impacts and these could be the focus of future regulatory efforts (311,643).

AMOUNTS OF EFFLUENT AND SLUDGE GENERATED AND DISPOSED

Effluent and Sludge Generation

Over 15,000 POTWs currently operate in the United States and each year they treat and discharge approximately 9.5 trillion gallons of wastewater. More than 2,200 POTWs are located in coastal counties, and they discharge about one-third of the Nation's municipal effluent (503,608). POTWs also produce increasing amounts of sewage sludge. The total amount generated by all POTWs more than doubled during the last decade, and almost 40 percent originates from POTWs located in coastal counties (table 22).

By the year 2000, total sludge production could increase to over 10 million dry metric tons (377, 503). The amount of effluent is expected to increase to between 13 and 16 trillion gallons per year. These increases will result from expanded use of secondary and advanced treatment processes, which produce more sludge, and increases in population, sewerage hookups, and numbers of POTWs (654).

Land-Based Management Technologies for Effluent and Sludge

Only about 2 percent of the Nation's total effluent is not discharged into surface waters (503). Instead, it is used to irrigate or fertilize agricultural and forest land, or for groundwater recharge, industrial uses, aquaculture, and underground injection to prevent saltwater intrusion. In Los Angeles County, for example, water reclaimed by five municipal treatment plants is used for landscaping, irrigation, groundwater recharge, and industrial processes (503).

The choice of disposal or treatment options generally is driven by site-specific factors such as POTW size and location, regulatory climate, State policies, qualitative assessments of impacts, and costs (502). Most land-based disposal or treatment of sludge involves land application, the use of the sludge as a commercial product for household or municipal use (known as distribution and market-

Table 22.—Amounts of Effluent and Sludge Generated by Municipal Treatment Plants (POTWs)

Product and treatment level[a]	All POTWs			POTWs discharging into estuaries			POTWs discharging into coastal waters		
	Number	Amount[b]	Percent[c]	Number	Amount	Percent	Number	Amount	Percent
Effluent:									
No discharge	1,577	0.49	1.9	0	0.00	0.0	0	0.00	0.0
Primary	1,023	2.35	9.1	55	0.94	17.1	11	0.15	18.2
Advanced primary	2,102	2.81	10.8	52	1.22	22.2	6	0.35	42.7
Secondary	8,005	10.47	40.3	272	2.43	44.2	46	0.31	38.4
Tertiary	2,775	9.84	37.9	121	0.91	16.5	7	0.01	0.7
Total	15,482	25.96	—	500[d]	5.50	—	70	0.82	—
Sludge:									
No discharge	1,577	0.00	0.0	0	0.00	0.0	0	0.00	0.0
Primary	1,023	0.40	5.9	55	0.18	11.8	11	0.03	15.0
Advanced primary	2,102	0.75	11.0	52	0.35	23.1	6	0.09	45.0
Secondary	8,005	2.85	41.5	272	0.71	46.7	46	0.08	40.0
Tertiary	2,775	2.86	41.6	121	0.28	18.4	7	0.002	<1.0
Total	15,482	6.86	—	500	1.52	—	70	0.20	—

[a]Treatment levels are defined in table 20; except no discharge = no discharge into waterbodies, and advanced primary = levels intermediate between primary and secondary levels.
[b]For effluent, amounts are in billion gallons per day. For sludge, amounts are in million *dry* metric tons per year.
[c]Percent of total amount (of effluent or sludge, as appropriate).
[d]The total of 570 POTWs (500 into estuaries and 70 into coastal waters) differs slightly from the total of 578 cited in ch. 3, because of differences in the way POTWs were classified during different analyses of data available from EPA.

SOURCE: Office of Technology Assessment, 1987; after Science Applications International Corp., *Overview of Sewage Sludge and Effluent Management*, prepared for U.S. Congress, Office of Technology Assessment (McLean, VA: 1986).

Photo credit: S.C. Delaney/U.S. Environmental Protection Agency

Sludge is a byproduct of municipal wastewater treatment processes. While composed primarily of water and solids, it also can contain pollutants such as metals, organic chemicals, and microorganisms. After treatment, sludge is often stored in large piles, as shown here. Depending on its composition, sludge can be incinerated, landfilled, applied on land, or dumped in marine waters.

ing), landfilling, and incineration (650). Other approaches used for significantly smaller amounts of sludge include incorporation into construction materials (e.g., brick, asphalt) and animal feed, and degradation by earthworms.

Land application usually involves spreading sludge on agricultural and forest land or using it in reclamation projects (e.g., on strip-mined land), and it has been the subject of much research (429, 525,591). Distributing and marketing sludge commercially involves using or selling composted or heat-dried sludge as a nutrient enricher and soil conditioner. Landfilling consists of placing sludge in an area dedicated solely to disposal and then covering it with soil. Incineration includes burning sludge alone, co-combustion with municipal waste

(to yield usable energy), and emerging technologies such as gasification and liquefaction (660). In general, it is used for sludges with a very high organic content where land is scarce (318); about 150 incinerators or co-combustion facilities now operate in the United States (660).

Together, these land-based options are used to treat or dispose of about 90 percent of the sludge generated in the United States. Up to one-half of all sludge is landfilled or put in surface impoundments, about one-fifth is incinerated, and about one-fourth is applied to the land (including distribution and marketing) (503,634).[7] The use of these

_____

[7]A more precise accounting of the prevalence of the various land-based options is not possible because of inconsistencies in how POTWs record and report data. Some POTWs do not report methods, some

options varies geographically and with POTW size. In 1980, for example, coastal POTWs landfilled 54 percent, ocean-dumped 22 percent, and land-applied 5 percent of their sludge. Small POTWs use land application or landfilling to dispose of almost three-fourths of their sludge, while larger POTWs use a greater variety of methods, including incineration.

Marine Disposal of Effluent

The most common way to dispose of effluent is to discharge it through pipelines into nearby waters. Almost 600 POTWs discharge effluent directly into estuaries or coastal waters (table 22). Although these POTWs represent only 4 percent of the Nations's total, they account for about one-fourth (2.3 trillion gallons annually) of all municipal effluent because many of them serve large urban areas. Of this, about 2.1 trillion gallons of effluent are discharged annually into estuaries (503). About three-fifths of the effluent discharged into marine waters receives secondary or greater treatment; effluent discharged into estuaries generally receives greater treatment than does effluent discharged into coastal waters.[8]

Pipelines can be designed so that discharges are more likely to meet specified water quality goals. For example, the use of very long pipelines, in combination with design features such as large multiport diffusers,[9] can result in effluent discharges that are highly diluted and far from shore (375,387).

---

report multiple methods without distinguishing amounts, and most do not list distribution and marketing as a separate option (R. Bastian, U.S. EPA, pers. comm., September 1985).

[8]In addition to treated sewage effluent, raw sewage also enters marine waters from routine discharges and combined sewer overflows (CSOs). For example, about 40 million gallons of raw, untreated sewage is discharged daily into the New York Bight. CSOs usually occur during storms and result in wastes flowing untreated into receiving waters, and they are a major problem in certain areas. In Seattle, for example, about 2 billion gallons of wastewater overflows annually and receives no treatment (460,503). Moreover, any industrial wastewater contained in these overflows never receives the "incidental" treatment provided by POTWs. In Boston, CSOs discharge about 9 billion gallons and are considered (along with sludge and industrial wastes) one of the major sources of pollutants in Boston Harbor.

[9]A "multiport diffuser" is a pipeline that has several ports or openings, located at various points along the pipeline, from which effluent can be discharged. As a result, effluent is discharged in multiple locations rather than only one, which allows greater mixing with surrounding water and greater dilution.

Marine Disposal of Sewage Sludge

Sludge is disposed of in marine environments in two ways, by dumping from barges or ships or by discharge from pipelines. Discharges of sludge into estuaries and coastal waters take place only in southern California and Boston and total about 110,000 dry metric tons of sludge or solids each year[10] (503). The discharge of sludge by the City of Los Angeles will cease when a sludge dehydration/incineration facility is completed in 1987. Under court order, Boston is developing secondary treatment facilities and alternative sludge management options (148,166). Sludge discharges into Boston Harbor could continue until these improvements are in place in the mid-1990s.

The amount of sludge that is dumped in marine waters has increased steadily, from over 2.5 million wet metric tons in 1959 to about 7.5 million wet metric tons in 1983 (648). This equals about 7 to 10 percent of all sludge generated in the United States (503). After the 1977 Clean Water Act amendments banned the dumping of sludge that would "unreasonably degrade" the ocean, over 100 small municipalities stopped dumping at sea, but these accounted for less than 5 percent of all dumped sludge. Because of a 1981 court decision (*City of New York* v. *United States Environmental Protection Agency*; see ch. 7), nine sewerage authorities in New York and New Jersey have continued to dump sludge at the 12-Mile Sewage Sludge Dump Site in the New York Bight, which has been used since 1924 (table 23) (306,503).

The use of the 12-Mile Sewage Sludge Dump Site is now being phased out and sludge dumping is being moved to the Deepwater Municipal Sludge Site (ch. 3). As of November 1986, Nassau and Westchester Counties had shifted all dumping to the deepwater site; New York City had shifted 10 percent; and the sewerage authorities in New Jersey had shifted 25 percent (F. Czulak, U.S. EPA Region II, pers. comm. November 1986). New York City has indicated that it will not be able to move all its dumping to the deepwater site prior to 1988 (502).

---

[10]The Los Angeles County Sanitation Districts mixes "centrate" (solid particles derived from centrifuge processes; these essentially are sludge particles but are not technically termed as such) with its effluent discharge.

Photo credit: National Oceanic and Atmospheric Administration

After sludge is produced at municipal treatment plants, it can be managed in several ways, including land application, landfilling, incineration, distribution and marketing, and marine dumping. When dumped, sludge is loaded onto barges (like those shown here, which are loaded with municipal waste and debris) or ships for transport to the dumping site.

Table 23.—Costs of Dumping Sludge at the 12-Mile Sewage Sludge Dump Site and the Deepwater Municipal Sludge Site

Permittee	Amount dumped in 1985[a]	Total cost, 12-mile site[b]	Total cost deepwater site[b]	Ratio of costs, deepwater/12-mile
New York City, NY	3.03	4.0	18.6	4.6
Middlesex County, NJ	0.94	3.3	11.5	3.5
Passaic Valley, NJ	0.80	2.4	13.2	5.5
Nassau County, NY	0.52	0.7	2.5	3.5
Westchester County, NY	0.43	1.3	3.5	2.7
Essex-Union Joint Meeting, NJ	0.31	0.8	2.1	2.7
Bergen County, NJ	0.28	0.9	2.6	2.9
Rahway Valley, NJ	0.17	0.8	2.3	3.0
Linden-Roselle, NJ	0.09	0.2	1.9	c
Total or average	6.57	14.4	58.2	4.0[d]

[a]In millions of wet metric tons.
[b]In millions of dollars, adjusted to 1982 dollar values.
[c]Not available.
[d]Weighted average ratio for entire New York Bight, excluding Linden-Roselle.

SOURCES: F. Czulak, U.S. Environmental Protection Agency Region 2, personal communication, Dec. 18, 1986; and T.M. Leschine and J.M. Broadus, "Economic and Operational Considerations of Offshore Disposal of Sewage Sludge," *Wastes in The Ocean, Volume 5,* D.R. Kester, et al. (eds.) (New York: John Wiley & Sons, 1985), pp. 287-315.

Future Demand for Marine Disposal of Sludge

It is difficult to predict the future demand for marine disposal of sludge. Many municipalities would probably be interested in dumping sludge in marine waters if the regulations are changed to allow increased dumping. Several large coastal municipalities (e.g., Baltimore, Boston, Washington, DC, Jacksonville, Philadelphia, San Diego, San Francisco, and Seattle) have expressed interest in maintaining dumping as a potential option should other options fail (32,532). In addition, Orange County, California, has proposed that it be allowed to discharge sludge into deep ocean waters on an experimental basis (see box W).

Estimating the amount of sludge that might be dumped in marine waters in the future is extremely difficult because:

- it is unclear whether current relatively restrictive Federal policies will change;
- it is unclear how many east coast communities would find it economically feasible to dump at the Deepwater Municipal Sludge Site;
- land-based disposal options could often be more attractive, especially if levels of toxic pollutants in sludge are reduced; and
- the granting of waivers from secondary treatment requirements (under Sec. 301(h)) could result in less sludge being produced by coastal municipalities, since lower treatment levels generate less sludge.

Costs of Sludge Disposal

The costs of sludge disposal and management options are determined primarily by sludge treatment requirements; economies of scale; land acquisition costs; capital, operating, and maintenance costs; transportation costs; and energy requirements (138, 504,635,639). Transportation alone can account for most of the costs associated with land application, landfilling, and ocean dumping.

The relative costs of dumping sludge at the 12-Mile Sewage Sludge Dump Site and of land-based disposal will vary. Dumping in marine waters generally is less costly, because it does not require the sludge to be dewatered, as is necessary for land application and incineration (306). Dumping will certainly be more costly at the Deepwater Municipal Sludge Site than at the 12-mile site. On average, dumping at the deepwater site is expected to cost four times more than dumping at the 12-mile site (table 23). These estimates largely reflect short-term transportation costs, but they may underemphasize the degree to which future capital investments reduce long-term costs. Dewatering, for example, would reduce the volume of sludge produced, which would reduce the number of trips to the deepwater site and decrease transportation costs, but it would also increase treatment costs prior to disposal (306). Dumping at the deepwater site generally could be less costly than land-based disposal for most municipalities currently dumping sludge (553).

GENERAL FATE OF SLUDGE AND EFFLUENT

Dumped Sludge

The potential for impacts from municipal waste disposal depends on what happens after disposal. When dumped from barges or ships, sludge is initially diluted[11] by currents and by turbulence from the wake, typically by a factor of several hundred within a few minutes and by a factor of 5,000 or

---

[11]"Initial dilution" is considered to occur until a discharge ceases rising in the water column (i.e., until it reaches water of equal density), or until dumped material either ceases moving in the water column or reaches the bottom (280).

more after 4 hours (385,387). Dilution is greater if the material is released in many smaller amounts rather than a few large ones (280). Subsequent dilution is much slower, so initial dilution greatly influences the concentration of sludge components to which marine organisms will be exposed (387).

After initial dilution, most of the particles in sludge are still denser than the surrounding seawater and tend to descend as a large "cloud" at a rate dependent on size and density. When it reaches water of equal density, the cloud tends to spread horizontally (385,387,503). Individual particles then slowly disperse and may settle toward the bottom.

The rate at which particles accumulate on the bottom is influenced by factors such as volume, dumping rate, type and size of particles, and physical and chemical processes. Large particles, for example, settle more rapidly than small ones; furthermore, they can be formed when small particles aggregate, a process enhanced when freshwater waste such as sludge mixes with saltwater. However, this tendency is decreased somewhat by dilution, which makes particles less likely to collide and aggregate (387).

The extent of settling—in terms of amount of material and area covered—varies significantly among different sites. In enclosed and shallow environments with little tidal action (e.g., many estuaries), material can accumulate on the bottom in the general vicinity of the dumpsite when sludge is dumped over a long period (85,87,400). These types of marine waters, however, are not used for sludge dumping in the United States.

In contrast, in more open and well-mixed waters such as are used in the United States, most particulate material (perhaps as much as 90 percent) is transported out of the immediate area by currents and may disperse over an area of several hundred square kilometers (387). In the New York Bight, for example, most particles disperse away from the immediate dumping area over the course of days or weeks; some particles and associated pollutants move into and accumulate in other areas of the Bight such as Christiaensen Basin (located northwest of the dumpsite).

The decomposition of the organic material in dumped sludge depends on the activity of microorganisms. Initially, much of the decomposition is performed by microorganisms that are present in the sludge before it enters marine waters. These microorganisms are adapted to survive in freshwater (the main component of sludge) and they may not survive when the sludge enters marine waters,

thereby reducing the initial decomposition of organic material. Some decomposition also occurs after the material settles on the bottom, but it tends to be slower in the conditions typical of bottom sediments and for sludges with low organic content. It also tends to be slower in waters that have low oxygen levels, since many or most of the decomposer microorganisms require oxygen. Some observers suggest that decomposition could be enhanced by "seeding" sludge with microorganisms developed (e.g., by genetic engineering) to survive in both fresh and marine water (R. Colwell, Univ. Maryland, pers. comm., October 1986; also see ref. 105).

Discharged Effluent

When discharged from a pipeline, effluent is primarily fresh, buoyant water which tends to form a "plume" that rises in the water column. The plume rises, entraining saltwater in the process, until it reaches either water of equal density or the surface; the plume can spread horizontally at either point. The particles in the effluent, already present at a concentration of less than 1 percent, are diluted as they mix with the denser saltwater, but the degree of initial dilution varies greatly. For pipelines that discharge into shallow water and that are not equipped with diffusers, dilution is only about a factor of 10 (280). In contrast, large outfalls that discharge in relatively deep water and that are equipped with long mulitport diffusers can achieve initial dilution of up to a thousandfold (280). Individual particles begin to sink slowly after this initial plume rise and dilution.

As with sludge, the fraction of particles in effluent that disperses from the discharge point varies markedly under different conditions. In the relatively dispersive conditions in the Southern California Bight, for example, only about 10 percent of the particles may settle in a well-defined zone around the discharge point (350), although accumulation of these particles can still result in significant impacts.

IMPACTS FROM EFFLUENT AND SLUDGE DISPOSAL

Marine Impacts From Particulate Material, Microorganisms, and Nutrients

Pollutants such as particulate material (suspended solids, organic material), fecal coliform bacteria and other microorganisms, and nutrients in sludge or effluent can cause a variety of beneficial and adverse impacts on marine environments.

Some observers argue that the beneficial nature of sludge and effluent disposal in marine waters has not been appreciated by the public (509,638). In estuaries and coastal waters, nutrients such as nitrogen and phosphorus can stimulate phytoplankton productivity, and in turn possibly enhance commercial fisheries. In the open ocean, nutrients from sludge dumping could stimulate increases in productivity, since lack of nutrients is a major constraint on productivity in most of these waters. Overall productivity, however, would still remain low relative to estuaries and coastal waters.

In contrast, the adverse impacts associated with dumping and discharges of sewage wastes have received considerable attention. One common problem is that particulate material can accumulate in the disposal area, especially if the activity is continuous or frequent, and alter bottom (i.e., benthic) habitats. This can lead to changes in population sizes or the diversity of marine organisms. The major change in species composition that typically occurs is a shift from communities dominated by suspension feeders, such as crabs and mollusks, to ones dominated by deposit feeders, such as worms (350).

These types of impacts are present in a range of sites around the country. In southern California coastal waters, for example, pollutants including suspended solids, and some metals and organic chemicals have affected about 5 percent of the benthic communities to some degree (350,387). One small area (less than 10 km²) around two outfalls was severely affected and up to about 85 km² of surrounding areas was moderately affected. At one dumpsite near Delaware Bay, once used by Philadelphia, gradual accumulation of material to the south and west of the site seems to have caused changes in benthic species abundance and diver-

sity (686). In the New York Bight, the most severely degraded areas (about 10 to 15 km²) occur just west of the dumpsite, on the margin of Christiaensen Basin (387).

Excessive inputs of nutrients and organic material (i.e., eutrophication) can lead to hypoxia—low dissolved oxygen levels—and other serious consequences. In some shallow and enclosed marine waters, these impacts have been caused at least in part by effluent discharges. Both problems, however, also are caused by other factors. Seasonally recurring episodes of extreme hypoxia in the New York Bight, for example, are caused by a combination of factors: natural stratification of the water prevents the mixing and reoxygenation of bottom waters; nutrients from a wide variety of sources, including raw sewage and municipal effluents carried by the Hudson and Raritan rivers, increase plant life, which can lead to reduced oxygen supplies when the plants die and are decomposed by microorganisms (416,548,632).[12]

Pathogenic microorganisms present in effluent and sludge can cause a variety of impacts, too, such as the contamination and closure of shellfish beds. Some cases of shellfish contamination have been unequivocally linked to sludge dumping, for example, at the old Philadelphia dumpsite (595). Such contamination may be partially reversible; 3 years after dumping ceased at the Philadelphia dumpsite certain pathogenic microorganisms were relatively rare, although still detectable (595). Although contamination by pathogenic microorganisms is common in the vicinity of effluent discharges, the microorganisms can also come from raw sewage, combined sewage overflows, and runoff. Some viral pathogens present in effluent discharges (e.g., enteric viruses) have high survival rates in marine waters and, if ingested, may adversely affect human health by causing gastrointestinal disorders and other diseases.

---

[12]In 1976, the public attributed the presence of fecal material on New Jersey and New York beaches to the dumping of sludge in the New York Bight. However, several other sources of material (including the raw sewage carried into the Bight from the Hudson and Raritan rivers), together with high river flows and various climatic factors, apparently were responsible for these episodes (377,547,632).

Marine Impacts From Metals and Organic Chemicals

The potential for *metals* in both sludge and effluent to cause adverse impacts depends on many chemical and physical factors (ch. 4). Organisms can ingest certain metals that sometimes cause immediate toxic effects (including death). Furthermore, because metals are persistent they can bioaccumulate within organisms and cause further impacts (e.g., impair growth or reproduction in fish and benthic invertebrates) (55). Most metals do not biomagnify in successive levels of the food chain. However, mercury—a common pollutant in sludge—can be converted by marine organisms to a form that has direct acute toxic effects on the organisms, biomagnifies in the food chain, and is toxic to humans.

Many *organic chemicals* also are persistent in the environment and often bioaccumulate. In contrast to metals, however, many also can biomagnify. These chemicals can cause severe sediment contamination problems, and a variety of short- and long-term effects on organisms.

Information on the potential impacts of metals and organic chemicals in effluent discharges has been summarized for 25 of the 30 largest coastal POTWs that applied for Section 301(h) waivers (503,649). If these POTWs were allowed to continue to provide less-than-secondary treatment, 12 were considered to have the potential to cause significant impacts because of large quantities of metals (e.g., copper, nickel, thallium, zinc) and organic chemicals (e.g., naphthalene, pentachlorophenol) in their effluents. Potential and observed impacts included contamination of sediments and organisms; fish disease and reproductive failure; degradation of benthic and plankton communities; and closures of shellfisheries and fisheries (503). Effluents from the other 13 applicants were considered to lack this potential (649).

Toxic pollutants in combined sewer overflows (CSOs) also have caused major impacts in marine waters (503,647).[13] For example, overflows into

Puget Sound have contributed to toxic "hotspots" in Elliot Bay, where the sediments have average concentrations of metals and organic chemicals greater than sediments from the deep central part of the Sound. Elevated levels of copper and lead also were found in fish exposed to the overflows and sediments. In San Francisco Bay, sediments located near CSOs had elevated concentrations of numerous pollutants, including many metals, and the sediments were considered unsuitable to support the normal diversity and abundance of organisms (233, 503). In contrast to Puget Sound, however, fish near the CSOs in San Francisco Bay did not exhibit elevated amounts of metals.

Impacts From Land-Based Disposal

Land applied sludge can be used beneficially as a fertilizer or soil conditioner on agricultural and forest lands (34,429,502,525). Seattle, for example, has applied sludge to forest lands and recently determined that revenues from the sale of sludge for forest application will at least partially offset the costs of sludge treatment and other management options (P. Machno, Seattle Metro, pers. comm., 1985; ref. 469).

Controversy surrounds many land application projects, however, because pathogens, metals, organic chemicals, and even nutrients in the sludge can cause adverse impacts (167). Nutrients such as organic nitrogen and ammonia, for example, can be converted by microorganisms into nitrates, which can leach into and contaminate surface water or groundwater.

Because of public health concerns, the presence of pathogens is a major factor limiting land application. Modern treatment processes can reduce the densities of most bacteria and some viruses, but not parasites, and sludges subjected to these processes are allowed to be land-applied in certain situations. Additional reduction of pathogens results from sunlight and drying after application. Despite this, bacteria, viruses, and parasites can survive in soil for months, depending on soil temperature, pH, organic content, and other factors. Viable pathogens have been found in runoff from fields that were subject to land application of sludge. No cases of human disease, however, have been documented to date from land application of treated sludge, al-

---

[13]The Water Quality Act of 1987 set aside some Construction Grants funding for the correction of CSOs that cause water quality problems in marine waters; the amount set aside is not to exceed 1 percent of the Construction Grants funding for fiscal years 1987 and 1988 and 1.5 percent for fiscal years 1989 and 1990.

though disease outbreaks have been linked with application of untreated sewage wastes (ch. 6).

Potential impacts from metals are another limiting factor. In general, metals adsorb strongly to particles or are not highly water soluble, so they tend to be retained in the soil. They are more mobile in sandy or acidic soils, however, from which they can leach into and contaminate surface runoff and groundwater. Some metals (e.g., cadmium, chromium, zinc, nickel) can be taken up by plants, sometimes affecting productivity, and excessive amounts of metals in plants can affect livestock or human health (118,387).

Some organic chemicals are lost from the soil through volatilization, but the fate of others depends on properties such as their solubility in water (503). They usually do not affect plant productivity significantly, but they can be ingested from soil or root surfaces by livestock, accumulate in animal tissues, and be consumed by humans.

The impacts associated with *landfilling* sludge are similar to those for land application. Anaerobic conditions are more common in landfills, however, which tends to retard the conversion of nitrogen and ammonia into nitrates, so there is generally less potential for leaching of nitrates into groundwater. Decomposition of organic material in landfills also can produce various gases; methane can be explosive, while carbon dioxide can acidify soils and increase the solubility of metals.

Sludge *incineration* significantly reduces the amount of material to be disposed, totally destroys pathogens, and can destroy more than 99 percent of organic chemicals under proper conditions. However, emissions of particulate material can affect ambient air quality, and emissions of volatilized metals or products of incomplete combustion can increase risks to human health. Incineration residuals, particularly metals, also remain in scrubber water or bottom ash, which also must be disposed of (usually in landfills, thereby adding to the potential risk of groundwater contamination).

Risks to Humans From Sludge Disposal Methods

Extensive research has been conducted on the potential risks to humans from different sludge disposal methods. EPA has developed a series of environmental and human health hazard indices for 50 pollutants found in sludge (656). Based on these indices, it appears that *contaminated* sludge applied to human food-chain cropland poses the greatest risk to humans, primarily because of the threat of PCBs and other nonvolatile, insoluble organic chemicals (503). In contrast, application of uncontaminated or even moderately contaminated sludge to non-food-chain croplands poses much less risk to humans. Evidence also suggests that risks to humans from land incineration and ocean dumping might be less than those from land application (503).

MAJOR ISSUES RELATED TO MARINE ENVIRONMENTS

The many issues that influence the management of sewage sludge and effluent in marine waters can be grouped into five broad categories:

1. compliance and enforcement,
2. how toxic pollutants and hazardous wastes affect sewage management,
3. new regulatory initiatives regarding sludge management,
4. the role of marine waters in waste management, and
5. the role of land-based disposal alternatives.[14]

Compliance and Enforcement

Municipal treatment plants have been slower than industrial facilities to respond to the original requirements of CWA. Originally, municipal plants were to achieve secondary treatment by 1977. Congress extended this date to 1988, and in 1984 EPA issued a National Municipal Policy statement affirming this goal. As of September 30, 1985, however, 37 percent of all major POTWs[15] still were not in compliance with secondary treatment requirements, in part because many have not com-

---

[14]The issue of ensuring funding of future municipal treatment plant construction, under the Construction Grants Program, is discussed in detail in ch. 1; indirect effects of the program on the above issues are discussed here when appropriate.

[15]"Major" POTWs include those discharging more than 1 million gallons per day or serving more than 10,000 people.

pleted the necessary construction (327). Substantial progress has been made during the last year in bringing POTWs onto a compliance schedule (designed to achieve compliance by mid-1988) or into actual compliance (table 24). As a result, under existing construction schedules, it appears that compliance could be achieved by the mid-1988 deadline by about 87 percent of major POTWs (table 24) (655). Almost 200 major POTWs, however, currently do not have compliance schedules and hence are likely to miss the deadline.

Even where required facilities have been built and are operational, some frequent and often serious violations of discharge standards have occurred. About 6 percent of the major POTWs that have completed construction have exhibited significant noncompliance. Noncompliance by these POTWs was attributed to inadequate facilities to provide required treatment levels; inadequate industrial pretreatment and treatment of combined sewer overflows; problems in maintaining sewer systems; and lack of appropriate local institutional structures to finance capital and operating costs and to efficiently manage facilities (327).

Implementing and enforcing CWA goals and requirements has been difficult, in part because of limited resources for monitoring and restrictions on the types of penalties that EPA can impose (327). Furthermore, these requirements focus largely on the removal of conventional pollutants. Quantities of metals and organic chemicals can be significant, however, and are likely to decline only if pretreatment is implemented and enforced.

Effect of Toxic Pollutants and Hazardous Wastes on Sewage Management

Many problems associated with municipal waste disposal stem from the presence of toxic metals and organic chemicals, which municipal treatment plants are not designed to treat. Hundreds of these pollutants enter municipal systems legally and illegally, although often at extremely low concentrations, and they are primarily contributed by industrial sources. A small portion is "incidentally" removed from wastewater, and some pollutants become incorporated in sludge. If POTWs continue to receive industrial wastes that contain these pollutants, questions will continue to arise regarding the ability of POTWs to produce clean effluent and sludge.

If levels of these pollutants in POTW influents were reduced, however, the feasibility of some disposal options, such as land application, would be enhanced.[16] Many constraints hamper the achievement of such a goal, however, including poor compliance with and enforcement of regulations, lack of standards for some management options, and lack of permit limits for some pollutants.

The lack of standards presents problems for both sludge and effluent disposal. Most sludge disposal options are not covered by regulations that limit metals and organic chemicals in sludge, leaving

_____

[16]Models developed for EPA suggest that full implementation of pretreatment regulations could reduce, in both sludge and effluent, the amount of CWA priority metals by about one-half and the amount of priority organic chemicals by about three-fourths (503).

Table 24.—Status of Major Municipal Facilities (POTWs) Not in Compliance With National Municipal Policy[a]

Status of POTW	Number of major POTWs (October 1985)	Number of major POTWs (July 1986)	Number of minor POTWs[b] (October 1985)
On final enforcement schedule or under referral[c]	835	1,066	586
Returned to compliance	162	234	—
Unresolved	581	191	2,775
Total subject to National Municipal Policy	1,578	1,491	3,361

[a]As of date indicated; data refer only to major POTWs (those designed to treat flows of 1 million gallons per day or more or to service a population of 10,000 or greater) not yet in compliance with the National Municipal Policy (which affirmed the goal of compliance by July 1, 1988; see box A).
[b]Represents only those minor POTWs that require further construction to achieve compliance; data are not available for minor POTWs that have completed construction but are not in compliance.
[c]Referral represents cases referred to Department of Justice or State Attorneys General for civil action; such cases usually result in the establishment of a final compliance schedule.

SOURCES: U.S. Environmental Protection Agency, Office of Water, "State Breakout of NMP Majors Construction Required, Status at End 3rd Quarter FY1986" (Washington, DC: data as of end of July 1, 1986); and Management Advisory Group to the EPA Construction Grants Program, Report to EPA: Municipal Compliance With the National Pollutant Discharge Elimination System (Washington, DC: June 1986).

POTWs without clearly defined goals for sludge quality. Among POTWs that receive significant quantities of industrial discharges, most have effluent discharge permits that contain limits on some metals but only a few organic chemicals (503). In part, this reflects the lack of State water quality standards for some metals and most organic chemicals (668). POTWs can also develop their own "local" limits on industrial discharges of metals and organic chemicals into sewers. Local limits on industries are generally developed, however, only if a POTW must meet a specific limit contained in its own discharge permit. Since most POTWs do not have limits on organic chemicals in their permits, there is little incentive for them to develop corresponding local limits on their industrial users.

Some of these problems are being addressed by EPA and the States. For example, EPA is developing comprehensive sludge disposal regulations and promoting the use of water quality-based permitting as a means of controlling toxic pollutants (49 FR 9016-9019, Mar. 9, 1984). EPA also has the statutory authority to develop regulations for potentially toxic substances that are currently unregulated by CWA but which may be present at high concentrations in municipal wastestreams. Under paragraph 4(c) of the 1976 Toxics Consent Decree, for example, EPA has identified six such pollutants, but no regulations have been developed (644).

These important issues—enforcement, local limits, additional national standards, and water quality criteria—are discussed in further detail in chapters 1 and 8. For most of these issues, improved implementation of existing programs *at all levels* of involvement is critical and will require much more rigorous enforcement. This will only be possible if funding for monitoring and enforcement programs is increased.

The issue of legal hazardous waste discharges into sewers is particularly vexing. Eliminating the Resource Conservation and Recovery Act's (RCRA) Domestic Sewage Exemption and regulating such discharges under RCRA is attractive because POTWs would receive fewer hazardous wastes. This option, however, could lead to increased illegal dumping into sewers and waterways, possibly making the problem worse (666). EPA proposed that the exemption be retained and that POTWs continue to develop and improve pretreatment programs to reduce the levels of hazardous and toxic pollutants that enter treatment plants. This approach thus will require effective implementation and enforcement of the pretreatment program, the likelihood of which is unclear. A related approach might be to develop regionalized waste treatment facilities, specifically designed to collect and treat hazardous or other industrial wastestreams (502).

New Regulatory Initiatives Regarding Sludge Management

The management of sludge is controlled by a patchwork of Federal, State, and local regulations, and no national sludge management program now exists. Instead, institutional arrangements among municipalities, counties, States, and EPA Regions are highly site-specific and complex, and often highly politicized (502).[17] Management also is complicated by a lack of comprehensive disposal standards, changing economic conditions, public opposition, and a relative lack of promotion of the idea that sludge can be used as a beneficial resource. As a result, most municipalities develop options haphazardly to take advantage of short-term opportunities.

Two sometimes antagonistic needs are key parts of the sludge management debate: the needs for stronger Federal guidance and regulation, and more flexibility to accommodate local conditions (502). Proponents of a minimal Federal role would let States develop their own regulations independently, with the Federal Government providing only technical assistance amd guidance. Because local sludge management decisions are highly site-specific and often difficult to implement (502), local managers need considerable flexibility in designing and implementing sludge disposal options.

---

[17]These and other institutional issues are discussed in a contract prepared for OTA which summarized case studies of sewage management and pretreatment programs in Boston, Hampton Roads, Houston, Los Angeles County, Miami, New York, Philadelphia, and Seattle (502). These localities were chosen to achieve diversity in geography and type of waterbody, use of sludge dumping, degree of industrialization, pretreatment program status, sludge management alternatives, and institutional or policy issues.

According to this perspective, however, Federal regulations do not allow sufficient flexibility or promote consideration of site-specific factors.

In contrast, proponents of an increased Federal role question whether current regulations for both sludge and effluent provide sufficient protection for the public and the environment. They contend that the Federal Government should continue to establish minimum national standards for most pollutants, conduct broad multimedia assessments, and possibly develop a large-scale, uniform national program with mandatory requirements for all States. Minimum standards, for example, could be included in the National Pollution Discharge Elimination System to provide performance goals for POTWs and promote the use of sludge as a resource.

In response, EPA has developed two new regulatory initiatives involving State sludge management programs and Federal regulations for sludge disposal. Under these initiatives, the Federal Government would play a stronger role in some areas: promoting the use of sludge as a resource, developing technical regulations for sludge disposal, and providing more technical assistance.[18]

First, EPA proposed a new rule to aid States in designing sludge management programs (51 FR 4458, Feb. 4, 1986); final action on the rule is scheduled for February 1987. Under the rule, States would develop plans for managing sludge (including promoting beneficial uses). To obtain EPA approval, a State would have to:

* demonstrate that it can ensure compliance with Federal regulations by overseeing how individual POTWs manage sludge;
* demonstrate that the State can monitor sludge to verify compliance and take enforcement actions against violators;
* possess legal authority to assess civil penalties for violations; and
* meet various reporting requirements on sludge inventories, noncompliance, and other aspects of sludge management (150,151).

These regulations would focus on improving sludge quality by implementing and enforcing pretreatment programs and sludge sampling and monitoring. According to EPA, the regulations would give States flexibility in using existing programs and institutional arrangements. Other than the loss of Federal funds for program development, there appear to be few penalties for States that do not submit plans for programs.

Second, EPA is developing technical regulations for five major sludge management methods—land application, landfilling, incineration, distribution and marketing, and ocean disposal. Federal regulations have never been promulgated for some of these options (e.g., for distribution and marketing), although nonbinding guidance has been issued. The new regulations would complement the regulations for State programs and would focus on quantifying the risks from and allowable concentrations of metals and organic chemicals in sludge; EPA identified pollutants that are candidates for regulation, and pollutants selected for actual regulation will be controlled either by numerical criteria for different disposal options or by technology-based management practices (311). These regulations would place sewage sludge management within a multimedia context; for a given situation, the risks of different options could be compared and the most environmentally acceptable option identified.

The regulations are scheduled for proposal in 1987. In a preliminary review, EPA's Science Advisory Board (SAB) indicated that the risk assessment methodologies being used by EPA to develop the regulations do not provide a clear way to compare the human health risks of different sludge management options (245); the SAB recommendations focused on improving these methodologies. In addition to this shortcoming, the regulations do not sufficiently address pathogens. Current regulations for pathogens are technology-based and focus on fecal coliform bacteria; they do not directly address other pathogens such as viruses and parasites.

Role of Marine Environments in Municipal Waste Management

If policy choices about waste disposal in different marine environments are made within the context of a waste management hierarchy that includes

---

[18]EPA also has drafted new regulations to establish conditions under which dumping in marine environments would be allowed; the regulations originally developed for ocean dumping of sewage sludge were overturned in court (see ch. 7).

other management options, then marine waste disposal may be acceptable in some cases and unacceptable in others. Furthermore, the particular policy choices made about disposal in estuaries and coastal waters could greatly influence decisions about land-based and open ocean disposal.

For example, a goal of maintaining or improving the quality of estuaries and coastal waters could preclude the dumping of sludge in coastal waters, where most of it currently occurs. This could increase the need for either land-based disposal or open ocean dumping of sludge. A comparison of the benefits and risks of all sludge disposal options could suggest that the best use of uncontaminated (i.e., containing minimal amounts of metals, organic chemicals, pathogens) sludge might be either on land or in open ocean waters. On the other hand, the choice of treatment or disposal options for contaminated sludge would be less clear because of the risks of groundwater contamination from land application or landfilling, air pollution from incineration, or marine impacts from dumping. EPA's new regulatory initiative will address some of these issues. Still, some basic questions regarding sludge and effluent disposal in marine environments remain unsettled.

Should Sludge Dumping Be Allowed in Marine Waters?

The basic choice to be made in any disposal operation is between dispersal or containment (387). Containment of sludge in marine waters is technically difficult, expensive, and can increase adverse impacts. Dispersal, on the other hand, is feasible and since its objective is to minimize the buildup of disposed material on the bottom, it reduces the probability of impacts.

Dispersal is generally greater in large and well-mixed water masses, where wastes are mixed rapidly into a large volume of water and dispersed over a wide area (132,387). These conditions are prevalent in open ocean environments and well-mixed coastal waters. In contrast, estuarine and calmer coastal wasters generally are less well-mixed or flushed. In addition, they receive large inputs of waste material from other sources. This section focuses on sludge *dumping*, while the proposed *discharge* of sludge by Orange County, California, into ocean waters is described in box W.

At the relatively deep Deepwater Municipal Sludge Site, for example, particles from dumped sludge are expected to be well-dispersed and result in little or no accumulation on the bottom (595). Although bottom-dwelling organisms at the site might be affected and monitoring should be conducted, the impacts are likely to be less severe than those seen near the present less dispersive site in the New York Bight Apex (416,548).

Relatively *uncontaminated* sludge thus could probably be dumped in open ocean waters without causing severe impacts, as long as a dispersal strategy and appropriate disposal technologies were used (51,87,338,387,402). In general, sludge should be dumped slowly in deep, dispersive waters to obtain the greatest mixing and dilution. Furthermore, impacts might be minimized by varying the location and frequency of disposal operations (509). Barges or ships could help achieve this goal, since the disposal location can be changed as needed. Shifting dumping to the Deepwater Municipal Sludge Site is one example of this strategy.

On the basis of these factors, there would be little rationale to eliminate dumping as a disposal option. In addition, controlled dumping under dispersive conditions also might be used beneficially to increase productivity in certain marine environments, for example midcontinental shelf areas with a naturally relatively barren benthic community (509).

On the other hand, however, most sludge dumped in marine waters will continue to be contaminated to some degree with microorganisms, metals, and organic chemicals. Furthermore, the likelihood that programs for reducing toxic pollutants in municipal wastes will be fully implemented and enforced is unclear. The uncertainties associated with our ability to detect microorganisms (including human pathogens) in marine waters, and to sufficiently reduce the amounts of metals and organic chemicals in sludge, thus argue in favor of a policy that call for restricting (at least to some degree) the dumping of sludge. For these reasons, many public groups remain adamantly opposed to dumping in any form.

If marine dumping of sludge is to continue, it seems prudent to use a dispersal strategy (e.g., dumping in well-mixed and deep waters) and to minimize the presence of metals, organic

Box W.—Proposed Discharge of Sludge by Orange County

The difficulty in accommodating environmental concerns, economic factors, and political considerations is exemplified by a recent proposal from Orange County, California, to discharge sludge through a pipeline into the ocean on an experimental basis. This type of discharge is currently illegal under the California Ocean Plan, but the Federal Water Quality Act of 1987 contained a provision allowing EPA to grant a permit for this project. The County's motivation for the proposal is both environmental and economic: ocean discharge would be environmentally preferable to land-based alternatives, and it would be about four times less costly than landfilling, the recommended alternative (294), with annual savings of up to $10 million.

Relatively uncontaminated sludge would be discharged about 8 miles offshore into the open ocean, at a depth of about 1,300 feet (50,116). This would be the only pipeline in the Southern California Bight that is located off the edge of the continental shelf and it would be four times deeper than any other discharge in the area. It would be designed with large diffusers to increase dispersion, and the project would include a long-term monitoring program. Analyses of potential impacts from such a discharge indicate that disposal of relatively uncontaminated sludge in appropriate waters (deep and open, with high dispersal capability) probably would cause relatively minor impacts, except on a small area around the discharge point (50,385). Orange County has proposed that the discharge be allowed only if inputs of metal and organic chemicals into its POTWs were sufficiently controlled, and that it be terminated if observed impacts were deemed unacceptable by the permitting authorities.

The proposal has been criticized by environmental groups for several reasons. The discharge would be located in the Bight, which already receives massive discharges of effluent from three large sewerage authorities (the City of Los Angeles, Los Angeles County, and Orange County) and several smaller ones, as well as some sludge from Los Angeles County. These discharges have been the source of intense and continuing public debate, particularly with regard to two issues: general degradation of the environment, and the presence of pollutants such as DDT in sediments and fish. EPA has already encountered great public opposition when the three large authorities applied for waivers from secondary treatment requirements for effluents.

Current disposal practices clearly have caused some localized degradation of a small portion of the benthic environment. Although the proposed discharge could be terminated if unacceptable impacts were observed, critics note that no scientific definition of "unacceptable impacts" currently exists; as a result, there is no clear point when termination would be deemed necessary (N.K. Tayor, Sierra Club Clean Coastal Water Task Force, pers. comm., January 1987). In addition, these observers contend that discharges into deep ocean waters may not be degraded as rapidly as in other environments because of low oxygen levels and that enhancement of productivity in these waters would be minimal at best.

DDT was legally discharged in relatively large amounts into the Bight until 1970; some apparently still remains in pipes and sewers because small amounts are present in some municipal effluents entering the Bight. It persists in the environment, has contaminated sediments and many organisms, and is known to have caused reproductive failure in fish-eating birds such as the brown pelican. Although the presence of DDT is primarily a result of past practices, its presence is still of concern because it continues to be detected in organisms from these coastal waters (52). Its presence also fuels the controversy about current municipal disposal practices, particularly the ability of municipal treatment plants to control industrial discharges into sewers and reduce the amounts of toxic metals and chemicals in effluent and sludge.

chemicals, and pathogens. This strategy would clearly preclude dumping in estuaries and poorly-mixed coastal waters.

In addition, if increased dumping were allowed, another legitimate concern is whether the magnitude of dumping could be sufficiently controlled. Economic pressures might force a substantial in-

crease in the amount of sludge being dumped in the ocean. The Marine Protection Research and Sanctuaries Act (MPRSA) permitting process may be adequate to temper these economic incentives; in addition, fees or taxes could be imposed on ocean dumpers so that the total cost of dumping is comparable to the cost of other disposal options. It may be difficult, however, to levy such a tax or fee on

POTWs using ocean dumping because the MPRSA only allows the collection of fees to process permit applications.

Treatment Levels for Effluent Discharges

Effluent cannot be readily contained and instead is generally discharged from pipelines. Since pipelines are fixed in one position and result in relatively low rates of initial dilution (280,509), their use can lead to the accumulation of particulate material in localized areas. Even so, these discharges can be suitable and dispersion can be enhanced if:

1. the amounts or concentrations of pollutants in effluents are reduced, and
2. pipelines are properly designed and placed in well-mixed and dispersive waters at appropriate distance from shore and depth (387).

Because of historical precedent and the current structure of municipal systems, few people question the need to discharge sewage effluent into estuaries and coastal waters.

At the same time, however, other options such as water conservation and reclamation could be used in some situations to reduce discharges into

Photo credit: Country Sanitation Districts of Los Angeles

Municipal wastewater can be treated at water reclamation facilities and then used for a variety of purposes, including irrigation. The golf course shown here is located next to a water reclamation facility and irrigated with water from the facility.

marine and fresh waters. For example, the development of small plants to treat and reclaim municipal wastewater might be environmentally and economically preferable to the continued development of larger and more expensive POTWs that discharge effluents into surface waters, especially as these larger plants begin to age and as Federal funding for the construction of municipal treatment plants declines (N.K. Taylor, Seirra Club Clean Coastal Water Task Force, pers. comm. 1987). The incentives to develop water reclamation and conservation plants are greatest in the more arid areas of the country.

For discharges into marine waters much disagreement exists about the acceptable levels of two conventional pollutants—suspended solids and biochemical oxygen demand—in such discharges and whether some POTWs should be allowed to provide less-than-secondary treatment. Under Section 301(h) of CWA, POTWs could apply for waivers from secondary treatment requirements in areas where environmental quality would not be harmed, primarily to reduce construction and operating costs. The implementation of the waiver program, as well as its merit, has been debated extensively (122,225,310,533,649).

There is little doubt that substantial cost savings, amounting to several billion dollars in construction costs and up to $100 million in annual operation and maintenance costs, could be achieved by allowing some waivers (504,575). Comparing cost savings to costs of subsequent changes in receiving water quality is difficult, however, because of the variety of other factors that can affect water quality.

From a technical perspective, the question of whether lower treatment levels should be allowed can only be determined on a case-by-case basis, after evaluating site-specific factors. These factors are evaluated as part of the 301(h) application process (40 CFR Part 125, Subpart G) and include:

- the quality of receiving water (i.e., ability to disperse material, degree of previous impacts);
- the sensitivity of indigenous organisms and communities; and
- the relative contributions of pollutants from other sources (e.g., nonpoint pollution, industrial effluents).

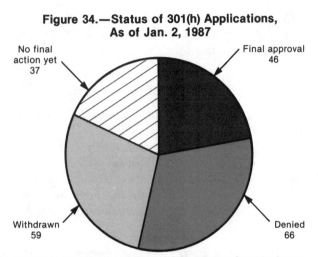

Figure 34.—Status of 301(h) Applications, As of Jan. 2, 1987

SOURCE: R. DeCesare, Office of Water, U.S. Environmental Protection Agency, personal communication, January 1987.

As of January 1987, EPA had decided that all relevant criteria appeared to be satisfied for 46 of 208 waiver applications (figure 34). Only a few large coastal POTWs, however, received approvals (e.g., Los Angeles County). Some municipalities (e.g., Seattle) withdrew their applications in part because of major public controversy.

From a policy perspective, prohibiting such waivers in the future could be justified because of the overall extent of pollutant inputs from many sources into estuaries and coastal waters and the expected trend of degradation in many of these waters. Indeed, some environmental groups have suggested that the Section 301(h) waiver provision be rescinded.

In one sense, this issue is largely moot, however, because decisions about most waivers have been made and no additional applications can be submitted. In addition, the National Municipal Policy calls for most POTWs to achieve secondary treatment by mid-1988. It is complicated, however, by uncertain future economic conditions. Many POTWs have not yet secured funding for building or upgrading plants to the secondary level, and Federal Construction Grant funds for such activities will be significantly reduced in the next few years.

In anticipation of reduced Federal funding, some States are developing revolving funds (through bond sales or initial capitalization by State appropriations) to meet future POTW construction costs (143,520). Some municipalities are turning to private developers in an attempt to finance necessary construction, although it is unclear whether incentives are sufficient for private developers to invest on a large scale in municipal treatment plants (143,685).

These economic conditions could lead to reconsideration of the issue of required treatment levels in the future, as municipal treatment needs in coastal areas increase and as older treatment plants require maintenance or expansion. At the same time, and in combination with general concerns about the quality of marine environments, they also could provide incentives to consider other options for managing effluent such as water reclamation and reuse.

Role of Land-Based Alternatives in Sludge Management

The availability of land-based sludge management options is a critical factor in decisions regarding marine disposal of sludge. In the context of the waste management hierarchy, use of sludge as a beneficial resource on land (or in marine waters) would generally be preferred to disposal. From a technical perspective, relatively uncontaminated sludge could be land-applied, under proper conditions (e.g., appropriate measures to control runoff), as a beneficial resource without causing significant impacts. In addition, destruction of uncontaminated sludge by incineration might also be preferred in many situations.

Implementation of land-based alternatives, however, often is difficult for several reasons. First, local public opposition to land application or incineration can be intense because sludge is often considered an undesirable waste and because of concerns about health risks arising from use of these methods. Second, long-term management arrangements often are difficult to maintain (502). Third, standards to compare the various land-based disposal alternatives have been lacking (502), but EPA is developing regulations to address this problem. Finally, most sludges are contaminated with pathogens and toxic pollutants, which limits the environmental acceptability of land-based (and marine) disposal options, especially those involving beneficial uses.

Chapter 10
Managing Dredged Material

CONTENTS

Managing Dredged Material

OVERVIEW

Dredging involves the removal of bottom sediment from rivers and harbors and its transportation to another location for disposal. "New work" dredged material is generated during the initial development of a port, harbor, or navigation channel, or the widening and deepening of existing navigational channels. In addition, "maintenance" dredging is required for most channels because fine-grained, river-borne sediment settles out of suspension and gradually accumulates in the channels (44), and because coarse-grained sediment is eroded from along shorelines and also begins to fill the channels. The U.S. Army Corps of Engineers (COE) considers a small fraction of this maintenance material, generally dredged from areas near highly industrialized ports and harbors, to contain high enough levels of pollutants to require special management during disposal.[1]

COE is responsible for maintaining over 25,000 miles of navigable waterways that service over 155 commercial ports and more than 400 small boat harbors; these ports and harbors are valuable for commercial, defense, and recreational purposes. Projects run by COE, other Federal and State agencies, and private efforts result in the dredging of about 550 million wet metric tons of sediment each year, at a cost of about $725 million (442).[2] Most dredged material originates from COE projects that have been authorized by Congress. Box X briefly describes the major Federal statutes that control the disposal of dredged material.

Dredged material accounts for about 80 to 90 percent by volume of the waste material that is dumped into marine environments each year. About one-third of all dredged material (180 million wet metric tons) is disposed of in marine environments: two-thirds of this material is disposed of in estuaries and the remainder is dumped in coastal waters or the open ocean. Two dozen sites receive about 95 percent of all dredged material dumped in coastal waters and the open ocean (442); an unknown but large number of sites are used for disposal in estuaries. Pressure to use marine environments for dredged material disposal will continue and possibly increase.

Disposal of *uncontaminated* dredged material does not appear to have had major negative impacts on organisms in most large estuaries or open ocean waters, although some temporary impacts have occurred. In some cases, uncontaminated dredged material can be used for beneficial purposes such as beach replenishment. In some smaller estuaries and some coastal waters, however, disposal of uncontaminated material can contribute to observable degradation. During disposal operations, most bottom dwelling (or "benthic") organisms that are covered by disposed material will die because of physical burial or suffocation. These physical impacts generally are short-term and restricted to the disposal sites; recolonization can take from several months to a few years after cessation of disposal activities. However, marine resources can be permanently damaged at disposal sites that are used regularly and/or that are used for large volumes of dredged material. For example, a quahog fishery off of Narragansett was totally lost after the disposal of several million cubic yards of dredged material.

The disposal of *contaminated* dredged material is generally of greater concern. According to COE, about 3 percent of the material dredged from estuaries and coastal areas is heavily contaminated with pollutants (metals and organic chemicals) derived from point and nonpoint sources. When this material is disposed of and settles on the bottom,

---

[1]COE uses a series of screening tests, discussed below, to determine when dredged material requires special handling. Quantitative national criteria that could be used to decide whether dredged material is contaminated, however, are currently lacking.

[2]Dredged material is usually measured by volume in cubic yards. To facilitate comparisons, where possible, with the amounts of other waste types, volumes of dredged material have been multiplied by the density of such material, which is approximately 1.18 metric tons per cubic yard, to give wet metric tons. The density of material from any given site may vary somewhat from this figure, however, so the resultant calculations should be considered estimates only.

Box X.—Statutory Framework

Several Federal statutes control dredging and disposal operations. The General Survey Act (1824) directed the Corps of Engineers (COE) to develop and improve harbors and navigation, and Section 10 of the River and Harbor Act (1899) required COE to issue permits for any work in navigable waters. Dredging and disposal operations have since been addressed more fully in the major environmental statutes passed during the 1970s, particularly the Marine Protection, Research, and Sanctuaries Act (MPRSA) and Clean Water Act (CWA).

Marine Protection, Research, and Sanctuaries Act

Under Section 103, COE must evaluate proposed projects that involve the transportation and dumping of dredged material in most coastal waters and in the open ocean. The evaluation of these activities is based on environmental impact criteria developed by the Environmental Protection Agency (EPA) in conjunction with COE; these criteria generally contain all the constraints set forth in the London Dumping Convention. Non-Federal projects that are approved receive an ocean dumping permit from COE. Federal projects performed by COE are evaluated in the same manner, but they do not receive permits.

EPA has the primary responsibility for designating ocean disposal sites, although COE may also initiate the site-selection process. EPA can prohibit the use of any disposal site, but has not done so. The National Marine Fisheries Service (NMFS), with assistance from the U.S. Fish and Wildlife Service (FWS), provides environmental input to COE during the review of Section 103 activities.

Clean Water Act

Under Section 404 of CWA, COE regulates discharges of dredged or fill material in "waters of the United States"—most freshwater areas and wetlands, estuaries, and coastal waters inside the territorial sea boundary. The permit applicant or project sponsor is responsible for finding appropriate disposal sites. Permit applications are evaluated by COE, using guidelines developed jointly by COE and EPA. Other agencies—EPA, FWS, and NMFS—provide comments and recommendations, and EPA may veto the use of a proposed disposal site. Where COE's jurisdictions under Section 103 and Section 404 of CWA overlap in the territorial sea, COE usually issues only an ocean dumping permit.

The States also review permit applications for disposal operations in estuaries and in the territorial sea. Under Section 401, these disposal operations must be certified by the affected State as complying with applicable water quality criteria.

Other Federal Statutes

The Comprehensive Environmental Response, Compensation, and Liability Act established the Superfund program to ensure the cleanup of sites that have been highly contaminated by improper disposal practices of the past (583). There are about 30 known freshwater and marine sites that may require dredging and disposal under the Superfund program (N. Francingues, pers. comm., 1986).

The Coastal Zone Management Act requires that States with federally approved coastal zone management programs review permit applications for dredging and disposal activities in estuaries and the territorial sea. Under Section 307, these activities must be certified by the applicant as complying to the maximum extent practicable with the State program. States must either concur with or object to the certification determination.

benthic organisms that recolonize the deposits may take up and bioaccumulate some of these pollutants. Although few potentially harmful pollutants appear to be released directly into the water column, the pollutants can be transferred from benthic organisms to predatory organisms. To date, no known human health impacts have been documented from the transfer of pollutants from dredged material up the food chain, although such impacts

would be difficult to detect and generally have not been investigated.

Disposal of contaminated material generally involves expensive techniques designed to isolate the material from the environment. In some marine environments, it can be covered or "capped" with a layer of uncontaminated sediment or special containment islands can be built. On land, it can be disposed of in specially managed upland containment areas. It is unlikely, however, that currently available disposal techniques will permanently isolate all pollutants.

No disposal option is categorically better than another, because operational, economic, environmental, and social factors vary greatly among dredging projects (496). As a result, the choice of disposal options usually requires site-specific and often subjective evaluations (169).

Finding suitable disposal sites is the overriding problem now facing COE and other sponsors of dredging projects. Although COE policy stresses the balanced consideration of all disposal options,

Federal regulatory requirements are generally stricter for disposal in the open ocean than for disposal in freshwater and estuarine environments. Federal and State requirements tend to be least restrictive for upland disposal, and policies that have attempted to curb pollution in freshwater and marine environments have indirectly encouraged dredged material disposal in upland containment areas. However, disposal in upland areas is generally costly, and finding upland sites is becoming more difficult. Thus, future decisions regarding disposal of dredged material in the Nation's estuaries and coastal waters will be greatly affected by policies regarding disposal on land and in open ocean waters.

DREDGED MATERIAL DISPOSAL: AMOUNTS AND SITES

Inventory of Dredged Material Amounts

Almost 60 percent—about 310 million wet metric tons—of all dredged material in the United States comes from estuarine and coastal areas. Of this material, an average of about 180 million wet metric tons is disposed of in marine environments: about two-thirds in estuaries, one-sixth in coastal waters, and one-sixth in the open ocean.[3] Of the remaining material dredged from estuarine and coastal areas, most is disposed of in upland containment areas (above the water table, but near the dredging site), but some is disposed of in intertidal areas, open freshwater areas, or containment islands.

The amount of material disposed of in coastal and open ocean waters has fluctuated greatly during the last 25 years, between 35 and 120 million wet metric tons per year, with much of the fluctuation resulting from varying dredging demands on the lower Mississippi River (figure 35; the figure and the discussion in this paragraph do not include disposal activities in estuaries). If the volumes of dredged material from the lower Mississippi River

are ignored, dredging volumes decreased from 1974 to 1981 and then began to increase in 1982.

Dredged material can be used in a variety of beneficial ways. About 15 to 20 percent of all dredged material is used for: beach nourishment; subaqueous mounds for shoreline protection; construction aggregate; fill material for commercial development or parks; cover material for sanitary landfills; construction material for dikes, levees, and roads; development of marshes and upland habitat; soil supplementation on agricultural land; and reclamation of strip mines (565).

Disposal Sites

Sites in Coastal and Open Ocean Waters

As of January 1987, about 126 disposal sites were located in coastal waters and the open ocean (D. Mathis, COE, pers. comm., January 1987). Most of these sites are distributed relatively evenly along the Atlantic, Gulf, and Pacific coasts, although a few are located in the Caribbean and the South Pacific (442). Although half of the sites receive some use each year, 95 percent of all coastal and open water disposal occurs at about two dozen sites. Some disposal sites are rarely used. Half of the sites have areas of 0.5 square miles or less; the remainder have areas of up to 4 square miles. About three-fourths are located in water less than 60 feet deep; only 18 are in water over 300 feet deep.

---

[3]These figures are based on an OTA survey of the 18 COE coastal districts with primary responsibility for maintaining ports and harbors. The amount of dredging and disposal fluctuates greatly from year to year and in different areas. Of the material that is disposed of in coastal and open ocean waters, almost half is disposed of in the Gulf of Mexico, and the remainder is split between the Atlantic and Pacific Oceans.

Figure 35.—Amounts of Dredged Material[a] Disposed of in Coastal Waters and the Open Ocean, 1974–84

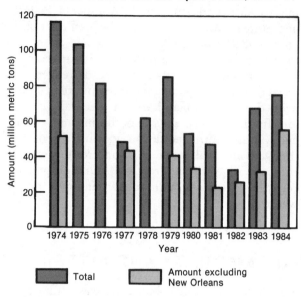

Total

Amount excluding New Orleans

[a]From U.S. Army Corps of Engineers' Projects.

NOTE: Missing second bar for 1975, 1976, and 1978 reflects lack of data.

SOURCES: U.S. Army Corps of Engineers, *1980 Report to Congress on Administration of Ocean Dumping Activities*, Pamphlet 82–P1 (Fort Belvoir, VA: Water Resources Support Center, May 1982); U.S. Army Corps of Engineers, *Ocean Dumping Report for Calendar Year 1981*, Summary Report 82–S02 (Fort Belvoir, VA: Water Resources Support Center, June 1982); U.S. Army Corps of Engineers, *Ocean Dumping Report for Calendar Year 1982*, Summary Report 83–SR1 (Fort Belvoir, VA: Water Resources Support Center, October 1983); J. Wilson, U.S. Army Corps of Engineers, personal communication, 1986.

Section 103 of the Marine Protection, Research, and Sanctuaries Act (MPRSA) established a formal process for designating coastal and open ocean disposal sites. In 1977, the Environmental Protection Agency (EPA) published comprehensive criteria for the designation process and granted 3-year "interim" designations for previously used disposal sites (630). Preparation of an environmental impact statement (EIS) for each site and "final" designation were to be completed by January 1981, but this process encountered many delays (579).

As of November 1986, EPA had granted final designation to 19 sites (Office of Marine and Estuarine Protection, EPA, pers. comm., November 1986). Although COE does not formally designate sites, as of January 1986 it had "selected" about 15 additional sites that it considered suitable for final designation (D. Mathis, COE, pers. comm., January 1987). Most of these sites were selected for one-time use, for example, for the disposal of material from channel deepening projects in Mobile and Norfolk, with disposal of any subsequent maintenance material occurring at other, EPA-designated sites. Most of the remaining 90 or so coastal and open ocean sites are being used under extensions of EPA's original interim designations.

No major dredging project has been canceled because a coastal or open ocean disposal site lacks final designation. However, a portion of the Tampa Harbor deepening was delayed, and the State of New York has prohibited the use of several undesignated disposal sites in coastal waters. COE, port authorities, and the dredging industry are concerned that future projects could be canceled or delayed if interim disposal sites cannot be used, for example, because of litigation over delays in the designation process (442).

Sites in Estuarine Waters

Defining the number of disposal sites in estuaries is difficult because material often is discharged along much of a navigation channel, although at some distance from the channel itself. Thus, it is difficult to judge whether this constitutes one or multiple disposal sites. Section 404 of the Clean Water Act (CWA) controls this kind of disposal through a permitting process, but it does not include provisions for formally designating sites. Generally, a permit for disposal in an estuary specifies that a site will be used for a given length of time; in many cases, the permit is only for one disposal operation, although the site could be used again under another permit.

In addition, about 30 sites have received multiple-use permits under Section 404 (D. Mathis, COE, pers. comm., January 1987). These sites—for example, Alcatraz Island in California and Puget Sound in Washington—tend to be controversial, and the permitting process often involves preparing an EIS, even if the site is not to be used on a continuing basis. Most dredged material disposal that occurs in estuaries, however, does not occur at these multiple-use sites.

Future Need for Marine Disposal

Dredged material disposal in marine waters could increase for several reasons. First, some projects

Photo credit: U.S. Army Corps of Engineers

After being excavated from a navigation channel, dredged material can be "sidecast" from the dredge and used to replenish beaches.

proposed for increasing commercial traffic would involve the development of harbors and deepening of channels. These projects would generate large amounts of material, and deeper channels, once created, generally would require increased maintenance dredging. Since most coastal ports have used marine waters heavily for disposal from similar projects in the past, such disposal is likely to continue; for example, about 90 percent of dredged material from COE's New York District is disposed of at the Mud Dump site in the New York Bight. Second, some observers argue that the United States must develop additional capacity to handle large, deep-draft vessels if it is to maintain or increase its present role in the international economy. Because of the high costs of port construction and uncertainties about the total required capacity for deep-draft ships, however, only a limited number of such ports are likely to be built (388). Third, some ports may need to be deepened to accommodate an increasing number of naval vessels.

Although all major U.S. ports have made some plans for expansion and channel deepening, new federally authorized port projects have faced many delays over the last 10 to 15 years. The Water Resources Development Act passed by Congress in 1986, however, authorized many of these projects, including 6 deep-draft projects and 35 general cargo and shallow water improvements. It also established cost-sharing between the Federal Government and local sponsors or port authorities. If all the authorized projects are completed, the Federal share would be $3 billion and the local or port share would be $2 billion. New dredging projects will now proceed, and marine disposal of some material is likely. The cost-sharing arrangement, however, could inspire local sponsors to reduce the amount of requested dredging, with a subsequent decrease in the amount of material requiring disposal (P. Johnson, Office of Technology Assessment, pers. comm., November 1986).

DREDGED MATERIAL CHARACTERISTICS

Sediment Composition

Sediment is composed of varying amounts of several natural substances: sand, gravel, silt, and/or clay; organic matter and humus (i.e., decomposed organic matter); and chemical compounds such as sulfides and hydrous iron oxide. Sediments also can be contaminated with various metals and organic chemicals, from both point and nonpoint pollution sources.

Pollutants commonly found in dredged material include metals (e.g., cadmium, chromium, copper, iron, lead, mercury, nickel, silver), chlorinated hydrocarbons (e.g., PCBs, DDT), polycyclic aromatic hydrocarbons, and other petroleum products. Most pollutants are adsorbed or tightly bound to the organic material and the clay particles in dredged material, thereby reducing their potential availability to marine organisms that inhabit the water column, but not necessarily to benthic organisms that dwell on or in the sediment (46,549).

No quantitative criteria exist for defining the degree to which dredged material is contaminated,[4] so it is difficult to estimate exactly how much dredged material is clean, somewhat contaminated, or highly contaminated. COE considers a large portion of dredged material to be relatively clean. For example, as much as 20 percent of all sediment dredged by COE is "new work" material, which generally is not contaminated because it was deposited long before the settlement and industrialization of North America. Sand, which does not readily adsorb pollutants, and sediments located some distance from present or past pollution sources usually are relatively clean.

To determine when dredged material is contaminated enough to require special management, either in upland or island containment areas or by capping, COE generally relies on a series of screening tests. Based on these tests, COE considers about 3 percent (approximately 7 million wet metric tons) of all material dredged in its coastal districts to be

highly contaminated and to require special management.[5] In addition, participants at an OTA workshop estimated that about 30 percent of all maintenance material might be contaminated to some degree, even though it is not managed specially (584). These participants and other observers consider highly contaminated dredged material to be less contaminated than material disposed of in sanitary landfills (103) and than some hazardous material (299).[6]

Most contaminated sediment is found in the immediate vicinity of ports and harbors, or in areas where direct municipal and industrial discharges into estuaries and coastal waters have contributed considerable amounts of pollutants. Riverborne clays that are "trapped" in estuaries and navigation channels also may have been contaminated when they were transported through upstream regions. Dredging and disposal operations do not introduce new pollutants into aquatic environments; they simply redistribute sediments that are already contaminated.

Effects of Dredging Equipment on Composition

The physical characteristics of dredged material are significantly influenced by the type of dredging equipment used to excavate and transport the sediment. Two major types of dredges are used: hydraulic and mechanical.

---

[4]EPA is studying the feasibility of developing sediment quality criteria for certain metals and organic chemicals (C. Zarba, EPA, pers. comm., November 1986). These criteria could be used, for example, to determine whether dredged material is sufficiently contaminated so as to pose undue risks to bottom-dwelling marine organisms.

[5]This figure is based on the survey of COE coastal districts; since quantitative nationwide criteria are lacking, responses about the degree of contamination were based on the best judgment of COE personnel and their experience with dredged material requiring special management.

[6]It is unclear whether the Resource Conservation and Recovery Act's toxic characteristic leachate procedure (TCLP) toxicity test, developed by EPA to evaluate the potential for leaching from land-based disposal sites, is legally applicable to dredged material. The previous test (the extraction procedure, or EP, toxicity test) specified that material should be tested in its field-collected form, which for dredged material would always be wet. According to COE, wet dredged material, even if highly contaminated, will generally pass the EP test (103,269,299). On the basis of the EP test, highly contaminated dredged material could be considered acceptable for upland disposal; subsequent oxidization of the material, however, could make it susceptible to leaching. In contrast, dried dredged material, even if not considered highly contaminated, might not pass the test.

HOPPER DREDGE
409'x78'x39'

Photo credit: U.S. Army Corps of Engineers

Much of the dredging in the United States is conducted by ships known as hopper dredges, which hydraulically excavate material from a channel bottom and place it in "hoppers" within the hull. After the ship moves to a designated disposal site, the material is either dumped into the water or discharged through a pipeline to a beach or a diked containment area. This diagram of a hopper dredge shows some of the equipment used to excavate material from the bottom.

Most of the dredging in the United States is done with *hydraulic dredges*. Hydraulic dredges add significant amounts of water to the material, which enables the resulting slurry—about 80 to 90 percent water—to be pumped through a pipeline to a disposal site. Hydraulic dredges include both pipeline dredges and hopper dredges. Pipeline dredges discharge the material through a pipeline leading away from the dredge; they are rarely used in the open ocean because the pipeline can be broken by sizable waves. Hopper dredges are ships that hydraulically fill their hulls with dredged material

slurry, cruise to the disposal site, and dump the dredged material through doors in the hull; they operate most effectively in the open ocean or in vessel traffic.

Mechanical dredges are used on small jobs (e.g., around piers). They have clamshells or buckets that remove sediment at its original, or "in situ," density without adding significant amounts of water. Mechanically dredged sediment is typically transported to a disposal site in barges (240).

PHYSICAL FATE OF DREDGED MATERIAL AFTER DISPOSAL

Disposal in Marine Environments

When a load of dredged material is dumped from a stationary hopper dredge in water depths of less than 200 feet, most of the material descends rap-

idly (a few feet per second) through the water column as a high-density mass. The remainder—about 1 to 5 percent of the released sediment—remains suspended in the water column as a plume of slowly settling particles. When the rapidly descending mass reaches the bottom, the largest blocks of material remain in the impact area. Fine-grained material (as opposed to coarser, sandy material) spreads several hundred feet radially along the bottom and away from the impact point, in the form of a layer of "fluid mud" that can measure several feet in thickness (234).

In estuaries and coastal waters, the accumulation of dredged material on the bottom can be significant if dumping occurs over long periods or involves large volumes. Ever since the late 1800s, for example, dredged material has been dumped in the New York Bight within a few miles of the present Mud Dump site (located 6 miles east of New Jersey and 11 miles south of Long Island). More than 85 percent of the dredged material has remained at the disposal site, forming three well-defined hills that rise 30 to 50 feet above the bottom (125,171). Other material has accumulated northwest of these mounds, evidence of shoreward transport and/or dumping of dredged material short of the disposal site.

In the deep ocean, where water may be thousands of feet deep, the rapidly descending mass of material will entrain water as it descends and lose its integrity. After this occurs the material continues to settle slowly through the water column as a widely dispersed plume of suspended sediment. This material eventually is distributed widely over the bottom, usually without any significant buildup (93,234,439). If the total volume of material dumped at an open ocean site exceeds a few hundred thousand metric tons, however, the deposition may be sufficient to bury benthic organisms.

When slurry made of dredged material is disposed of in rivers, estuaries, and enclosed coastal areas via pipeline dredges, an estimated 97 to 99 percent of the slurry descends rapidly to the bottom of the disposal area. Once on the bottom, the slurry typically accumulates in the form of a low gradient mound of fluid mud. About 1 to 3 percent of the slurry remains suspended in the water column in the form of a turbidity plume. Turbidity plumes usually spread a few thousand feet from the discharge point and dissipate within several hours after the dredging is completed (500).

Once deposited, the long-term fate of dredged material depends on many factors, including bottom topography, the nature of the sediment, and erosion and transport by currents (234). Several models can be used to predict long-term transport of dredged material (235).

Disposal in Intertidal and Upland Containment Areas

To dispose of dredged material in intertidal or upland areas, it is usually pumped by a pipeline dredge into a diked containment area. The sediment will settle inside the area, and the ponded surface water is generally drained off as "effluent." If the ponded water and any rainfall on top of the settled material is drained completely, a dried crust will form over the surface of the area (217). However, the dredged material under the crust will remain almost indefinitely at a grease-like consistency with a solids content of about 30 to 40 percent (431).

IMPACTS OF DREDGED MATERIAL DISPOSAL

Disposal in Marine Environments

In marine environments, a number of adverse impacts can occur as dredged material descends through the water column or when it settles on the bottom. Impacts can occur in two areas: 1) on the water column and pelagic organisms, and 2) on bottom-dwelling organisms. No impacts on human health are known to have occurred from the transfer of pollutants in dredged material through the food chain; even if they were detected, however, it would be difficult to attribute them solely to

dredged material disposal because other sources of pollutants are present in most areas.

Impacts on the Water Column and Pelagic Organisms

Levels of suspended solids in the water column during dredging and disposal operations are usually low enough to cause few, if any, detectable physical impacts on pelagic organisms (i.e., those in the water column) (232). If surrounding waters become too turbid, several control techniques can be used (14). Another approach is to schedule projects to avoid seasons, such as during spawning, when potential impacts to marine organisms are great (320).

As dredged material descends through the water column, some pollutants (e.g., hydrogen sulfide, manganese, iron, ammonia, and phosphorus) may be released and their concentration in the water column may increase. In dispersive waters these increases are usually diluted rapidly (62,232,708). In small estuaries and sheltered coastal waters, however, such releases may adversely affect organisms in the water column. It appears to be rare, however, for pelagic organisms to bioaccumulate metals and organic chemicals released from contaminated dredged material, although detecting such impacts and attributing them to a particular waste type is difficult.

Impacts on Bottom-Dwelling Organisms

Physical Impacts.—Most benthic organisms that are covered by more than a foot or so of dredged material will die from suffocation within minutes or hours after disposal (232,708). Some burrowing organisms may be able to move vertically through the deposited material, especially if the dredged material is similar to the original bottom sediment. Bottom-dwelling fish typically leave a dumpsite during disposal operations, but they may return later if the habitat is not severely altered.

Physical impacts to bottom-dwelling organisms are generally restricted to the dumpsites. In large estuaries and coastal waters such as Chesapeake Bay or Puget Sound, disposal of uncontaminated material probably has an insignificant overall impact on the ecosystem. Physical impacts also are likely to be less significant (although possibly more widespread) at deeper sites because the material is more dispersed over the ocean bottom (440). In smaller estuaries or embayments, however, ecosystem impacts can be more significant. In general, disposal sites that are used once a year or more frequently will not fully recover as long as disposal continues. In addition, particularly sensitive benthic communities such as undersea grasses, coral reefs, and oyster beds may never recover.

In most areas that have been covered by uncontaminated dredged material, recolonization by benthic organisms begins within a period of weeks; extensive recolonization can take from several months to a few years after cessation of dumping activities (232,708). If the dredged material is similar to the original sediment, recolonization will occur more rapidly and the new community will more closely resemble the original community. Otherwise, the new community may show changes in the types and distribution of species, or changes in the total biomass of organisms present.

Some disposal of uncontaminated dredged material has harmed fisheries and other marine resources. For example, a quahog fishery was totally lost when several million wet metric tons of dredged material were disposed of at the Brenton Reef disposal site near Narragansett in 1969 and 1970 (442). In another instance, a coral reef off southern Florida was smothered by a disposal operation (J. Wilson, COE, pers. comm., 1986). On the other hand, some sites adversely affected by dredged material disposal may later be colonized by other organisms and become important commercial and sport fishing sites. The Brenton Reef site, for instance, was colonized by lobsters and is now a favorite location for lobster fishermen (442).

Chemical Impacts.—In addition to physical impacts, some pollutants in dredged material may be released to the lower water column over a period of months or years because of: the expulsion of interstitial (or pore) water as the material consolidates on the bottom (125), the burrowing of organisms in the upper layers of sediment (124), or the resuspension of material during storms (709). Benthic organisms that recolonize the deposit area may take up and sometimes bioaccumulate some pollutants from the lower water column or the sediment itself (46,232,260). Pollutants of particular importance include methyl mercury, cadmium, and many chlorinated and polynuclear aromatic hydrocarbons (442).

Potential and actual adverse effects from pollutant uptake are discussed in chapters 5 and 6. In many situations, it is difficult to discern how bioaccumulation affects benthic organisms or the ecosystem in general (437), or where the pollutants came from. In the New York Bight, for example, dumped sewage sludge and dredged material are major contributors of many metals, suspended solids, phosphorus, and PCBs, but it is difficult to ascertain how much is contributed by the dumping of dredged material alone (125). In addition, metals often tend to be tightly bound to the clays and organic matter in dredged material; in such cases, they are less likely to be released to the water column.

Disposal in Intertidal Areas and Upland Containment Areas

In *intertidal areas*, impacts depend largely on the characteristics of the disposal site. If a 3- to 6-inch layer of clean dredged material is placed over a marsh, most plants will regrow within several years. Once vegetated, these marshes are quickly colonized by various invertebrates (e.g., mussels, snails, and crabs) and birds. Thicker layers of dredged material will probably destroy the marsh grass and significantly change the elevation of the area so that marsh grass may not regrow (319). Some metals and organic chemicals can be taken up by plants, especially if the dredged material is exposed to oxidizing conditions during low tide (179,265,479). Once taken up by plants, pollutants can be transferred to wildlife or to estuarine organisms that feed on the plants or plant detritus (301).

Upland disposal in diked containment areas has been favored by many Federal and State regulatory agencies during the last decade because of concerns about potential impacts associated with marine and freshwater disposal (259). Several problems, however, can occur in upland situations. For instance, if ponded water on top of the settled material is drained, pollutants can escape in the drained effluent (15,289,431,449). In addition, the material remaining in the containment area often oxidizes, increasing the acidity of the material. Under these conditions, metals that were formerly bound to the dredged material (e.g., cadmium, lead, mercury, nickel) can be released in runoff from the containment area (179,300). Most organic chemicals tend to remain with the fine-grained dredged material in the containment area or volatilize into the atmosphere (R. Peddicord, COE, pers. comm., 1986). Upland containment areas have also been identified as sources of saltwater intrusion and other groundwater contamination (179,442; R.M. Engler, COE, pers. comm., 1986).

PREDICTING ENVIRONMENTAL IMPACTS

Predictions about project-specific impacts can only be made if there is adequate information about the proposed operation, the material that will be disposed of, and the disposal site environment. In some cases, predictions can be generalized for situations in which the sediments, dredging equipment, and disposal environments are similar.

Predicting Physical Fate in Marine Environments

The short-term physical fate of disposed dredged material is generally quite predictable and computer and empirical models are available for making more detailed assessments. Predicting long-term sediment transport, however, is usually more difficult (234). For example, the accuracy of long-term predictions is directly related to the availability of data on currents and other hydrographic conditions in a particular area (235). If this information is not available, detailed and costly predictions of long-term physical fate for single disposal operations may not be worthwhile, unless a large volume of material is involved. Using information from past disposal operations at similar sites is often a more effective way of predicting long-term transport.

Predicting Chemical Impacts on Marine Organisms

The COE uses a number of laboratory tests to evaluate the potential short-term toxicity of pollutants in dredged material to marine organisms:

- Bulk sediment composition analyses indicate the concentrations of pollutants that are present in the dredged material.
- Elutriate tests indicate the degree to which different pollutants might be released to the water column during disposal.
- Laboratory bioassays indicate the potential "acute" toxicity (or lethality) of pollutants to organisms, over a period of 2 to 60 days (436).

• Bioaccumulation tests indicate which pollutants might be taken up by marine organisms over the short-term (436).

These standard tests generally require several months to a year to complete; costs can range from $1,000 to $30,000 per sample (table 25). The number of samples required to evaluate a particular disposal operation depends on several factors, including size of the dredging project, sensitivity of the disposal environment, and similarity of the project to past projects that have been monitored for impacts.

Federal and State regulators have tried to use bulk sediment composition analyses as the primary (and sometimes only) method for determining whether a particular sediment is contaminated enough to require special management, in part because these analyses do not require expensive and time-consuming tests. In addition, the potential toxicity to marine organisms of a given sample of dredged material is usually lower than indicated by the tests, because metals and some organic chemicals tend to be tightly bound to clay particles and humus in the material and to be less available to the organisms (259). Because of this factor, COE has assumed that sediment passing these tests can be disposed of safely in any environment without concern about potential chemical effects; if the sediment fails these tests, it is considered contaminated enough to be unsafe for unrestricted disposal in marine environments.

The use of these types of tests to determine whether dredged material is contaminated enough to require special management has been criticized. For instance, bulk sediment composition analyses do not necessarily indicate the likelihood that pollutants will be released from the sediment to the

water column or whether they will be taken up by marine organisms. Laboratory bioassays do not indicate which pollutants are responsible for any observed effects nor can they detect long-term, chronic effects, and bioaccumulation tests do not necessarily indicate whether further effects will occur (436,437).

In addition, the lack of standardized, quantitative "sediment quality criteria" is a major problem, especially for sediments that are neither extremely contaminated nor extremely clean (C. Zarba, U.S. EPA, pers. comm., November 1986; K. Kamlet, A.T. Kearney, Inc., pers. comm., November 1986). Regulatory agencies could use sediment quality criteria—designed to indicate when pollutant levels in dredged material are likely to cause impacts on marine organisms—to assess the degree of contamination of dredged material. EPA is considering developing quantitative sediment quality criteria for some metals and organic chemicals (C. Zarba, U.S. EPA, pers. comm., November 1986). Without such criteria, the results of any qualitative tests are subject to varying interpretation. Thus, tests such as the bulk sediment composition analyses are probably most useful as screening tools to identify clean sediments, but not to assess the degree of contamination of sediments.

Predicting Impacts From Other Disposal Options

In upland containment areas, the consolidation of dredged material slurry can be predicted with reasonable accuracy, based on empirical observations and measurements (555). Chemical impacts, however, are more difficult to predict.

COE has developed several tests to predict the chemical impacts that might be associated with upland disposal. These tests show the quality of ponded water that might be discharged during the disposal operation, the quality of surface runoff from the disposal area, the potential for long-term leaching of pollutants into adjacent aquifers or surface waters, and the potential uptake of pollutants by plants and animals that might eventually inhabit the area. These tests also require several months to complete, but they are generally more expensive than tests for marine disposal (table 26). EPA has developed other tests under the Resource Con-

Table 25.—Approximate Costs of Laboratory Tests for Predicting Chemical Impacts of Marine Disposal, Per Sample

Bulk sediment analyses	$ 5,000
Elutriate test	$ 1,000 to $ 5,000
Bioassay	$ 1,000 to $ 5,000
Bioaccumulation tests	$15,000 to $30,000

SOURCE: Office of Technology Assessment, 1987; based on W.E. Pequegnat, *An Overview of the Scientific and Technical Aspects of Dredged Material Disposal in the Marine Environment,* contract prepared for U.S. Congress, Office of Technology Assessment (College Station, TX: January 1986).

Table 26.—Approximate Costs of Laboratory Tests for Predicting Chemical Impacts of Upland Disposal, Per Sample

Bulk sediment analyses	$ 5,000
Quality of ponded water tests	$ 7,500
Runoff quality tests	$20,000
Leachate quality tests[a]	$75,000 to $100,000
Plant/animal uptake tests	$30,000 to $ 40,000

[a]Leachate quality tests address the potential for contaminating groundwater. Standard tests are being developed and costs for routine testing may be lower than values cited.

SOURCE: Office of Technology Assessment, 1987; based on W.E. Pequegnat, *An Overview of the Scientific and Technical Aspects of Dredged Material Disposal in the Marine Environment*, contract prepared for U.S. Congress, Office of Technology Assessment (College Station, TX: January 1986).

servation and Recovery Act (RCRA) to assess the potential for hazardous wastes to leach from land disposal sites. Whether these tests are legally applicable to the disposal of dredged material is unclear.

DISPOSAL TECHNIQUES FOR CONTAMINATED DREDGED MATERIAL

If dredged material is determined, by whatever method, to be highly contaminated, it generally requires special management during disposal to isolate it from the environment. These special techniques include disposal in:

1. water, under a layer or "cap" of uncontaminated sediment,
2. an upland containment area, or
3. a containment island.

These disposal options all require long-term maintenance to maximize the degree of sediment isolation. Even then, it is unlikely that all pollutants will be permanently isolated.

Capping in Marine Environments

In relatively quiescent marine (and freshwater) environments, contaminated dredged material can sometimes be isolated from aquatic organisms by covering it with a layer of uncontaminated material (47,261,363,511). About 3 feet of cover material is usually required to minimize the possibility of organisms burrowing into the contaminated material. Capping does not change the nature of the material under the cap: it still remains contaminated.

Capping has several advantages, especially in relatively quiescent marine environments. First, caps appear to be stable and subject to little erosion in these environments; it is conceivable that additional capping material might not be needed

for several decades. Second, as long as the cap is not disturbed the contaminated material remains in a relatively unoxidized state, thereby minimizing the upward migration of pollutants from the dredged material (47). Finally, costs for capping can be much lower than costs for disposal in upland containment areas or containment islands. The London Dumping Convention considers capping in quiescent marine environments to be an acceptable method, if subject to careful monitoring (136).

Capping also has several disadvantages, however. First, the water column may be exposed to some pollutants as the dredged material descends to the bottom, although releases to the water column tend to be small. Second, if water depth at the site increases beyond 100 feet, it becomes difficult to restrict lateral spreading when dredged material is placed on the bottom.[7] Third, capping requires a large volume of clean cover material, leading to increased costs. This can be minimized if contaminated material from one project is covered with uncontaminated material from another

---

[7]When contaminated sediment is removed with a pipeline dredge, the slurry can be pumped through a "diffuser" mounted at the end of the pipeline (390). By reducing the velocity of the slurry as it leaves the pipeline, a diffuser system minimizes the exposure of the water column to the contaminated dredged material, allows more precise placement of the material, and minimizes the lateral movement of the material along the bottom. First designed in the United States in the late 1970s, diffusers have been used to cap contaminated sediment in Rotterdam Harbor and recently were tested on contaminated dredged material from Calumet Harbor, Chicago, IL (R. Montgomery, pers. comm., 1986).

project. Fourth, once the dredged material is on the bottom, contaminated water may be released as the sediment consolidates prior to capping or after capping if the cap "sinks" into the contaminated sediment. Finally, storms or currents can erode the cap, thus requiring the expense of additional clean material.

Capping of contaminated material has been used over the last decade in the United States, Japan, Canada, and the Netherlands (442). In this country it has been used in water depths of 100 feet or less in Long Island Sound, the New York Bight, off the coast of Maine, the Duwamish Waterway near Seattle, and Alaska.

In some cases, natural or artificial subaqueous pits can be filled with contaminated dredged material and capped (37). Pits restrict the lateral spreading of the dredged material, and a cap that is level with the bottom is less susceptible to erosion than a mounded surface. Approximately 125 natural and artificial pits have been identified in rivers, estuaries, and coastal areas of the United States, many near ports and harbors (39).

Upland Containment Areas

Upland containment areas are used to physically, and presumably chemically, contain contaminated dredged material. As with capping, material disposed of in upland containment areas still remains contaminated.

The primary advantage of upland disposal is that it allows greater management and control than is possible in marine environments (259,317,431). For instance, the area could be lined with clay or synthetic materials to reduce the potential for groundwater contamination, water discharged from the site could be controlled and treated, lime could be applied to increase pH and minimize the leaching of metals, or the area could be covered with clean sediment to isolate the contaminated sediment from runoff. Such management techniques, however, can greatly increase the overall costs of disposal.

The major disadvantage of upland disposal is the potential for release of metals (e.g., cadmium, lead,

mercury, nickel). Drying the dredged material to enhance its consolidation and increase the area's capacity significantly increases the potential for mobilization and subsequent release of most metals to the environment. Oxidizing conditions, which are more common in upland than aquatic environments, increase the likelihood that metals will be taken up by plants and transferred to animals, released in runoff from the site, or leach into groundwater.

Containment Islands

Another option to dispose of contaminated sediment is to build containment "islands" in relatively protected areas close to a port. These islands consist of an outer perimeter that is gradually filled with contaminated dredged material over a period of a few decades. The primary advantage of containment islands over upland containment areas is that the contaminated material is maintained in a saturated and unoxidized chemical environment, thereby minimizing the potential for migration and release of pollutants. These islands can cause undesirable changes in water circulation or benthic communities or become navigational hazards, however, unless they are located away from navigation channels or biologically important resources.

Several containment islands have either been built or proposed. For example, Hart-Miller Island in the Chesapeake Bay was designed to accept contaminated dredged material from Baltimore Harbor for the next two decades. It has a capacity of 53 million cubic yards (about 63 million wet metric tons), covers approximately 1,100 acres of shallow bottom, and cost $59 million to build (442). The New Jersey Department of Environmental Protection has proposed that the New York and New Jersey Port Authority build a containment island in Raritan Bay. After a containment island has been filled and capped, the island can be developed commercially or used for recreation or wildlife habitat if continued isolation of the contaminated material can be ensured.

Photo credit: U.S. Army Corps of Engineers

In some cases, dredged material is disposed of in diked containment islands that are built in relatively protected areas. The containment island shown here, located in North Carolina, is being filled with dredged material pumped through a pipeline from the dredge in the foreground.

COSTS OF DISPOSAL OPERATIONS

Dredging and disposal costs vary significantly from one project to another. In 1986, uncontaminated material averaged about $1.50 per cubic yard for disposal. For marine disposal, the costs of operations using pipeline dredges ranged from about $0.50 to $2.00 per cubic yard (table 27). Costs for ocean disposal using a hopper dredge or dumping barge are usually somewhat greater, reflecting the transport distance to the disposal sites. The use of upland containment areas is considerably more expensive than disposal in marine environments. Costs for disposing highly contaminated dredged material may be 2 to 10 times higher than ordinary disposal costs.

Table 27.—Approximate Costs Per Cubic Yard for Dredged Material Disposal

	Uncontaminated	Contaminated
Marine environments[a]	$0.50 to $2	$3 to $5
Upland containment	$5 to $20	$30 to $60

[a]Using pipeline dredge with disposal in adjacent waters.

SOURCE: Office of Technology Assessment, 1987; based on W.E. Pequegnat, *An Overview of the Scientific and Technical Aspects of Dredged Material Disposal in the Marine Environment,* contract prepared for U.S. Congress, Office of Technology Assessment (College Station, TX: January 1986).

POLICY ISSUES

Dealing With the Shortage of Disposal Areas

Estuaries and coastal waters are among the most important aquatic environments, but many are severely stressed by pollutants from a variety of waste disposal activities and from runoff. Because of concerns about the immediate physical impacts and potential long-term chemical impacts of dredged material disposal in these waters, coastal States (e.g., California, Delaware, Maryland, North Carolina, Florida, and others) have increasingly restricted waste disposal during the last decade. For example, Maryland does not allow dredged material from Baltimore Harbor to be disposed of in its marine waters without special management, even though some of the material (that generated by any new work, as opposed to maintenance work) would not be contaminated (442).

Despite this trend, two-thirds of all marine dredged material disposal still takes place in estuaries. One reason for this is that most of the material is dredged from these waters. Another is that disposal in estuaries generally is less costly than disposal in upland areas or in waters more distant from shore. Third, according to COE, Federal regulatory policies have made it easier to obtain permits for disposal in estuaries than in coastal and open ocean waters. COE considers the testing requirements under CWA for disposal in estuaries and rivers (or for pipeline discharge of dredged material within the territorial sea) to be less stringent than MPRSA requirements for dumping in coastal and open ocean waters.[8]

Policies about disposal of wastes in estuaries and coastal waters are influenced by policies about disposal in open ocean waters and on land. If a policy to provide greater protection to estuaries and coastal waters is pursued, it may become difficult to decide where else to put dredged material. As

discussed, disposal in the open ocean already appears to be regulated more stringently than in other marine environments. In addition, open ocean disposal usually increases transportation costs significantly and it may not be practical in most situations.

Although the regulatory requirements for upland disposal are less comprehensive (and probably less stringent) than requirements for disposal in aquatic environments, finding new upland containment areas can be difficult and costly. Disposal in upland containment areas is controlled indirectly through Section 404 of CWA and State coastal zone management programs, and some States have imposed standards on the effluent discharged from upland containment areas. Upland disposal may also be subject to State or local land-use requirements. Finally, a shortage of upland containment areas is developing as available areas reach their designed capacity and as concerns about the effects of upland disposal increase.

At the same time, it is not clear that dredged material disposal should be prohibited totally in estuaries and coastal waters. Disposal in some instances causes only short-term and reversible impacts, especially when the volume of material is small, dumping does not occur frequently or continuously, and the dredged material is not a significant source of pollutants. In addition, COE contends that dredged material should receive less stringent regulatory treatment than other wastes because comparable concentrations of pollutants in dredged material may be less available to organisms than the same pollutants in other wastes. COE also notes that dredged material that is disposed of in estuaries usually is not ''added'' to estuaries, but simply moved from one location to another.

In general, finding a disposal site for dredged material in *any* environment is becoming increasingly difficult, partly because of public attitudes. Sometimes public attitudes reflect real uncertainties about the impacts of dredged material disposal. For example, it is true that the long-term impacts of disposal in marine waters and on land are not well understood. Similarly, the relative importance of different sources of pollutants are also uncertain and in some cases may never be resolved. (For instance,

---

[8]It is unclear whether the decrease in material disposed of in coastal and open ocean waters that occurred prior to 1982 (other than in the Mississippi River area) reflects differences in regulatory requirements or simply normal fluctuations in dredging operations. Similarly, it is unclear whether the increase beginning in 1982 reflects an easing of the requirements for ocean disposal permits, perhaps in response to the Federal District Court decision that overturned a ban on the disposal of sewage sludge in the ocean (ch. 7).

despite millions of dollars of research, the importance of dredged material relative to other pollutant sources in the New York Bight is still unclear.) Public attitudes, however, also can reflect a lack of awareness of the facts. For example, many people think that dredged material from the Mud Dump site in the New York Bight adversely affected New Jersey beaches (441); sediment transport studies, however, show that less than 0.1 percent of the sediment near these beaches is derived from dredged material disposed of at the site (709).

Siting decisions can also be affected by questions about the credibility of COE statements regarding the impacts of dredged material disposal. These questions arise in part because COE both regulates dredging activities and conducts many of them, and because it traditionally has been managed as a development agency. In addition, it does not need Section 404 or Section 103 permits to conduct its own federally authorized dredging projects, although it must comply with all applicable Federal laws and regulations and obtain appropriate State permits. COE also has sponsored most of the technical research on the impacts of marine and land-based disposal, having spent over $100 million since the early 1970s. To avoid bias, however, much of the research has been conducted and/or reviewed by non-COE groups. Recent State-supported research has tended to support the findings of COE-sponsored research.

Since disposal sites are becoming increasingly difficult to develop and use, one management approach is to minimize the amount of required dredging, and thus the impacts generated, by developing long-term management plans for dredged material disposal. Revised regulations proposed by COE (for 33 CFR Part 337.9) recognize the value of such plans and suggest their use (51 FR 104: 19693-19706, May 30, 1986). In addition, since most marine disposal consists of material dredged from ports and harbors, it makes sense to link dredging plans with broader management plans for surrounding estuaries and coastal waters. Several examples of such planning efforts exist. The Grays Harbor (Washington) Estuary Management Plan provides a blueprint for the port's future dredging and development (441). In the Chesapeake Bay, EPA is conducting a $27 million study of water quality and resources, including all different causes

of degradation, and as a part of this effort COE's Baltimore District is studying the effects of dredging on the Bay (441). Finally, the Puget Sound Water Quality Authority was formed by the State of Washington to provide a broad framework for cleaning up Puget Sound, and as part of this initiative COE's Seattle District is developing a management plan for dredged material disposal (441).

Long-term management plans for dredged material disposal have several advantages. First, they can be used to guide future decisions about port development while protecting the long-term productivity of an estuary. Such plans can consider the physical, biological, aesthetic, social, and economic resources associated with each proposed project, as well as interrelationships among proposed projects (61,273,496). Second, management plans can provide consistency and predictability for regulators, developers, environmentalists, and all State and Federal agencies. To ensure broad consensus, the planning process can involve all relevant State and Federal agencies, as well as local special interest groups with broad public representation. Finally, long-term plans could help streamline the regulatory review process, thereby reducing the time required to approve future projects and allowing more efficient scheduling of dredging and disposal operations.

Long-term planning efforts, however, are not without potential drawbacks. First, they are initially time-consuming and expensive. Second, unless these plans are incorporated into long-term compliance documents or permits, they may be subject to shifting agency policies. Finally, planning can be difficult because the long-term need for disposal sites is often hard to predict, given uncertainties about future port development.

In addition to developing long-term management plans where feasible, it will be important to conduct more peer-reviewed research about long-term impacts. Both laboratory and long-term field studies need to continue addressing several areas, including: how to assess sediment contamination; the long-term effects of bioaccumulation on individual organisms and community structure; the quality of effluent water discharged from upland containment areas; and procedures to minimize adverse impacts from the dredging and disposal of contaminated sediment in different environments (296). To increase the usefulness of research results to Federal and State decisionmakers, COE and EPA could jointly summarize and periodically update the state-of-knowledge on dredging and disposal operations. Short, readable publications would also help to explain to the public the necessity for, and the impacts associated with, dredging and disposal operations (e.g., see ref. 563).

Providing Balanced Consideration of All Disposal Options

As discussed in chapter 1, specific decisions or general policies about the disposal of any material should be based on a comprehensive evaluation of all available options. Evaluations are becoming more important, especially for large dredging projects or ones that involve contaminated material, as disposal sites become scarcer. One example of this type of evaluation is the Dredged Material Disposal Management Plan, prepared by the Port of New York and New Jersey. The plan compares eight disposal alternatives that might be used when the Mud Dump site in the New York Bight is filled to capacity in 15 to 20 years (562).

Regulations proposed by COE in May 1986 would provide general guidelines to ensure that all dredging and disposal alternatives are compared comprehensively by all COE districts. Such comparisons might indicate, for example, that dredged material disposal in the open ocean is more favorable in a specific situation than disposal in estuaries, coastal waters, or on land, even though regulatory requirements appear to be more stringent for the open ocean.

COE has developed a technical framework, with laboratory testing procedures, for evaluating the potential impacts of different disposal options (169). This framework is being tested on contaminated sediment from Commencement Bay, Washington (438). A comprehensive implementation manual describing the different laboratory procedures, however, does not yet exist. This kind of manual, which would need periodic updating to incorporate newly developed techniques, could provide guidance so planners could routinely evaluate the potential impacts of waste disposal. However, it also could impose detailed and costly analyses all disposal operations, regardless of their size or characteristics. The development of national sediment quality criteria also would aid the evaluation of potential impacts.

Chapter 11
Managing Industrial Wastes Dumped in Marine Waters

CONTENTS

Chapter 11
Managing Industrial Wastes Dumped in Marine Waters

INTRODUCTION

Marine disposal of industrial waste involves two primary modes of delivering wastes to marine waters. *Dumping* typically involves the use of barges to deliver industrial sludges and slurries directly to surface waters at designated marine dumping sites. In general, dumping of such wastes is not as significant in causing environmental impacts as are the far greater quantities of industrial waste directly *discharged* through pipelines or outfalls into marine waters within a short distance of the coastline (ch. 8).

Marine dumping of industrial wastes has been greatly reduced in the United States in the last decade, with respect to both number of permittees and quantities of waste. Prior to 1973, over 300 firms used marine waters for dumping; by 1979 the number had fallen to 13 (6), and currently only 3 firms are dumping wastes in marine waters (139). The quantity of dumped industrial wastes has steadily declined from 4.6 million metric tons in 1973 to about 200,000 metric tons in 1985 (figure 36).

Numerous sites have been used by the United States for marine dumping of industrial wastes, but only three have received significant amounts since 1977. These sites, which are administered by EPA Region II, are: 1) the New York Bight Acid Waste Disposal Site, 2) Deepwater Industrial Waste Disposal Site, and 3) the Pharmaceutical Waste Site off Puerto Rico. Only the first two are in current use (see figure 3 in ch. 3), receiving waste from three firms (139). Use of the Puerto Rico site was discontinued in 1981 (594).

Industrial waste disposal planning involves considerable capital investment, and the decreasing availability of marine waters as a viable disposal option caused many firms to make long-term investments in land-based disposal or treatment options (140). However, some previously attractive land-based options are now subject to much stricter

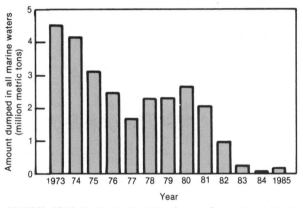

Figure 36.—Quantities of Industrial Wastes Dumped in Marine Waters, 1973-85

SOURCES: EG&G Washington Analytical Services Center, Inc., *Industrial Waste Disposal in Marine Environments*, contract prepared for U.S. Congress, Office of Technology Assessment (Waltham, MA: 1986); R. Schwer, E.I. du Pont de Nemours & Co., personal communication, November 1986; L. Mattioli, Allied Chemical, personal communication, December 1986; R. DeCesare, Office of Water, U.S. Environmental Protection Agency, personal communication, January 1987.

regulation and this may increase pressure to consider marine dumping. Indeed, for particular waste types in some regions of the country and for new generators, the marine option may be very attractive. It is difficult to predict the extent of future pressures to use marine environments for dumping or to gauge what effect a change in regulations might have on marine dumping.

This chapter discusses the marine dumping of drilling fluids, acid and alkaline wastes, pharmaceutical wastes, fish processing wastes, and coal ash and flue gas desulfurization (FGD) sludges.[1] For each waste type considered, the following topics are covered wherever data are available: waste composition; quantities generated and marine-disposed;

---

[1]Waste effluents and sludges resulting from the treatment of industrial process wastewaters prior to direct discharge into navigable waters or into sewerage systems are considered in ch. 8.

management and disposal practices currently used; the fate, availability, and impacts of waste constituents or contaminants; and the current regulatory framework. Table 28 provides a summary of current management practices for the various wastes considered in this chapter. Table 29 summarizes the regulatory framework governing their marine disposal.

Table 28.—Current Management Practices for Industrial Wastes

Waste type	Drilling fluids	Acid/alkaline	Pharmaceuticals	Fish processing	Coal ash/sludge
Marine pipeline	1[a]		X	X	
Marine dumping	2	3		2	
Land disposal[b]	X	2		X	1
Land application				X	
Physical, chemical, or biological treatment		1	1	X	
Incineration			X		
Recycling/reuse.............				1	2

[a]Numbers indicate relative prevalence of use of an option; X indicates an option used to a lesser extent than the numbered options, but to an unknown extent relative to the other options.
[b]Land disposal includes landfilling, surface impoundment, and deep-well injection.
SOURCE: Office of Technology Assessment, 1987.

Table 29.—Regulatory Framework for Marine Disposal of Industrial Wastes

Waste type	Statute	Agency	Program or regulations
Drilling muds	CWA	EPA	NPDES
	CWA	EPA	New Source Performance Standards
	Outer Continental Shelf (OCS) Lands Act	MMS	OCS
Acid/alkaline	MPRSA	EPA	Ocean dumping regulations
Pharmaceutical..............	MPRSA (dumping)	EPA	Ocean dumping regulations
	CWA (pipeline)	EPA	NPDES
	CWA (sewer)	EPA	Pretreatment regulations
Fish processing	MPRSA	EPA	Ocean dumping regulations
Coal ash/FGD sludge	MPRSA	EPA	Ocean dumping regulations

KEY: CWA = Clean Water Act
   MPRSA = Marine Protection, Research, and Sanctuaries Act
   MMS = Minerals Management Service, U.S. Department of the Interior
NPDES = National Pollutant Discharge Elimination System
FGD = flue gas desulfurization.
SOURCE: Office of Technology Assessment, 1987.

DRILLING FLUIDS

Composition

The discharge of spent drilling fluids accompanies the exploration and development phases, but not the production phase, of offshore oil and gas activities. Hundreds of compounds are used in drilling fluids formulations, depending on the particular needs of each well. However, four materials account for over 90 percent of the mass of essentially all drilling fluids: barium sulfate (or barite), clays, lignosulfates, and lignites (50 FR 34592, Aug. 26, 1985).

Drilling fluids can be water- or oil-based. Water-based fluids are more commonly used for offshore operations, for both regulatory and technical reasons. EPA's chemical analysis of 8 generic water-based fluids detected no priority pollutant organic chemicals, but 10 priority pollutant metals were detected. In particular, mercury and cadmium were found in all formulations tested (50 FR 34592, Aug. 26, 1985; ref. 384).

---

[2]Drilling fluids are one of several types of waste created by offshore drilling. Others include brine and sand brought up along with oil or gas; drill cuttings (the solids resulting from drilling); well treatment wastes (resulting from operations to enhance oil or gas recovery); and deck drainage and sanitary wastes. Most regulatory attention has focused on drilling fluids, and thus the discussion here is limited to this waste type. Although drilling fluids are often discharged through pipelines, the offshore location of such discharges justifies their consideration here with other dumping activities.

Quantities of Waste Generated and Marine-Disposed

About 2 million dry metric tons of drilling fluid components are used annually in offshore drilling activities and discharged directly into marine waters. About 3,900 offshore platforms currently produce oil and gas, accounting for roughly 20 percent of all domestic production (50 FR 34592, Aug. 26, 1985). Almost all (98 percent) such operations are located in the Gulf of Mexico, and over 90 percent of drilling fluid discharges occur there (384); however, increasing exploration in the waters of southern California and Alaska (the only other major sites for drilling fluid discharges) is expected to alter this distribution. EPA estimates that about 800 new platforms will be built between 1986 and 2000, a rate greatly reduced from that between 1972 to 1982, when an average of 1,100 new wells were drilled each year.

Management and Disposal Practices

Under existing regulations, used *oil-based* drilling fluids (or water-based fluids significantly contaminated with oil) are prohibited from marine discharge. Such fluids must be transported to land for disposal in a facility permitted under the Resource Recovery and Conservation Act or for reconditioning and reuse (40 CFR 435, Subpart A). In contrast, used *water-based* drilling fluids typically are dumped overboard or discharged from a pipe; both of these practices require a National Pollution Discharge Elimination System (NPDES) permit. Most coastal EPA Regions have prohibited direct offshore discharge near certain drilling sites because of ecological sensitivity at the sites (482). In addition, some State laws require land disposal for spent fluids generated in coastal waters (384). Using barges to transport used drilling fluids to shore or to distant ocean sites can be expensive and logistically difficult.

Fate, Availability, and Impact of Waste Constituents

In most areas on the continental shelf, the majority of particles present in drilling fluids and cuttings settle rapidly to the seabed, generally within 1,000 meters of the point of discharge (384).

Further dispersion may occur because of bottom currents or tidal action. A plume of particles, however, remains in suspension and is subject to relatively more rapid dispersion and dilution.

The main environmental concerns related to marine disposal of drilling fluids include potential toxicity of various chemical additives or trace metals, increased turbidity in the water column, physical burial of benthic organisms or alteration of physical substrates available to these organisms, and possible long-term accumulation of metals in sediments and marine organisms. The primary acute effects appear to be physical and limited to the benthic environment.

Evidence indicates that most water-based drilling fluids are relatively nontoxic to marine organisms at the concentrations that are achieved shortly after discharge (384). For those fluids exhibiting significant toxicity, it appears to be primarily attributable to the presence of diesel fuel, which can comprise as much as 2 to 4 percent of the total volume.

While most research has focused on acute effects, the concentrations of potentially toxic constituents present at most sites are in a range that is more likely to induce chronic or sublethal effects. Data are limited on such impacts, however, so considerable uncertainty remains regarding the long-term environmental impacts associated with marine disposal of drilling fluids.

Regulatory Framework

The principal authority to regulate marine disposal of drilling discharges lies with EPA, through the NPDES program of the Clean Water Act (Section 402). Such discharges are subject to the "best practicable control technology" (BPT) and "best available technology" (BAT) limitations of the Clean Water Act (ch. 7). Offshore oil and gas operations do not receive individual NPDES permits and instead are covered by a *general* NPDES permit (ch. 8), unless a facility requests its own permit. Prior to issuance of an individual NPDES permit, EPA must determine that the discharges will not unreasonably degrade the marine environment, in compliance with the Ocean Discharge Criteria of the Clean Water Act (40 CFR 125, Subpart M).

Using this regulatory authority, EPA has imposed a variety of permit conditions, including: limits on the amount of toxic substances or total hydrocarbons in drilling fluids, requirements to conduct toxicity testing of drilling fluid formulations prior to use, seasonal or zonal restrictions on discharges, and special monitoring and reporting requirements (482).

EPA recently proposed BAT effluent limitation guidelines and new source performance standards for offshore oil and gas facilities (50 FR 34592, Aug. 26, 1985). In addition to maintaining the prohibitions against the discharge of oil-based drilling fluids and water-based fluids containing oil, EPA proposed two further controls: a limit on the acute toxicity of drilling fluid discharges and a limit on the discharge of cadmium and mercury to a maximum of 1 part per million (ppm).

A recently renewed general permit covering drilling operations in the Gulf of Mexico (51 FR 24897, July 9, 1986) incorporates the proposed BAT limits, but not the proposed limit for mercury and cadmium. General permits incorporating the BAT limits on these two metals have been issued for Alaska (51 FR 35460, Oct. 3, 1986) and proposed for southern California (50 FR 34036, Aug. 22, 1985).

The Minerals Management Service of the Department of the Interior regulates drilling discharges through lease stipulations and Outer Continental Shelf operating orders, some of which prohibit the use of certain additives. In addition, individual States may impose further requirements on drilling discharges for operations taking place in their territorial waters.

While drilling fluids are not classified as hazardous, their disposal on land is regulated under the solid waste provisions of RCRA.

ACID AND ALKALINE WASTES

Quantities of Waste Generated

Most liquid acid and alkaline wastes are classified generically as corrosive wastes, a RCRA category that also includes sludges and solids. Corrosive wastes comprise almost half of the total hazardous waste generated in the United States (140,690), but it is not known what fraction of these are relatively uncontaminated acids and alkalis with potential for marine disposal.

Management and Disposal Practices

Most corrosive wastes are disposed or treated on-site, using methods such as deep-well injection and neutralization. Only 1 to 2 percent of corrosive wastes are disposed off-site (140), including the acid and alkaline wastes that are currently dumped in marine waters. When disposed of in marine waters, these wastes are barged to the disposal site and then dumped in bulk at a permitted rate into the wake of the vessel.

Composition and Quantities of Waste Disposed of in Marine Waters

Acid and alkaline wastes from three industrial firms are presently dumped in marine waters (table 30).[3] Under current permit schedules, about 200,000 metric tons from Allied Chemical, DuPont-Edge Moor, and DuPont-Grasselli will continue to be dumped annually after 1986 (139).

Quantities of waste dumped at the New York Bight Acid Waste Site by Allied Chemical have decreased from a high of 60,000 metric tons in 1973 to the current level of about 40,000 metric tons annually; no change is anticipated in the amount of waste to be dumped during the next several years (L. Mattioli, Allied Chemical, pers. comm., Dec. 1986). The current permit expires September 30, 1988; application for renewal of the permit is due

---

[3]Unless otherwise noted the following discussion is based on material from references 139 and 140.

Table 30.—Origin and Quantities of Acid and Alkaline Wastes Currently Dumped in Marine Waters

Company	Type of waste/process	Composition	Dumpsite	Annual quantity[a]
Allied Chemical	Hydrochloric acid from fluorocarbon refrigerants and polymer manufacturing[b]	About 30% HCl 1 to 2.5% fluoride Suspended solids and total organic carbon in 10s ppm range Petroleum hydrocarbons in 1-10 ppm range Chromium, nickel, zinc in <0.1 to 3 ppm range Arsenic, cadmium, copper, lead, mercury in <0.01 to 1 ppm range pH <1.0	New York Bight Acid Waste Site	30,000 mt
Du Pont-Edge Moor	Iron and other acidic metal chlorides from titanium dioxide production	Chromium in 100s ppm range Zinc and lead in 10s ppm range Copper and nickel in 1-10 ppm range Cadmium in 0.001 ppm range pH 0.1 to 1.0	Deepwater Industrial Waste Site	≤50,000 mt
Du Pont-Grasselli	Sodium sulfate from agricultural chemical production	10% sodium sulfate Low molecular weight organics in 10s-1000s ppm range Chromium, copper, nickel, lead in 0.01 to 0.1 ppm range Cadmium in 0.001 ppm range pH 10 to 12.5	Deepwater Industrial Waste Site	110,000 mt

[a]Based on present permit limits.
[b]The fluorocarbon polymer manufacturing facility was recently sold to Ausimont U.S.A., Inc. (L. Mattioli, Allied Chemical, personal communication, December 1986).

SOURCES: R. Schwer, E.I. du Pont de Nemours & Co., personal communication, November 1986; L. Mattioli, Allied Chemical, personal communication, December 1986; and EG&G Washington Analytical Services Center, Inc., *Industrial Waste Disposal in Marine Environments*, prepared for U.S. Congress, Office of Technology Assessment (Waltham, MA: 1986).

by April 3, 1988. The only alternative to marine disposal currently being used is to sell the waste for use as hydrochloric acid; about 10 percent of the waste was sold in 1984 and about 6 percent was sold in 1985 (L. Mattioli, Allied Chemical, pers. comm., Dec. 1986).

DuPont-Edge Moor has dumped acid wastes at the Deepwater Industrial Waste Site since 1968. Since 1973, its permits have contained provisions for the cessation of ocean dumping and the development of feasible alternatives. By 1984, such changes had reduced ocean dumping by 70 percent, to less than 115,000 wet metric tons. The current permit expires June 30, 1987, but it can be renewed provided that, despite good-faith efforts, DuPont-Edge Moor has been unable to develop sufficient land-based alternatives to completely replace marine disposal; an application for a new 3-year permit is being submitted on this basis (R. Schwer, E.I. du Pont de Nemours & Co., pers. comm., Nov., 1986).

DuPont-Grasselli has dumped alkaline wastes from 1968 to 1973 at the New York Bight Acid Waste Site and from 1968 to the present at the Deepwater Industrial Waste Site. Between 1973 and 1983, amounts dumped ranged from 118,000 to 290,000 wet metric tons per year. The last permit, valid through the end of 1986, required the development of alternative treatment methods, but contained no deadline for halting dumping.

Fate, Availability, and Impact of Waste Constituents

The rationale for allowing marine disposal of acid and alkaline wastes is that they rapidly (within 1 to 4 hours) neutralize after coming into contact with seawater, which has a high natural buffering capacity (157). In addition, discharge into the turbulent wake of the vessel provides rapid mixing and a several thousand-fold dilution. Acute impacts due to transient changes in acidity may occur prior to neutralization. Such effects, however, would generally be limited to the immediate vicinity of the discharge, and past monitoring has not detected any observable effects on marine life (83,157). Trace pollutants, primarily toxic metals, may pose longer term risks, although such pollutants are generally rapidly and extensively diluted under typical disposal conditions. Some metals (primarily iron and magnesium) precipitate upon entering marine waters and can remain in suspension for several days or more.

Regulatory Framework

Most acid and alkaline wastes are generically classified under RCRA as hazardous due to their corrosivity. However, wastewater treatment sludges from the DuPont-Edge Moor facility, which are usually disposed of by landfilling or ocean dumping, have been delisted and excluded from regulation as a hazardous waste (45 FR 72037, Oct. 30, 1980), and wastewater from the DuPont-Grasselli facility is not considered hazardous because it does not exceed the upper pH limit specified for corrosivity by RCRA (R. Schwer, E.I. du Pont de Nemours & Co., pers. comm., Nov. 1986). In all cases, however, dumping of acid and alkaline wastes in marine waters is allowed only after a case-by-case determination of compliance with the Ocean Dumping Criteria of the Marine Protection, Research, and Sanctuaries Act (MPRSA). Permits issued by EPA under authority of MPRSA do not contain requirements for specific treatment of acid and alkaline wastes that are to be dumped; rather they require that dilution to background levels be achieved in a specified length of time.

PHARMACEUTICAL WASTES

Composition

Pharmaceutical wastes originate from the biological production of antibiotics and the chemical production of pharmaceuticals. They are typically aqueous suspensions near neutral pH; have densities similar to seawater; and contain about 1 percent suspended solids, 2 percent organic carbon, and very low (<1 ppm) levels of metals or high molecular weight organochlorines (relative to other industrial wastes). However, they can contain high concentrations (10 to 100 ppm) of any of several common industrial solvents (139,594).

Management and Disposal Practices

Most U.S. pharmaceutical companies responded to EPA regulations by investing in land-based disposal, including onsite incineration and secondary wastewater treatment (e.g., anaerobic digestion, activated sludge treatment). Marine dumping generally is only attractive to those pharmaceutical companies not yet having secondary treatment (138).

Quantities of Waste Marine-Disposed

Seven pharmaceutical companies in Puerto Rico dumped pharmaceutical wastes at the designated Pharmaceutical Waste Site (74 km north of Puerto Rico) from at least 1973 until 1981. The amounts dumped increased from 38,000 wet metric tons in 1973 to over 300,000 metric tons in the late 1970s and early 1980s (594).

Because of lower disposal costs and EPA mandates, these wastes are now discharged into a secondary sewage treatment plant completed at Barceloneta in 1981 (Black and Veatch, cited in ref. 594). Almost half of the wastewater entering the plant is pharmaceutical wastes. Effluent from the plant is not disinfected and is discharged through a pipeline 800 meters offshore, in waters less than 30 meters deep (208). The volume of pharmaceutical wastes entering the plant is about fourfold higher than that previously marine dumped, but is about tenfold less concentrated in suspended solids, organic carbon, and nitrogen (594).

Fate, Availability, and Impact of Waste Constituents

The dispersive high-energy conditions at the Puerto Rico dumpsite diluted dumped wastes by a factor of up to 100,000-fold soon after dumping and by a factor of about 10,000,000-fold over a long-term period (594). However, one clearly demonstrated change at the dumpsite was the almost complete replacement, within 7 years after disposal began, of resident bacteria with other bacterial species (including several human pathogens—e.g., *Salmonella and Vibrio*) (207). Some experts attribute the change to a selection for the other species able to degrade particular pharmaceutical waste components, but this conclusion has been questioned (594). Shifts also occurred in the composition and size of the phytoplankton community in the vicinity of the dumpsite (594).

Other potentially more serious impacts have been attributed to the discharge of pharmaceutical wastes into the Barceloneta secondary treatment plant. These include frequent disruption of the treatment process, reduced removal of bacterial pathogens prior to effluent discharge, and significantly elevated levels of fecal bacteria (including human pathogens) in coastal waters (208). Because discharge

now occurs close to shore in shallow waters, concerns have been raised that human health impacts under the current disposal system may be greater than those associated with the previous ocean dumping of untreated pharmaceutical waste. In particular, currents and wave action in the area of discharge have been demonstrated to carry waste constituents back to shore (208,594).

Regulatory Framework

Dumping of pharmaceutical wastes at the Puerto Rico dumpsite was regulated under the Ocean Dumping Criteria of MPRSA. The present discharge of effluent from the Barceloneta treatment plant is regulated under an NPDES permit.

FISH PROCESSING WASTES

Composition and Quantities of Waste Generated

These wastes arise from the processing of seafood for a variety of products.[4] The large tuna and fishmeal industries engage in year-round operations; most other food waste generators are small, seasonal, specialized facilities. Prior to treatment, these wastes are composed entirely of organic matter and "conventional" pollutants: oil, grease, and solids (139). In 1980, 1.4 to 2.0 million metric tons of seafood processing material was produced, an increase of 35 to 41 percent since 1970 (138).

Management and Disposal Practices

Most seafood processing material is converted to marketable meal and fertilizer or is recycled (140). The remaining waste can be treated prior to NPDES-permitted discharge, or directly disposed of by landfilling, land application, or marine dumping. Data on the relative use of these options are not available.

Primary treatment using dissolved air flotation (DAF) systems is typically employed by large proc-

essing facilities (140). These systems use coagulants to remove solids from wastewater, thereby generating DAF sludges which are not exempt from ocean dumping regulations. The small quantity of DAF sludge currently generated is disposed of in landfills.

EPA is considering requiring DAF systems for virtually all seafood processing waste generators. If applied throughout the industry, about 2000 metric tons of DAF sludge would be produced annually (140).

Quantities of Waste Marine-Disposed

The only site designated for dumping of fish processing wastes is in the Pacific Ocean near American Samoa (45 FR 77435, Nov. 24, 1980; ref. 86). This interim site, which is administered by EPA Region IX, was approved for up to 118,000 wet metric tons per year and for a period of 3 years, pending completion of further studies. Actual quantities of waste dumped at the site were 17,000 wet metric tons in 1982 and 19,500 wet metric tons in 1983 (648).

Region IX expects an increase in requests to dump seafood processing wastes in marine waters, primarily DAF treatment sludge from tuna canneries (140). EPA has proposed to designate a dump-

---

[4]These wastes are distinguished from unprocessed wastes arising from seafood cleaning, which are exempted from ocean dumping regulations (40 CFR 220.1(c)(1)).

site in the Southern California Bight for cannery wastes generated by Star-Kist Foods, Inc. (102).

Fate, Availability, and Impact of Waste Constituents

Disposal of these wastes in poorly mixed or relatively enclosed coastal marine environments could cause nutrient overloading. Proper disposal in well-mixed ocean environments appears to be essentially nonhazardous, although some concerns have been raised over potential impacts of the chemical coagulants introduced during DAF treatment (140).

COAL ASH AND FLUE GAS DESULFURIZATION SLUDGES

Composition

Coal ash is the incombustible, inorganic, or mineral fraction of coal remaining after its combustion in industrial and public-owned boilers. It includes *fly ash* captured by stack scrubbers and *bottom ash/slag* that is left behind in the boiler. FGD sludges are produced when sulfur-containing flue gases react with air pollution control scrubber reagents (usually calcium carbonate or limestone).

The chemical composition of coal ash varies considerably with the type of coal from which it is derived (288). Primary constituents include the salts and oxides of silicon, aluminum, iron, calcium, magnesium, sodium, potassium, titanium, and sulfur. Other significant trace (<1 percent) constituents include barium, strontium, manganese, and boron. Coal ash leachate has a pH between 6 to 11.5. The composition and quantity of FGD sludge depends on the type of air pollution controls that are employed. It typically contains 5 to 15 percent solids initially and 30 to 80 percent solids after dewatering, and has a pH of 3 to 13 (139,141). Primary components are calcium sulfites, sulfates, and carbonate; major trace elements include barium, boron, copper, fluorine, manganese, molybdenum, nickel, and zinc (288).

Currently, coal ash and FGD sludge from electric utilities are characterized as nonhazardous under RCRA (40 CFR 261.4(b)(4)). However, EPA has proposed new procedures for testing the toxicity of leachates from wastes, which could lead to the classification of some utility wastes as hazardous. The original exemption of such wastes from regulation as hazardous wastes was intended to allow EPA to complete environmental impact studies and determine a course of action (141).

Quantities of Waste Generated

Estimates of the amount of coal ash and FGD sludge generated annually in the United States vary considerably, but they are clearly very high-volume wastes that will continue to increase in quantity at least through the end of this century (table 31). Coal-burning utilities are estimated to produce about 95 percent of total utility ash; the remainder arises from burning of other fossil fuels.[5] Rapid increases in the generation of these wastes are occurring, due to wider application and greater efficiency of sulfur dioxide removal technology, as well as increased coal consumption.[6]

The East Coast States, from Pennsylvania south, and the Gulf States of Florida, Alabama, and Louisiana, are the major sources of utility coal ash and FGD sludge. These States are also expected to exhibit the largest increase in waste generation through 2000 (141; Tobin, 1982, cited in ref. 140).

Management and Disposal Practices[7]

Most utility coal waste is presently disposed of onsite, in unlined landfills and impoundments (table 31); however, the use of unlined impoundments is declining because of concerns about ground-

---

[5]Nonutility industries generate much larger quantities of air pollution control dusts and sludges (257). Many of these are listed as hazardous under RCRA, however, and have not been discussed as candidates for dumping. Nonutility boilers generate an additional 8 million metric tons of coal wastes (257), similar in composition to that generated by utilities.

[6]Fly ash and FGD sludge together constitute 5 to 15 percent of the mass of coal from which they are generated. A typical, 1000-megawatt plant will consume 2.4 million metric tons of coal annually, and generate about 650,000 metric tons of ash and sludge (131).

[7]This discussion is limited to practices used for utility coal waste. Data on the disposition of coal wastes from nonutility industrial boilers are scant (257).

Table 31.—Quantities and Current Management of Utility Coal Ash and FGD Sludge

	Coal ash	FGD sludge	Reference[a]
Annual quantities: [b]			
1985	85 mmt	18 mmt	USWAG, 1982; EG&G, 1986
2000	155 mmt	52 mmt	
Current management practices			
Percent disposed onsite	75%	82%	JRB, 1983
Percent disposed offsite	16%	18%	
Percent sold	9%	—	
Percent in landfills............	40%	30%	EG&G, 1983; EPRI, 1985
Percent in wet ponds	39%	67%	
Percent in mines	—	2%	
Percent used commercially	21%	<1%	

[a]See list of references at end of report.
[b]Quantities are in million metric tons (wet weight).

SOURCE: Office of Technology Assessment, 1987.

water contamination and land reclamation (141). These wastes are distinguished from unprocessed wastes arising from seafood cleaning, which are exempted from ocean dumping regulations (40 CFR 220.1(c)(1)).

Some 10 to 20 percent of utility coal wastes are currently recycled or reused in cement manufacture, building materials, road surfacing, sand blasting, roofing, and ice and snow control. While recycling and reuse will increase somewhat, it is not expected to keep up with increases in waste generation. By the year 2000, only an estimated 16 percent of utility coal ash and a much smaller fraction of FGD sludge will be recycled or reused (680). Thus, the vast majority of coal wastes will continue to require disposal.

In contrast to present practices, most future disposal—at least 80 percent—is expected to take place offsite because of insufficient onsite disposal capacity (Tobin, 1982, cited in ref. 140).

Potential for Marine Disposal

No coal ash or FGD sludge is currently disposed of in marine waters, except for research purposes. However, with land for disposal becoming increasingly scarce, and with accelerating coal conversion taking place in New England, the use of the ocean for such disposal has recently received attention: in particular, ConEdison of New York has requested permission to dump fly ash at the 106-mile deepwater dumpsite (488). In the absence of regulatory restrictions on the ocean disposal of coal ash, over one-fourth of the total ash generated in the

coastal regions (representing over one-tenth of the national total) might be economically disposed of in the ocean by 2000. Estimates for FGD sludge are even higher: about 40 percent of all the FGD sludge generated in coastal regions in 2000, representing over one-fourth of the national total (140). Estimated costs for ocean disposal are about $5 per metric ton, compared to about $16 per metric ton for landfilling (1982 dollars).[8]

Utility coal ash has been dumped in marine waters for many years by the United Kingdom (131,401). In the United States, two research projects involving dumping of consolidated coal ash in the New York Bight have been conducted (131), and it is possible that coal-waste blocks could be used to construct artificial reefs in both marine and freshwater environments (131,213).

Fate, Availability, and Impact of Waste Constituents

The principal problems associated with marine disposal of untreated coal ash or FGD sludge include dissolved oxygen depletion, increased turbidity, sulfite toxicity to marine organisms, smothering of benthic organisms, and release of metals. These concerns particularly limit the potential for disposing of FGD wastes in marine waters (131).

A promising potential option to address these problems involves the consolidation of a mixture

---

[8]This estimate, derived by ConEdison, includes tug and barge leasing and monitoring costs, and assumes that at least 500,000 metric tons per year are disposed at the Deepwater Industrial Waste Disposal Site (140).

of FGD sludge, fly ash, and lime into solid forms (433). The resulting material exhibits significantly decreased release rates for sulfite and metals, and resulting leachates show reduced toxicity to marine organisms. The consolidated material maintains structural integrity over periods of years in the marine environment, and may therefore be an appropriate material for building artificial reefs to provide substrates for colonization by marine organisms and enhance fisheries.

Appendixes

Appendix A
Acknowledgments

Contractors

Arthur D. Little, Inc.
Brookhaven National Laboratory
EG&G Washington Analytical Services Center, Inc.
Chris Elfring, *Editor*
Florida Institute of Technology
Jay Grimes, University of Maryland
Joseph O'Connor, New York University Medical Center
Willis Pequegnat
Resources For the Future
Science and Public Policy Program, University of Oklahoma
Science Applications International Corp.

OTA Services

The unsung heroes of every OTA assessment:

* Information Center
* Publishing Office
* Service Center

Reviewers and Other Contributors

Many individuals and organizations contributed valuable information, guidance, and substantive reviews of draft material. OTA wishes to extend its gratitude to the following for their help (and apologizes to any contributors inadvertently omitted from this list):

Talibah Adisa, Dade County Department of Environmental Resources Management
Ken Adler, Environmental Protection Agency
Lorenzo Alonzo, Houston Public Works Department
Blake Anderson, County Sanitation Districts of Orange County
Vyto Andreliunas, Corps of Engineers New England Division
Forest Arnold, National Oceanic and Atmospheric Administration
Judith Ayres, Environmental Protection Agency Region IX
James Bajek, Corps of Engineers New England Division
Dru Barrineau, Corps of Engineers Mobile District
Dan Basta, National Oceanic and Atmospheric Administration

Bob Bastian, Environmental Protection Agency
Warren Baxter, Corps of Engineers Seattle District
Ed Bender, Environmental Protection Agency
D.W. Bennett, American Littoral Society
Harvey Beverly, Corps of Engineers Los Angeles District
Rosina Bierbaum, Office of Technology Assessment
Allen Blume, Center for Environmental Education
Donald Boesch, Louisiana Universities Marine Consortium
Suzanne Bolton, Battelle Columbus Laboratories
Charles Bookman, National Academy of Sciences
James Borberg, Hampton Roads Sanitation District
Robert Borchardt, Houston Public Works Division
David Brailey, Environmental Defense Fund
William Brandt, Dade County Department of Environmental Resources Management
N.H. Brooks, California Institute of Technology
David Brown, Southern California Costal Water Research Project
Robert Burd, Environmental Protection Agency Region X
Jack Campbell, South Dade Soil and Water Conservation District
Judy Capuzzo, Woods Hole Oceanographic Institution
Charles Carry, Sanitation Districts of Los Angeles County
Mike Champ, Science Applications International Corp.
Frank Covington, Environmental Protection Agency Region IX
Jerome Cura, Goldberg-Zoino & Associates, Inc.
Clif Curtis, The Oceanic Society
Phil Dagatano, New York State Department of Environmental Conservation
Tudor Davies, Environmental Protection Agency
Jack Davis, Gulf Coast Waste Disposal Authority
Ron DeCesare, Environmental Protection Agency
Stephen C. Delaney, Environmental Protection Agency
D.A. Dennis, Corps of Engineers Sacramento District
Harriet Diamond, Massachusetts Coastal Zone Management
Iver Duedall, Florida Institute of Technology
Al Dufour, Environmental Protection Agency
Robert Dyer, Environmental Protection Agency
Charles Ehler, National Oceanic and Atmospheric Administration

Robert Engler, Waterways Experiment Station

Dan Farrow, National Oceanic and Atmospheric Administration

Robert Fergen, Miami-Dade Water and Sewer Authority

David Fierra, Environmental Protection Agency Region I

Larry Fink, Environmental Protection Agency Region V

David Fluharty, University of Washington

Rod Frederick, Environmental Protection Agency

Jim Gallup, Environmental Protection Agency

Charles Ganze, Gulf Coast Waste Disposal Authority

William Garber, Management Consultant

Mary Jo Garreis, Maryland Department of Health and Mental Hygiene

Jack Gentile, Environmental Protection Agency

Roger Gerth, Corps of Engineers Mobile District

Leonard Gianessi, Resources For the Future

Tim Goodger, National Marine Fisheries Service

Court Greenfield, Soil Conservation Service

Henry Gregory, Houston Public Works Department

Jay Grimes, University of Maryland

Philip Gschwend, Massachusetts Institute of Technology

LeeAnn Hammer, Environmental Protection Agency

Mary Hamrick, Science Applications International Corp.

Rixie Hardy, Corps of Engineers New Orleans District

Nan Harllee, Maritime Administration

Jake Harari, Corps of Engineers San Francisco District

Fred Harper, Management Consultant

Irwin Haydock, Sanitation Districts of Los Angeles County

Thomas Healey, Philadelphia Water Department

Doug Hilderbrand, Seattle METRO

Charles Hollister, Woods Hole Oceanographic Institution

John Hoornbeek, Environmental Protection Agency

Mercedes Hsu, California Regional Water Quality Control Board

Fran Irwin, Conservation Foundation

Robert Johns, Dade County Department of Environmental Resources Management

Dennis Johnson, Agency for International Development

Peter Johnson, Office of Technology Assessment

Galen Jones, University of New Hampshire

David Kern, Corps of Engineers Pacific Ocean Division

Ken Kirk, Association of Metropolitan Sewerage Agencies

Jay Kremer, Sanitation Districts of Los Angeles County

Braxton Kyzer, Corps of Engineers Charleston District

William Lahey, Palmer & Dodge

John Lampe, Seattle METRO

Ron Landy, National Oceanic and Atmospheric Administration

Dick Lee, Waterways Experiment Station

Tom Leschine, University of Washington

Leonard Levine, Gulf Coast Waste Disposal Authority

David Lim, Environmental Protection Agency Region I

Steve Lipman, Massachusetts Department of Environmental Quality Engineering

Carol Litchfield, E.I. du Pont de Nemours & Co., Inc.

John Lyon, Environmental Protection Agency

David Lyons, Environmental Protection Agency

Peter Machno, Seattle METRO

Gary Magnuson, Coastal States Organization

Donald Malins, National Marine Fisheries Service

Samual Maloof, Management Consultant

Bruce Mannheim, Environmental Defense Fund

Deborah Martin, Environmental Protection Agency

Dave Mathis, Corps of Engineers

Leon Mattioli, Allied Chemical Corp.

Gordon McFeters, Montana State University

Jeffrey McKee, Corps of Engineers Baltimore District

John McKern, Corps of Engineers Walla Walla District

Terrence McKiernan, Corps of Engineers North Pacific Division

Alan Mearns, National Oceanic and Atmospheric Administration

Richard Medina, Corps of Engineers Galveston District

Bob Miele, Sanitation Districts of Los Angeles County

Kathy Minsch, Environmental Protection Agency

Ralph Mitchell, Harvard University

Robert Mitchell, Resources For the Future

Joe Moran, Environmental Protection Agency

Robert Morton, Science Applications International Corp.

Donald Mount, Environmental Protection Agency

William Murden, Corps of Engineers

Leonard Nash, Environmental Protection Agency Region III

Joel O'Connor, National Oceanic and Atmospheric Administration

Tom O'Connor, National Oceanic and Atmospheric Administration

Tom O'Farrell, Environmental Protection Agency

Linda O'Leary, Port Authority of New York
Robert Ottesen, Corps of Engineers Jacksonville
  District
Henry Peskin, Resources For the Future
Carolyn Smith Pravlik, Terris & Sunderland
Andrew Robertson, National Oceanic and
  Atmospheric Administration
Robert Robichaud, Environmental Protection
  Agency Region X
Gilbert Rowe, Brookhaven National Laboratory
Marvin Rubin, Environmental Protection Agency
Thomas Schine, Corps of Engineers Philadelphia
  District
Lewis Schinazi, California Regional Water Quality
  Control Board
Moira Schoen, Environmental Protection Agency
Doug Segar, Aquatic Habitat Institute
Frank Senske, Philadelphia Water Department
Peter Shelley, Conservation Law Foundation
Diane Sheridan, League of Women Voters of Texas
Keith Silva, Environmental Protection Agency
  Region IX
Jerome Simpson, Corps of Engineers Portland
  District
Carl Sindermann, University of Miami
Kirvil Skinnarland, Puget Sound Water Quality
  Authority
John Spencer, Seattle METRO
Elisa Speranza, Massachusetts Water Resources
  Authority
Frank Steimle, National Oceanic and Atmospheric
  Administration
William Stelle, Jr., House Committee on Merchant
  Marine and Fisheries
Harold Stevenson, McNeese State University
Frank Sweeney, Science Applications International
  Corp.
Fern Summer, Water & Wastewater Equipment
  Manufacturers Association
Dennis Suszkowski, Environmental Protection
  Agency Region II
Wayne Sylvester, Sanitation Districts of Orange
  County
John Tavolaro, Corps of Engineers New York
  District
Nancy Taylor, Sierra Club Clean Coastal Waters
  Task Force
Joe Teller, Gulf Coast Waste Disposal Authority
Thomas Thornton, Environmental Protection
  Agency Region I

Peter Trick, Science Applications International
  Corp.
Glenn Unterburger, Environmental Protection
  Agency
Ron Vann, Corps of Engineers Norfolk District
John Waesche, Jardine Insurance Brokers, Inc.
Eric Washburn, Science Applications International
  Corp.
Thomas Walton, Philadelphia Water Department
Al Wastler, Environmental Protection Agency
Ginger Webster, Environmental Protection Agency
Judith Weis, Rutgers University
Bill Westermeyer, Office of Technology Assessment
Lee White, White, Fine, & Verville
Gene Whitehurst, Corps of Engineers Norfolk
  District
Richard Whitsel, California Regional Water Quality
  Control Board
Joe Wilson, Corps of Engineers
Douglas Wolfe, National Oceanic and Atmospheric
  Administration
Thomas Yourk, Corps of Engineers Savannah
  District
Chris Zarba, Environmental Protection Agency

Participants in Dredged Material Workshop, July 23, 1985

The following individuals participated in a workshop
held by OTA on the disposal of dredged material:

Bill Conner, National Oceanic and Atmospheric
  Administration
Robert Engler, Waterways Experiment Station
Dan Farrow, National Oceanic and Atmospheric
  Administration
Tim Goodger, National Marine Fisheries Service
Frank Hamons, Maryland Port Administration
David Hankla, Fish and Wildlife Service
George Hanson, Fish and Wildlife Service
Jim Houston, Waterways Experiment Station
Ken Kamlet, A.T. Kearney, Inc. (formerly with
  National Wildlife Federation)
Bob Morton, Science Applications International
  Corp.
Tom O'Connor, National Oceanic and Atmospheric
  Administration
Willis Pequegnat, Consultant
Rónald Vann, Waterways Experiment Station
Al Wastler, Environmental Protection Agency

Appendix B
Acronyms and Abbreviations

AO — Administrative Order
BAT — Best Available Technology (Economically Achievable)
BCT — Best Conventional Pollutant Control Technology
BMR — Baseline Monitoring Report
BOD — Biochemical Oxygen Demand
BPJ — Best Professional Judgment
BPT — Best Practicable Control Technology (Currently Available)
CBP — Chesapeake Bay Program
CEQ — Council on Environmental Quality
CERCLA — Comprehensive Environmental Response, Compensation, and Liability Act
CFR — Code of Federal Regulations
COE — U.S. Army Corps of Engineers
CSI — Compliance Sampling Inspection
CSO — Combined Sewer Overflow
CWA — Clean Water Act
CZMA — Coastal Zone Management Act
DAF — Dissolved Air Flotation
DDT — Dichlorodiphenyl Trichloroethane
DMR — Discharge Monitoring Report
DOJ — Department of Justice
EDF — Environmental Defense Fund
EIL — Environmental Impairment Liability
EIS — Environmental Impact Statement
EMS — Enforcement Management System
EP — Extraction Procedure
EPA — U.S. Environmental Protection Agency
FDA — Food and Drug Administration
FDF — Fundamentally Different Factor
FGD — Flue Gas Desulfurization
FR — Federal Register
FWS — Fish and Wildlife Service
GAO — General Accounting Office
GCWDA — Gulf Coast Waste Disposal Authority
LDC — London Dumping Convention
LLW — Low-level Radioactive Waste
MDSD — Monitoring and Data Support Division
MPRSA — Marine Protection, Research, and Sanctuaries Act
NACOA — National Advisory Committee on Oceans and Atmosphere
NEA — Nuclear Energy Agency
NEMP — Northeast Monitoring Program

NEP — National Estuary Program
NEPA — National Environmental Policy Act
NIMBY — Not In My Backyard
NMFS — National Marine Fisheries Service
NMPP — National Marine Pollution Program
NOAA — National Oceanic and Atmospheric Administration
NPDES — National Pollutant Discharge Elimination System
NRDC — Natural Resource Defense Council
NSPS — New Source Performance Standards
OAD — Ocean Assessments Division
OCPSF — Organic Chemicals, Plastics, and Synthetic Fibers
OMEP — Office of Marine and Estuarine Protection
PAH — Polycyclic Aromatic Hydrocarbon
PCB — Polychlorinated Biphenyl
PCS — Permit Compliance System
PFRP — Process to Further Reduce Pathogens
PIRT — Pretreatment Implementation Review Task Force
POTW — Publicly Owned Treatment Work
PSES — Pretreatment Standards for Existing Sources
PSNS — Pretreatment Standards for New Sources
PSRP — Process to Significantly Reduce Pathogens
PSWQA — Puget Sound Water Quality Authority
QNCR — Quarterly Noncompliance Report
RCRA — Resource Conservation and Recovery Act
RFF — Resources For the Future
SAB — Science Advisory Board
SASSR — Semi-Annual Statistical Summary Report
SAV — Submerged Aquatic Vegetation
SCCWRP — Southern California Coastal Water Research Project
SNC — Significant Noncompliance
SPMS — Strategic Planning and Management System
TBT — Tributyltin
TCDD — Tetrachlorodioxin
TCDF — Tetrachlorodibenzofuran
TCLP — Toxic Characteristic Leachate Procedure
TSCA — Toxic Substances Control Act
TSS — Total Suspended Solids
U.S.C. — United States Code
USGS — U.S. Geological Survey
VRAC — Violation Review Action Criteria

References

References

1. Adamkus, V., "Restoring the Great Lakes," *EPA Journal* 11(2):2-4, March 1985.
2. Alabama Department of Environmental Management, *Water Quality Report to Congress for Calendar Year 1984 & 1985* (Montgomery, AL: April 1986).
3. Almeida, E.G., "Coastal Systems Program of the Brazilian 1st Sectional Plan for Marine Resources," in *Proceedings of the International Symposium on Utilization of Coastal Ecosystems: Planning, Pollution and Productivity*, vol. 1, N.L. Chao and W. Kirby-Smith (eds.) (Beaufort, NC: Duke University Marine Laboratory, 1985), pp. 17-26.
4. Anderson, D.M., "Paralytic Shellfish Poisoning," in *Natural Toxins and Human Pathogens in the Marine Environment*, R.R. Colwell (ed.), UM-SG-TS-85-01 (College Park, MD: Maryland Sea Grant College, 1985).
5. Anderson, D.W., et al., "Brown Pelicans: Improved Reproduction Off the Southern California Coast," *Science* 190:806-808, Nov. 21, 1975.
6. Anderson, P.W., and Dewling, R.T., "Industrial Ocean Dumping in EPA Region II—Regulatory Aspects," in *Ocean Dumping of Industrial Wastes*, B.H. Ketchum, et al. (eds.) (New York: Plenum Press, 1981), pp. 25-38.
7. Anderson, T., "State of the Sound," *Long Island Sound Report* 3(2):1,8-9, fall 1986.
8. Associated Press, "Rayonier Worst Polluter Among Paper Mills," *Associated Press*, Fernandina Beach, FL, Bjt/460, Aug. 28, 1986.
9. Associated Press, "Shellfish," *Associated Press*, Georgia, Bjt/340, Oct. 27, 1986.
10. Associated Press, "NJ Brief," *Associated Press*, Avalon, NJ, Dec. 30, 1986.
11. Ayres, R.U., and Rod, S.R., "Patterns of Pollution in the Hudson-Raritan Basin," *Environment* 28(4):14-43, May 1986.
12. Bakalian, A., "Regulation and Control of United States Ocean Dumping: A Decade of Progress, An Appraisal for the Future," *Harvard Environmental Law Review* 8(1):193-256, 1984.
13. Barile, D. (ed.), *Proceedings of the Indian River Resources Symposium, The Indian River Lagoon*, Sea Grant Project No. IR 84-28 (Melbourne, FL: Marine Resources Council of East Central Florida, 1985).
14. Barnard, W.D., *Prediction and Control of Dredged Material Dispersion Around Dredging and Open-Water Pipeline Disposal Operations*, Technical Report DS-78-13 (Vicksburg, MS: U.S. Army Engineer Waterways Experiment Station, 1978).
15. Barnard, W.D., and Hand, T.D., *Treatment of Contaminated Dredged Material*, Technical Report DS-78-14 (Vicksburg, MS: U.S. Army Engineer Waterways Experiment Station, 1978).
16. Baross, J.A., Hanus, F.J., and Morita, R.Y., "Survival of Human Enteric and Other Sewage Microorganisms Under Simulated Deep-Sea Conditions," *Applied Microbiology* 30:309-318, 1975.
17. Bascom, W. (ed.), *The Effects of Waste Disposal on Kelp Communities* (La Jolla, CA: University of California Institute of Marine Resources, 1983).
18. Bascom, W. (ed.), *Coastal Water Research Project Biennial Report, 1983-1984* (Long Beach, CA: Southern California Coastal Water Research Project, 1984).
19. Bascom, W. (ed.), "Director's Statement," in *Southern California Coastal Water Research Project Biennel Report, 1983-1984* (Long Beach, CA: Southern California Coastal Water Research Project, 1984), pp. 1-3.
20. Bascom, W. (ed.), "The Concept of Assimilative Capacity," in *Southern California Coastal Water Research Project Biennial Report, 1983-1984* (Long Beach, CA: Southern California Coastal Water Research Project, 1984), pp. 171-177.
21. Battelle Human Affairs Research Center, *Institutional Options for Improved Water Quality Management: Policy Direction*, prepared for U.S. Environmental Protection Agency Region X, EPA 901/9-83-116a (Seattle, WA: May 16, 1986).
22. Bean, M., "United States and International Authorities Applicable to Entanglement of Marine Mammals and Other Organisms in Lost or Discarded Fishing Gear and Other Debris, A Report to the Mammal Commission" (Springfield, VA: National Technical Information Service, Oct. 30, 1984).
23. Beccasio, A.D., Fotheringham, N., Redfield, A.E., et al., *Gulf Coast Ecological Inventory, User's Guide and Information Base*, FWS/OBS-82/55 (Washington, DC: U.S. Fish and Wildlife Service, 1982).
24. Beccasio, A.D., Isakson, J.S., Redfield, A.E., et al., *Pacific Coast Ecological Inventory, User's Guide and Information Base*, FWS/OBS-81/30 (Washington, DC: U.S. Fish and Wildlife Service, 1981).
25. Beccasio, A.D., Weissberg, G.H., et al., *Atlantic Coast Ecological Inventory, User's Guide and Information Base*, FWS/OBS-80/51 (Washington, DC: U.S. Fish and Wildlife Service, 1980).
26. Beier, K., and Jerelov, A., "Methylation of Mer-

cury in Aquatic Environments," in *The Biochemistry of Mercury in the Environment*, J. Nriaqu (ed.) (New York: Elsevier, 1979), pp. 203-210.

27. Bell, F.W., and Leeworthy, V.R., *An Economic Analysis of the Importance of Saltwater Beaches in Florida*, FLSGP-T-86-003 (Gainesville, FL: University of Florida Sea Grant Program, 1986).

28. Belton, T., Hazan, R., Ruppel, B., et al., *A Study of Dioxin (2,3,7,8-tetrachlorodibenzo-p-dioxin) Contamination in Select Finfish, Crustaceans, and Sediments of New Jersey Waterways* (Trenton, NJ: New Jersey Department of Environmental Protection, 1985).

29. Belton, T., Roundy, R., and Weinstein, N., "Urban Fishermen: Managing the Risks of Toxic Exposure," *Environment* 28(9):19-20,30-37, November 1986.

30. Belton, T., Ruppel, B., and Lockwood, K., *Polychlorinated Biphenyls (Aroclor 1254) in Fish Tissue Throughout the State of New Jersey: A Comprehensive Survey* (Trenton, NJ: New Jersey Department of Environmental Protection, 1983).

31. Belton, T., Ruppel, B., Lockwood, K., et al., *A Study of Toxic Hazards to Urban Recreational Fishermen and Crabbers*, Technical Report 51 (Trenton, NJ: New Jersey Department of Environmental Protection, 1985).

32. Bick, T., and Kamlet, K., "The Ocean Dumping Debate—Continued," *Environmental Law Reporter* 13:10034-10035, February 1983.

33. Biddinger, G.R., and Gloss, S.P., "The Importance of Trophic Transfer in the Bioaccumulation of Chemical Contaminants in Aquatic Ecosystems," *Residue Reviews* 91:103-145, 1984.

34. Biocycle, "Land Application Gains Wider Acceptance," *Biocycle* 26(8):25-27, November/December 1985.

35. Blake, P.A., et al., "Cholera—A Possible Endemic Focus in the United States," *The New England Journal of Medicine* 302:305-309, Feb. 7, 1980.

36. Boehnke, C.A., et al., "Diseased Fish From the St. Johns River, Preliminary Report" (Jacksonville, FL: Jacksonville University Charter Marine Science Center, 1986).

37. Bokuniewicz, H.J., "Submarine Borrow Pits as Containment Sites for Dredged Sediment," in *Wastes in the Ocean*, vol. 2, D.R. Kester, et al. (eds.) (New York: John Wiley & Sons, 1983), pp. 215-227.

38. Bonzek, C.F., Myers, W.L., Parolari, B.W., and Patil, G.P., "Sources of Bias in Harvest Surveys for Marine Fisheries," in *Oceans '86 Conference Record*, vol. 3, 86CH2363-0 (Washington, DC:

Marine Technology Society and IEEE Ocean Engineering Society, Sept. 23-25, 1986), pp. 908-913.

39. Boughton, J.D., *Investigation of Subaqueous Borrow Pits as Potential Sites for Dredged Material Disposal*, Technical Report D-77-5 (Vicksburg, MS: U.S. Army Engineer Waterways Experiment Station, 1977).

40. Bouwer, E.J., and McCarty, P.L., "Transformations of 1-Carbon and 2-Carbon Halogenated Aliphatic Organic Compounds Under Methanogenic Conditions," *Applied and Environmental Microbiology* 45(4):1286-1294, 1983.

41. Bouwer, E.J., and McCarty, P.L., "Transformations of Halogenated Organic Compounds Under Denitrification Conditions," *Applied and Environmental Microbiology* 45(4):1295-1299, 1983.

42. Bowden, W.B., "Nutrients in Albemarle Sound" (Raleigh, NC: University of North Carolina Sea Grant, 1977).

43. Bower, B., Barre, R., Kuhner, J., and Russell, C., "Incentives in Water Quality Management: France and the Ruhr Area," Research Paper R-24 (Washington, DC: Resources For the Future, 1981).

44. Boyd, M.B., Saucier, R.T., Keeley, J.W., et al., *Disposal of Dredged Spoil: Problem Identification and Assessment and Research Program Development*, Technical Report H-72-8 (Vicksburg, MS: U.S. Army Engineer Waterways Experiment Station, 1972).

45. Boyer, B., and Meidinger, E., "Privatizing Regulatory Enforcement: A Preliminary Assessment of Citizen Suits Under Federal Environmental Laws," in *Administrative Conference of the United States Recommendations and Reports* (Washington, DC: U.S. Government Printing Office, 1986).

46. Brannon, J.M., *Evaluation of Dredged Material Pollution Potential*, Technical Report DS-78-6 (Vicksburg, MS: U.S. Army Engineer Waterways Experiment Station, 1978).

47. Brannon, J.M., Hoeppel, R.E., Sturgis, T.C., et al., *Effectiveness of Capping in Isolating Contaminated Dredged Material From Biota and the Overlying Water*, Technical Report D-85-10 (Vicksburg, MS: U.S. Army Engineer Waterways Experiment Station, 1985).

48. Brookhaven National Laboratory, *Technologies For and Impacts of Low-Level Radioactive Waste in the Deep Sea*, contract prepared for U.S. Congress, Office of Technology Assessment (Upton, NY: 1986).

49. Brooks, N.H., "Evaluation of Key Issues and Alternative Strategies," in *Ocean Disposal of Municipal Wastewater: Impacts on the Coastal Envi-*

ronment, E.P. Myers (ed.) (Cambridge, MA: MIT Sea Grant College Program, 1983), pp. 707-759.

50. Brooks, N.H., Arnold, R.G., Koh, R.C.Y., et al., *Deep Ocean Disposal of Sewage Sludge Off Orange County, California: A Research Plan*, Environmental Quality Laboratory Report No. 21 (Pasadena, CA: California Institute of Technology, November 1982).

51. Brooks, N.H., and Krier, J.E., "Alternative Strategies for Ocean Disposal of Municipal Wastewater and Sludge," background paper for *Symposium on Engineering Aspects of Using the Assimilative Capacity of the Oceans*, Lewes, DE, June 23-24, 1981.

52. Brown, D.A., "Testimony on Contaminant Studies in Southern California Coastal Waters, As Presented to the SCCWRP Scientific Review Panel on May 28, 1985" (Long Beach, CA: May 28, 1985).

53. Brown, D.A., "Overview of the Status of California's Oceans and Bays," paper presented at the *Ocean Pollution Conference*, Santa Barbara, CA, Laurel Springs Canyon Ranch, Sept. 12-13, 1986.

54. Brown, D.A., Testimony Before the Subcommittee on Health and the Environment, House Committee on Energy and Commerce, U.S. Congress, "Hearings on Health Implications of Toxic Chemical Contamination of the Santa Monica Bay," 99th Cong., 2d sess., Serial No. 99-74 (Washington, DC: U.S. Government Printing Office, 1986), pp. 117-133.

55. Brown, D.A., Jenkins, K.D., et al., "Detoxification of Metals and Organic Compounds in White Croakers," in *Coastal Water Research Project Biennial Report for the Years 1981-1982*, W. Bascom (ed.) (Long Beach, CA: Southern California Coastal Water Research Project, 1982), pp. 157-172.

56. Brown, G.M., Jr., and Johnson, R., "Pollution Control by Effluent Charges: It Works in the Federal Republic, Why Not in the U.S.?" *Natural Resources Journal* 24:929-966, October 1984.

57. Brown, M., Werner, M., Sloan, R., et al., "Recent Trends in the Distribution of Polychlorinated Biphenyls in the Hudson River System," *Environmental Science and Technology* 19:656-661, 1985.

58. Brungs, W.A., and Mount, D.I., "Introduction to a Discussion of the Use of Aquatic Toxicity Tests for Evaluation of the Effects of Toxic Substances," in *Estimating the Hazard of Chemical Substances to Aquatic Life*, Special Technical Publication 657, J. Cairns, Jr., et al. (eds.) (Philadelphia, PA: American Society for Testing and Materials, 1978), pp. 15-26.

59. Bryan, E.H., "Innovative Management of Resid-

uals—Research Needs," in *Public Waste Management and the Ocean Choice*, MITSG 85-36, K.D. Stolzenbach, et al. (eds.) (Cambridge, MA: MIT Sea Grant College Program, April 1986), pp. 85-93.

60. Burge, W.D., and Marsh, P.B., "Infectious Disease Hazards of Landspreading Sewage Wastes," *Journal of Environmental Quality* 7(1):1-9, 1978.

61. Burke, R.A., and McDonald, G.T., "Long-Term Planning for Dredged Material Disposal," in *Dredging and Dredged Material Disposal*, Conference Proceedings (New York: American Society of Civil Engineers, November 1984), pp. 999-1005.

62. Burks, S.A., and Engler, R.M., *Water Quality Impacts of Aquatic Dredged Material Disposal (Laboratory Investigations)*, Technical Report DS-78-4 (Vicksburg, MS: U.S. Army Engineer Waterways Experiment Station, 1978).

63. Cabelli, V.J., Dufour, A.P., et al., "Swimming-Associated Gastroenteritis and Water Quality," *American Journal of Epidemiology* 115:606-616, 1982.

64. Cairns, J., Jr., "The Myth of the Most Sensitive Species," *BioScience* 36(10):670-672, November 1986.

65. Calambokidis, J.A., et al., *Biology of Puget Sound Marine Mammals and Marine Birds: Population Health and Evidence of Pollution Effects*, Technical Memorandum NOS OMA 18 (Seattle, WA: National Oceanic and Atmospheric Administration, 1985).

66. California Regional Water Quality Control Board, San Francisco Bay Region, *Aquatic Habitat Program*, information sheet (Oakland, CA: undated).

67. California Regional Water Quality Control Board, San Francisco Bay Region, "Second Draft Basin Review Document," File No. 1537.00(SRR) (Oakland, CA: Aug. 29, 1986).

68. California State Water Resources Control Board, *Water Quality Control Plan, Ocean Waters of California* (Sacramento, CA: November 1983).

69. California State Water Resources Control Board, *Water Quality Inventory for Years 1982 & 1983*, Water Quality Monitoring Report No. 84-3TS (Sacramento, CA: 1984).

70. California State Water Resources Control Board, "Plan for Assessing the Effects of Pollutants in the San Francisco Bay-Delta Estuary," Aquatic Habitat Program Pub. No. 82-7SP (Sacramento, CA: 1982, reprinted October 1984).

71. California State Water Resources Control Board, *California State Mussel Watch 1983-84*, Water Quality Monitoring Report No. 85-2WQ (Sacramento, CA: 1985).

72. Callahan, M., et al., *Water-Related Fate of 129*

Priority Pollutants, vol. 2, EPA 440/4-79-029B (Washington, DC: U.S. Environmental Protection Agency, 1979).

73. Camp, Dresser, & McKee, *A Comparison of Studies of Toxic Substances in POTW Effluents*, report to U.S. Environmental Protection Agency, Office of Water Regulations and Standards (Annandale, VA: 1984).

74. Capuzzo, J.M., "Predicting Pollution Effects in the Marine Environment," *Oceanus* 24(1):25-33, 1981.

75. Capuzzo, J.M., Burt, W.V., et al., "Future Strategies for Nearshore Waste Disposal," in *Wastes in the Ocean*, vol. 6, B.H. Ketchum, et al. (eds.) (New York: John Wiley & Sons, 1985), pp. 491-512.

76. Capuzzo, J.M., Farrington, J.W., et al., "Bioaccumulation and Biological Effects of PCBs on Marine Bivalve Molluscs in New Bedford Harbor, Massachusetts," paper presented at Sixth International Ocean Disposal Symposium, Pacific Grove, CA, Apr. 21-25, 1986.

77. CE Maguire, Inc., *Boston Harbor Water Quality Baseline for the SDEIS on Boston Harbor Wastewater Facilities Siting* (Providence, RI: December 1984).

78. Centers for Disease Control, "Follow-up on Vibrio Cholerae Infection—Louisiana," *Morbidity & Mortality Weekly Report* 27:365-367, 1978.

79. Centers for Disease Control, "Water-Related Disease Outbreaks in the United States—1980," *Morbidity & Mortality Weekly Report* 30:623-634, 1982.

80. Centers for Disease Control, "Enteric Illness Associated with Raw Clam Consumption—New York," *Morbidity & Mortality Weekly Report* 31:449-451, 1982.

81. Centers for Disease Control, "Toxigenic *Vibrio cholerae O1* Infections—Louisiana and Florida," *Morbidity & Mortality Weekly Report* 35:606-607, 1986.

82. Centers for Disease Control, "Cholerae in Louisiana —Update," *Morbidity & Mortality Weekly Report* 35:687-688, 1986.

83. Champ, M.A., *Operation SAMS: Sludge Acid Monitoring Survey*, CERES Program Publication No. 1 (Washington, DC: American University Center for Earth Resources and Environmental Studies, 1973).

84. Champ, M.A., "Monitoring—Painting a Moving Train," *Sea Technology* 27(9):73, 1986.

85. Champ, M.A., Norton, M.G., and Devine, M.F., "A Semi-Quantitative Model for the Assessment of Dispersion at Near-Shore Ocean Dumpsites," paper presented at Contaminant Fluxes Through the Coastal Zone Conference, Nantes, France, May 1984.

86. Champ, M.A., O'Connor, T.P., and Park, P.K., "Ocean Dumping of Seafood Wastes in the United States," *Marine Pollution Bulletin* 12(7):241-244, 1981.

87. Champ, M.A., and Park, P.K., "Ocean Dumping of Sewage Sludge: A Global Review," *Sea Technology* 22(2):18-24, 1981.

88. Champney, L., "A Case Study in the Implementation of the Federal Water Pollution Control Act Amendments," *Water Resources Bulletin* 15(6): 1602-1607, 1979.

89. Chapman, P.M., et al., *Survey of Biological Effects of Toxicants Upon Puget Sound Biota, I: Broad-Scale Toxicity Survey*, NOAA Technical Memorandum OMPA-25 (Rockville, MD: National Oceanic and Atmospheric Administration, December 1982).

90. Chapman, P.M., et al., *A Field Trial of the Sediment Quality Triad in San Francisco Bay*, NOAA Technical Memorandum NOS OMA 25 (Rockville, MD: National Oceanic and Atmospheric Administration, 1986).

91. Chapman, W., Fisher, H., and Pratt, M., *Concentration Factors of Chemical Elements in Edible Aquatic Vertebrates*, UCRL 50564 (Berkeley, CA: Lawrence Livermore Labs, 1968).

92. Chartrand, A.B., et al., *Ocean Dumping Under Los Angeles Regional Water Quality Control Board Permit: A Review of Past Practices, Potential Adverse Impacts, and Recommendations for Future Action* (Los Angeles, CA: California Regional Water Quality Control Board, March 1985).

93. Chave, K.E., and Miller, J.N., "Pearl Harbor Dredged-Material Disposal," in *Wastes in the Ocean*, vol. 2, D.R. Kester, et al. (eds.) (New York: John Wiley & Sons, 1983), pp. 91-98.

94. Chemical Manufacturers Association, "Comments of the Chemical Manufacturers Association on EPA's July and October 1985 Notices of Availability and Request for Comments on the Effluent Limitations Guidelines and Standards for the Organic Chemicals and Plastics and Synthetic Fibers Category, Volume II" (Washington, DC: Dec. 23, 1985).

95. Chesapeake Bay Executive Council, *Chesapeake Bay Restoration and Protection Plan* (Washington, DC: September 1985).

96. Chesapeake Bay Executive Council, *First Annual Progress Report Under the Chesapeake Bay Agreement* (Washington, DC: December 1985).

97. Citizens for a Better Environment, *Toxics Control at the Source: Industrial Pretreatment of*

Waste Water Prior to Disposal to Municipal Sewer Systems (San Francisco, CA: May 1984).

98. Citizens for a Better Environment, *Toxics Down the Drain: A Review of Industrial Compliance With Federal Pretreatment Standards in California* (San Francisco, CA: November 1985).

99. Clark, J., *Coastal Ecosystems: Ecological Considerations for Management of the Coastal Zone* (Washington, DC: The Conservation Foundation, 1974).

100. Clark, J.R., *Coastal Ecosystem Management, A Technical Manual for the Conservation of Coastal Zone Resources* (New York: John Wiley & Sons, 1977).

101. Clevenger, T.E., et al., "Chemical Composition and Possible Mutagenicity of Municipal Sludges," *Water Pollution Control Federation Journal* 55(12): 1470-1475, 1983.

102. Coastal Zone Management, "Ocean Science News," *Coastal Zone Management*, June 7, 1984, p. 5.

103. Coch, C.A., and Mansky, J.M., "Feasibility of Use of Dredged Material as Sanitary Landfill Cover in the New York/New Jersey Harbor Area," in *Dredging and Dredged Material Disposal (Conference Proceedings)* (New York: American Society of Civil Engineers, November 1984), pp. 830-838.

104. Coleman, F.C., and Wehle, D.H.S., "Plastic Pollution: A Worldwide Oceanic Problem," *Parks* 9(1):9-12, April/May/June 1984.

105. Colwell, R.R., "Biotechnology in the Marine Sciences," in *Biotechnology in the Marine Sciences, Proceedings of the First Annual MIT Sea Grant Lecture and Seminar*, R.R. Colwell, et al. (eds.) (New York: John Wiley & Sons, 1984), pp. 3-36.

106. Colwell R.R., et al., "Viable, But Non-culturable *Vibrio cholerae* and Related Pathogens in the Environment: Implications for Release of Genetically Engineered Microorganisms," *Biotechnology* 3: 817-823, 1985.

107. Commission on Marine Science, Engineering and Resources, *Our Nation and the Sea—A Plan for National Action* (Washington, DC: 1970).

108. Connecticut Department of Environmental Protection, Water Compliance Unit, *State of Connecticut 1984 Water Quality Report to Congress* (Hartford, CT: 1984).

109. Connor, M.S., "Comparison of the Carcinogenic Risks From Fish vs. Groundwater Contamination by Organic Compounds," *Environmental Science & Technology* 18:628-631, 1984.

110. Conservation Foundation, "Controlling Cross-Media Pollutants," Issue Report (Washington, DC: 1984).

111. Conservation Foundation, "New Perspectives on Pollution Control, Cross-Media Problems," Issue Report (Washington, DC: 1985).

112. Conservation Foundation, "National Conference on Risk Communication" (Washington, DC: Jan. 29-31, 1986).

113. Coolbaugh, J.C., et al., "Bacterial Contamination of Divers During Training Exercises in Coastal Waters," *Marine Technology Society Journal* 15(2):15-21, 1981.

114. Cormick, G.W., "Siting New Hazardous Waste Management Facilities Using Mediated Negotiations," paper presented at Conference on Meeting the New RCRA Requirements on Hazardous Waste, Alexandria, VA, Oct. 8, 1985.

115. Council on Environmental Quality, *Ocean Dumping: A National Policy* (Washington, DC: U.S. Government Printing Office, 1970).

116. County Sanitation Districts of Orange County, "Statement for the Hearing Record of the Senate Environment and Public Works Committee, Subcommittee on Environmental Pollution, Regarding Reauthorization of the Clean Water Act," Mar. 27, 1985.

117. Cousteau, J., "Statement of Captain Jacques-Yves Cousteau, Chairman, The Cousteau Society and the Foundation Cousteau," Hearings Before the Subcommittee on Oceanography and the Subcommittee on Fisheries and Wildlife Conservation and the Environment of the Committee on Merchant Marine and Fisheries, House of Representatives, 97th Cong., Mar. 23, 1982, Serial No. 97-40 (Washington, DC: U.S Government Printing Office, 1982).

118. Crites, R.W., "Land Use of Wastewater and Sludge," *Environmental Science & Technology* 18(5):140A-147A, 1984.

119. Cross, J.N., "Fin Erosion Among Fishes Collected Near a Southern California Municipal Wastewater Outfall (1971-1982)," *Fishery Bulletin* 83(2): 195-206, 1985.

120. Csanady, G.T., "Long-Term Mixing Processes in Slopewater," in *Wastes in the Ocean*, vol. 1, I.W. Duedall, et al. (eds.) (New York: John Wiley & Sons, 1983), pp. 103-116.

121. Curtis, C., Testimony Before the Subcommittee on Oceanography and the Subcommittee on Fisheries and Wildlife Conservation and the Environment, House Committee on Merchant Marine and Fisheries, "Hearings on Ocean Dumping," 96th Cong., Committee Serial No. 96-40 (Washington, DC: U.S Government Printing Office, 1980).

122. Dalpra, C., "Secondary Treatment Waivers Issued, Controversy Continues," *Water Pollution*

Control Federation Journal 53(11):1554-1557, 1981.

123. Davis, C., "Availability of Environmental Impairment Liability Insurance," paper presented to the Annual Meeting of the American Political Science Association, Washington, DC, Aug. 28-31, 1986.

124. Davis, W.R., "Sediment-Copper Reservoir Formation by the Burrowing Polychaete Nephtys Incisa," in Wastes in the Ocean, vol. 2, D.R. Kester, et al. (eds.) (New York: John Wiley & Sons, 1983), pp. 173-184.

125. Dayal, R., Heaton, M.G., Fuhrmann, M., et al., A Geochemical and Sedimentological Study of the Dredged Material Deposit in the New York Bight, NOAA Technical Memorandum OMPA-3 (Rockville, MD: National Oceanic and Atmospheric Administration, February 1981).

126. DeLong, R.L., Gilmartin, W.G., and Simpson, J.G., "Premature Births in California Sea-Lions: Association With High Organochlorine Pollutants Residue Levels," Science 181:1168-70, 1973.

127. Dillon, T.M., Biological Consequences of Bioaccumulation in Aquatic Animals: An Assessment of the Current Literature, Technical Report D-84-2 (Vicksburg, MS: U.S. Army Engineer Waterways Experiment Station, June 1984).

128. D'Itri, P., and D'Itri, F., Mercury Contamination: A Human Tragedy (New York: John Wiley & Sons, 1977).

129. Doull, J., Klaassen, C., and Amdur, M., Toxicology—The Basic Science of Poisons (New York: MacMillan, 1980).

130. Downing, D., and Sessions, S., "Innovative Water Quality-Based Permitting: A Policy Perspective," Water Pollution Control Federation Journal 57(5): 358-365, May, 1985.

131. Duedall, I.W., Kester, D.R., et al. (eds.), "Energy Wastes in the Marine Environment: An Overview," in Wastes in the Ocean, vol. 4 (New York: John Wiley & Sons, 1985), pp. 3-44.

132. Duedall, I.W., Ketchum, B.H., et al. (eds.), "Global Inputs, Characteristics, and Fates of Ocean-Dumped Industrial and Sewage Wastes: An Overview," in Wastes in the Ocean, vol. 1 (New York: John Wiley & Sons, 1983), pp. 3-45.

133. Duedall, I.W., Ketchum, B.H., et al. (eds.), "Scientific Strategy for Industrial and Sewage Waste Disposal in the Ocean," in Wastes in the Ocean, vol. 1 (New York: John Wiley & Sons, 1983), pp. 399-413.

134. Dybdahl, D.J., "Availability of Environmental Impairment Liability Insurance," paper presented to Federation of Environmental Technologies Seminar on Environmental Liability, Pewaukee, WI, Jan. 16, 1986.

135. Dykstra, M.J., et al., "Characterization of the Aphanomyces Species Involved With Ulcerative Mycosis (UM) in Menhaden," Mycologia 78(4): 664-672, 1986.

136. Edgar, C.E., and Engler, R.M., "The London Dumping Convention and Its Role in Regulating Dredged Material: An Update," in Dredging and Dredged Material Disposal (Conference Proceedings) (New York: American Society of Civil Engineers, November 1984), pp. 140-149.

137. Edwards, M.L., "Mobile Bay's Miracle Is Not Without a Price," National Fisherman: 12-15, September 1986.

138. EG&G Environmental Consultants, Assessment for Future Environmental Problems—Ocean Dumping, prepared for U.S. Environmental Protection Agency, Office of Strategic Assessment and Special Studies (Waltham, MA: September 1983).

139. EG&G Washington Analytical Services Center, Inc., Industrial Waste Disposal in Marine Environments, contract prepared for U.S. Congress, Office of Technology Assessment (Waltham, MA: 1986).

140. EG&G Washington Analytical Services Center, Inc., Oceanographic Services, Projected Ocean Dumping Rates for Municipal and Industrial Wastes in the Year 2000, prepared for National Oceanic and Atmospheric Administration, National Marine Pollution Office (Waltham, MA: 1986).

141. Electric Power Research Institute, "Utility Solid Waste: Managing the Byproducts of Coal Combustion," EPRI Journal 10(8):20-35, October 1985.

142. Engineering News Record, "Second Ozone Plant Started," Engineering News Record 215(8):15-16, Aug. 22, 1985.

143. Engineering News Record, "Questions Shroud Privatization," Engineering News Record 216(2): 26-27, Jan. 9, 1986.

144. Environmental Defense Fund, Approaches to Source Reduction, Practical Guidance From Existing Policies and Programs (Berkeley, CA: June, 1986).

145. Environmental Defense Fund, "Environmental Defense Fund Seeks Action To Ban the Use of TBT To Protect Virginia Shellfish Industry," News Release (Washington, DC: June 20, 1986).

146. Environmental and Energy Study Conference, "Oceans," Outlook (Washington, DC: Jan. 20, 1986), p. 29.

147. Environmental and Energy Study Conference, "Water Quality Act of 1987," House-Senate Floor Brief, Jan. 6, 1987.

148. Environmental Policy Alert, "Boston Harbor Dis-

charges—Court Orders Cleanup," *Environmental Policy Alert* 3(2):9, Jan. 15, 1986.

149. Environmental Policy Alert, "EPA FY-87 Budget Request," *Environmental Policy Alert* 3(4):17, Feb. 12, 1986.

150. Environmental Policy Alert, "Sewage Sludge—EPA Proposes State Program Requirements," *Environmental Policy Alert* 3(5):5, Feb. 26, 1986.

151. Environment Reporter, "States Would Be Required To Adopt Plans for Managing Sludge, Under EPA Proposal," *Environment Reporter* 16(40): 1805-1806, Jan. 31, 1986.

152. Environment Reporter, "RCRA May Cover 700 Plants Receiving Hazardous Wastes, Hedeman Tells AMSA," *Environment Reporter* 16(42): 1880-1881, Feb. 14, 1986.

153. Environment Reporter, "Massachusetts Plan To Use 106-Mile Site for Boston Sludge Dropped Due to Objections," *Environment Reporter* 17(14): 491-492, Aug. 1, 1986.

154. Environment Reporter, "New EPA Survey Shows Reduced POTW Needs; Drop Attributed to Documentation Requirement," *Environment Reporter* 17(42):1733, Feb. 13, 1987.

155. Erdheim, E., "United States Marine Waste Disposal Policy," in *Wastes in the Ocean*, vol. 6, B. Ketchum, et al. (eds.) (New York: John Wiley & Sons, 1985), pp. 421-461.

156. Fanning, K.A., and Bell, L.M., "Nutrients in Tampa Bay," in *Proceedings Tampa Bay Area Scientific Information Symposium*, S.F. Treat, et al. (eds) (New York: Bellwether, 1985), pp. 109-129.

157. Fay, R.R., and Wastler, T.A., "Ocean Incineration: Contaminant Loading and Monitoring," in *Wastes in the Ocean*, vol. 5, D.R. Kester, et al. (eds.) (New York: John Wiley & Sons, 1985), pp. 73-90.

158. Feldman, D.L., "Towards a Political Theory of Environmental Values: Water Resources Development and the Problems of a Policy Meta-Ethic," paper presented at 1986 Annual Meeting of the American Political Science Association, Washington, DC, Aug. 28-31, 1986.

159. Feliciano, D.V., "Fact Sheet for Wastewater Treatment," *Water Pollution Control Federation Journal* 54(10):1346-1348, 1982.

160. Fields, R.W., Mueller, R.T., and McGeorge, L.J., "The Occurrence and Fate of Toxic Substances in New Jersey Sewage Treatment Facilities," CN-409 (Trenton, NJ: Department of Environmental Protection, September 1986).

161. Flemer, D.A., Duke, T.W., and Mayer, F.L., Jr., "Integration of Monitoring and Research in Coastal Waters: Issues for Consideration From a Regulatory Point of View," in *Oceans '86 Conference Record*, vol. 3, 86CH2363-0 (Washington, DC: Marine Technology Society and IEEE Ocean Engineering Society, Sept. 23-25, 1986), pp. 980-992.

162. Fleming, W.J., Clark, D.R., Jr., and Henny, C.J., "Organochlorine Pesticides and PCBs: A Continuing Problem for the 1980's," *Transactions North American Wildlife Conference* 48:186-199, 1983.

163. Florida Department of Environmental Regulation, Biology Section, *Bioassays of Frost Seafoods (Scallop Processing Facility), St·Augustine, St. Johns County, Florida* (Tallahassee, FL: July 1985.)

164. Florida Department of Environmental Regulation, "Tables and Figures Derived From STORET Water Quality Data (1970-1979) on Reach 17 of the St. Johns River Basin (Lower)" (Tallahassee, FL: 1986).

165. Florida Institute of Technology, *Definition of the Oceanic, Atmospheric, and Terrestrial Media, an Overview of Resources and Processes in These Media, and a Discussion of the Concepts of Assimilative Capacity and Resiliency*, contract prepared for U.S. Congress, Office of Technology Assessment (Melbourne, FL: 1986).

166. Flynn, K.C., "Turning the Tide in Boston Harbor," *Water Pollution Control Federation Journal* 57(11):1048-1054, November 1985.

167. Forster, D.L., and Southgate, D.D., "Social Institutions Influencing Land Application of Wastewater and Sludge," *Water Pollution Control Federation Journal* 56(5):399-404, May 1984.

168. Fox, J.C., Testimony Before the Subcommittee on Governmental Efficiency and the District of Columbia, Committee on Governmental Affairs, U.S. Congress, "Hearing on Chesapeake Bay Cleanup Program," 99th Cong., 2d sess., Senate Hearing 99-950, June 24, 1986 (Washington, DC: U.S. Government Printing Office, 1986).

169. Francingues, N.R., Jr., et al., *Management Strategy for Disposal of Dredged Material: Contaminant Testing and Controls*, Miscellaneous Paper D-85-1 (Vicksburg, MS: U.S. Army Engineer Waterways Experiment Station, 1985).

170. Fredette, T.J., et al., "Biological Monitoring of Open-Water Dredged Material Disposal Sites," in *Oceans '86 Conference Record*, vol. 3, 86CH2363-0 (Washington, DC: Marine Technology Society and IEEE Ocean Engineering Society, Sept. 23-25, 1986), pp. 764-769.

171. Freeland, G.L., Swift, D., and Young, R.A., "Mud Deposits Near the New York Bight Dumpsites: Origin and Behavior," in *Ocean Dumping and Marine Pollution* (New York: Van Nostrand Reinhold, 1979), pp. 73-95.

172. Freeman, A.M., "The Ethical Basis of the Eco-

nomic View of the Environment," in *People, Penguins, and Plastic Trees*, D. VanDeVeer and C. Pierce (eds.) (Belmont, CA: Wadsworth Publishing Co., 1986), pp. 218-227.

173. Fricke, C., et al., "Comparing Priority Pollutants in Municipal Sludges," *Biocycle* 26(1):35-37, 1985.

174. Fried, J.P., "Illegal Taking of Clams From Tainted Bay Rises," *New York Times*, Sept. 21, 1986, p. 50.

175. Fuhrman, R.E., "History of Water Pollution Control," *Water Pollution Control Federation Journal* 56(4):306-313, 1984.

176. Funk, W., "The Exception That Approves The Rule: FDF Variances Under the Clean Water Act," *Environmental Affairs* 13(1):1-60, fall 1985.

177. Gaba, J., "Regulation of Toxic Pollutants Under the Clean Water Act: NPDES Toxics Control Strategies," *Journal of Air and Law Commerce* 50(3/4):761-791, 1985.

178. Gadbois, D.F., and Maney, R.S., "Survey of Polychlorinated Biphenyls in Selected Finfish Species From United States Coastal Waters," *Fishery Bulletin*, 81(2):389-396, 1983.

179. Gambrell, R.P., Khalid, R.A., and Patrick, W.H., Jr., *Disposal Alternatives of Contaminated Dredged Material as a Management Tool to Minimize Adverse Environmental Effects*, Technical Report DS-78-8 (Vicksburg, MS: U.S. Army Engineer Waterways Experiment Station, 1978).

180. Garber, W.F., and Wada, F.F., "Water Quality in Santa Monica Bay, as Indicated by Measurements of Total Coliform," in *Oceanic Processes in Marine Pollution, Volume 5, Urban Wastes in Coastal Marine Environments*, D.A. Wolfe and T.P. O'Connor (eds.) (Malabar, FL: Robert E. Krieger Publishing Co., 1987).

181. Garrity, R.D., McCann, N., and Murdoch, J.D., "A Review of Environmental Impacts of Municipal Services in Tampa, Florida," in *Proceedings Tampa Bay Area Scientific Information Symposium*, S.F. Treat, et al. (eds.) (New York: Bellwether, 1985), pp. 526-550.

182. Georgia Department of Natural Resources, Environmental Protection Division, *Water Quality Control in Georgia, 1982-1983* (Atlanta, GA: 1984).

183. Gerba, C.P., Smith, E.M., and Melnick, J.L., "Development of a Quantitative Method for Detecting Enteroviruses in Estuarine Sediments," *Applied Environmental Microbiology* 34:158-163, 1977.

184. Giam, C.S., et al., "Phthalate Ester Plasticizers: A New Class of Marine Pollutant," *Science* 199:419-421, 1978.

185. Giam, C.S., et al., "Aquatic Pollution Problems, Southeastern U.S. Coasts: I. Chemical Aspects," paper presented at Toxic Chemicals and Aquatic Life Symposium, sponsored by National Marine Fisheries Service, Seattle, WA, Sept. 16-18, 1986.

186. Gill, O.N., et al., "Epidemic of Gastroenteritis Caused by Oysters Contaminated With Small Round Structured Viruses," *British Medical Journal* 287:1532, 1983.

187. Gilmartin, W.G., et al., "Premature Parturition in the California Sea-Lion, *Journal of Wildlife Diseases* 12:104-115, 1976.

188. Goldberg, E.D. (ed.), *Proceedings of a Workshop on Scientific Problems Relating to Ocean Pollution, July 10-14, 1978* (Boulder, CO: National Oceanic and Atmospheric Administration, Environmental Research Laboratory, 1979).

189. Goldberg, E.D., "The Oceans as Waste Space: The Argument," *Oceanus* 24(1):2-9, 1981.

190. Goldberg, E.D., "TBT, An Environmental Dilemma," *Environment* 28(8):17-20,42-44, October 1986.

191. Goldstein, J., et al., "Effects of Pentachlorophenol on Hepatic Drug Metabolizing Enzymes and Porphyria Related to Contamination With Chlorinated Dibenzo-p-dioxins and Dibenzofurans," *Biochemical Pharmacology* 26:1549-1557, 1977.

192. Goodman, A., et al., "Gastrointestinal Illness Among Scuba Divers—New York City," *Morbidity & Mortality Weekly Report* 32(44):576-577, 1983.

193. Gordon, W.G., "NMFS Position Statement on 12-106 Mile Ocean Dump Sites," Letter to T.A. Wastler, Marine Protection Branch, U.S. Environmental Protection Agency, Aug. 24, 1983.

194. Gossett, R.W., et al., "DDT, PCB and Benzo(a)pyrene Levels in White Croaker (*Genyonemus lineatus*) From Southern California," *Marine Pollution Bulletin* 14(2):60-65, February 1983.

195. Gossett, R.W., Brown, D.A., and Young, D.R., "Predicting the Bioaccumulation and Toxicity of Organic Compounds," in *Coastal Water Research Project Biennial Report for the Years 1981-1982*, W. Bascom (ed.) (Long Beach, CA: Southern California Coastal Water Research Project, 1982), pp. 149-156.

196. Government of the State of Sao Paulo, *Environmental Pollution Control in Cubatao, Results Report—July/83 to July/86* (Sao Paulo, Brazil: 1986).

197. Goyal, S.M., "Viral Pollution of the Marine Environment," *CRC Critical Reviews in Environmental Control* 14(1):1-32, 1984.

198. Grassle, J.F., et al., *Contaminant Levels and Relative Sensitivities to Contamination in the Deep-Ocean Communities*, NOAA Technical

Memorandum NOS OMA 26 (Rockville, MD: National Oceanic and Atmospheric Administration, March 1986).

199. Greaney, D. "How Clean Is Your Beach?" *Save the Bay* 16(4):1, August/September 1986.

200. Great Lakes Water Quality Board, *Great Lakes Water Quality, Report to the International Joint Commission* (Kingston, Ontario: June 1985).

201. Green, R., "Another Kesterson?" *Sierra* 70(5):29-32, September/October 1985.

202. Grieg, R., et al., "Trace Metals in Organisms from Ocean Disposal Sites of the Middle-eastern United States," *Archives of Environmental Contamination and Toxicology* 6:395-409, 1979.

203. Grimes, D.J., "Release of Sediment-bound Fecal Coliforms by Dredging," *Applied Microbiology* 29:109, 1975.

204. Grimes, D.J., "Bacteriological Water Quality Effects of Hydraulically Dredging Contaminated Upper Mississippi River Bottom Sediment," *Applied and Environmental Microbiology* 39:782-789, 1980.

205. Grimes, D.J., *Human Health Impacts of Waste Constituents, Pathogens and Antibiotic- and Heavy Metal-Resistant Bacteria*, contract prepared for U.S. Congress, Office of Technology Assessment (College Park, MD: 1986).

206. Grimes, D.J., Attwell, R.W., et al., "Fate of Enteric Pathogenic Bacteria in Estuarine and Marine Environments," *Microbiological Sciences*, in press.

207. Grimes, D.J., Singleton, F.L., and Colwell, R.R., "Allogenic Succession of Marine Bacterial Communities in Response to Pharmaceutical Waste," *Journal of Applied Bacteriology* 57:247-261, 1984.

208. Grimes, D.J., Singleton, F.L., Stemmler, J., et al., "Microbiological Effects of Wastewater Effluent Discharge Into Coastal Waters of Puerto Rico," *Water Research* 18(5):613-619, 1984.

209. Grohman, G.S., et al., "Norwalk Virus Gastroenteritis in Volunteers Consuming Depurated Oysters," *Australian Journal of Experimental Biology and Medical Science* 59:219, 1981.

210. Gross, M.G., *Oceanography: A View of the Earth* (Englewood Cliffs, NJ: Prentice-Hall, Inc., 1972).

211. Gross, M.G., "Sedimentation and Waste Deposition in New York Harbor," *Annals of the New York Academy of Sciences* 250:112-128, 1974.

212. Gross, M.G. (ed.), *Middle Atlantic Continental Shelf and the New York Bight* (Lawrence, KS: Allen Press, Inc, 1976).

213. Grove, R.S., et al., "Artificial-Reef Development as a Means of Disposal of Coal Wastes Off Southern California," in *Wastes in the Ocean*, vol. 4, I.W. Duedall, et al. (eds.) (New York: John Wiley & Sons, 1985), pp. 515-534.

214. Guarascio, J., "The Regulation of Ocean Dumping After *City of New York v. Environmental Protection Agency*," *Environmental Affairs* 12:701-741, 1985.

215. Gulf Coast Waste Disposal Authority, *Annual Report* (Houston, TX: 1984).

216. Gusman, S., and Huser, V., "Mediation in the Estuary," *Coastal Zone Management Journal* 11(4): 273-295, 1984.

217. Haliburton, T.A., *Guidelines for Dewatering/ Densifying Confined Dredged Material*, Technical Report DS-78-11 (Vicksburg, MS: U.S. Army Engineer Waterways Experiment Station, 1978).

218. Hall, L.W., Jr., et al., "Mortality of Striped Bass Larvae in Relation to Contaminants and Water Quality in a Chesapeake Bay Tributary," *Transactions American Fisheries Society* 114(6):861-868, November 1985.

219. Hall, L.W., Jr., Hall, W.S., and Bushong, S.J., *In-Situ Investigations for Assessing Striped Bass, Morone saxatilis, Prolarval and Yearling Survival as Related to Contaminants and Water Quality Parameters in the Potomac River—Contaminant and Water Quality Evaluations in East Coast Striped Bass Habitats* (Shady Side, MD: Johns Hopkins University, Applied Physics Laboratory, November 1986).

220. Hand, J., et al., *Water Quality Inventory for the State of Florida* (Tallahassee, FL: Florida Department of Environmental Regulation, June 1986).

221. Hand, J., et al., *Water Quality Inventory for the State of Florida: Technical Appendix* (Tallahassee, FL: Florida Department of Environmental Regulation, June 1986).

222. Hardin, G., "The Tragedy of the Commons," *Science* 162:1243-1245, 1968.

223. Harrington, M., et al., "A Survey of a Population Exposed to High Concentrations of Arsenic in Well Water in Fairbanks, Alaska," *American Journal of Epidemiology* 108:377-385, 1978.

224. Harrison, P., and Sewell, W., "Water Pollution Control by Agreement: The French System of Contracts," *National Resources Journal* 20:765-786, October 1980.

225. Harwell, C., "Analysis of Clean Water Act Section 403—Ocean Discharge Criteria" (Ithaca, NY: Cornell University Ecosystems Research Center, February 1984).

226. Harwell, M.A., Harwell, C.C., and Kelly, J.R., "Regulatory Endpoints, Ecological Uncertainties, and Environmental Decision-Making," in *Oceans '86 Conference Record*, vol. 3, 86CH2363-0 (Washington, DC: Marine Technology Society and IEEE Ocean Engineering Society, Sept. 23-25, 1986), pp. 993-999.

227. Hasit, Y., "Sludge Treatment, Utilization, and Disposal," *Water Pollution Control Federation Journal* 58(6):510-518, June 1986.

228. Hawaii Department of Health, *305(b) Report on Water Quality* (Honolulu, HA: April 1986).

229. Hazen, R., and Kneip, T., "Biogeochemical Cycling of Cadmium in a Marsh Ecosystem," in *Cadmium in the Environment*, part 1, J. Nriagu (ed.) (New York: Wiley-Interscience, 1981).

230. Heltz, G., Huggett, R., and Hill, J., "Behavior of Mn, Fe, Cu, Zn, Cd, and Pb Discharged From a Wastewater Treatment Plant Into an Estuarine Environment," *Water Research* 9:631-636, 1975.

231. Hermens, J., Leeuwangh, P., and Husch, A., "Quantitative Structure-Activity Relationships and Mixture Toxicity Studies of Chloro- and Alkyl Anilines at an Acute Lethal Toxicity Level to the Guppy (*Poecilia reticulata*)," *Ecotoxicology and Environmental Safety* 8:388-394, 1984.

232. Hirsch, N.D., DiSalvo, L.H., and Peddicord, R., *Effects of Dredging and Disposal on Aquatic Organisms*, Technical Report DS-78-5 (Vicksburg, MS: U.S. Army Engineer Waterways Experiment Station, 1978).

233. Hoffman, R.W., and Meighan, R.B., "The Impact of Combined Sewer Overflows From San Francisco on the Western Shore of Central San Francisco Bay," *Water Pollution Control Federation Journal* 56(12):1277-1285, 1984.

234. Holliday, B.W., *Processes Affecting the Fate of Dredged Material*, Technical Report DS-78-2 (Vicksburg, MS: U.S. Army Engineer Waterways Experiment Station, 1978).

235. Holliday, B.W., et al., *Predicting and Monitoring Dredged Material Movement*, Technical Report DS-78-3 (Vicksburg, MS: U.S. Army Engineer Waterways Experiment Station, 1978).

236. Holznagel, B., "Negotiation and Mediation: The Newest Approach to Hazardous Waste Facility Siting," *Boston College Environmental Affairs Law Review* 13(3):329-378, 1986.

237. Horn, E., and Skinner, L., *Final Environmental Impact Statement for Policy on Contaminants in Fish* (Albany, NY: State Department of Environmental Conservation, Division of Fish and Wildlife, 1985).

238. Hudson River Sloop Clearwater, Inc., *Polluting the Hudson: A Gentlemen's Agreement, A Report on the SPDES Permit Program for the Period October 1979-September 1981* (Poughkeepsie, NY: February 1983).

239. Huggett, R.J., Nichols, M.M., and Bender, M.E., "Kepone Contamination of the James River Estuary," in *Contaminants and Sediments*, vol. 1, R.A. Baker (ed.) (Ann Arbor, MI: Ann Arbor Science, 1980), pp. 33-52.

240. Huston, J.W., *Hydraulic Dredging, Theoretical and Applied* (Cambridge, MD: Cornell Maritime Press, 1970).

241. ICF, Inc., *Assessment of Impacts of Land Disposal Restrictions on Ocean Dumping and Ocean Incineration of Solvents, Dioxins, and California List Wastes*, draft report prepared for U.S. Environmental Protection Agency, Office of Solid Waste (Washington, DC: October 1986).

242. Inside EPA, "EPA Agrees to OMB Call For Construction Grants Phaseout By 1990," *Inside EPA* 6(11):1,9-12, Mar. 15, 1985.

243. Inside EPA, "EPA Eyes 'Flexible' RCRA Cleanups for POTWs Trucking in Hazardous Wastes," *Inside EPA* 6(36):1,12, Sept. 6, 1987.

244. Inside EPA, "SAB R&D Push on Nonpoint Water Toxics May Pose Policy Choice For Thomas," *Inside EPA* 8(1):12-13, Jan. 2, 1987.

245. Inside EPA, "SAB Pushes Revisions to Allow More Risk Comparison of Sludge Disposal Options," *Inside EPA* 8(3):5-6, Jan. 16, 1987.

246. Inside EPA, "EPA To Launch Major New 'Nonpoint Source' Plan to Clean Up Coastal Waters," *Inside EPA* 8(8):1,9-10, Feb. 20, 1987.

247. Inter-Governmental Maritime Consultative Organization, "Scientific Criteria for the Selection of Waste Disposal Sites at Sea," *Reports and Studies*, No. 16 (London: 1982).

248. International Joint Commission, *Great Lakes Water Quality Agreement of 1978, With Annexes and Terms of Reference, Between the United States of America and Canada* (Ottawa, Canada: Nov. 22, 1978).

249. Jelinek, C., and Corneliussen, P., "Levels of PCBs in the U.S. Food Supply," *Proceedings of the National Conference on PCBs*, EPA-560/6-75-004 (Washington, DC: 1976), pp. 147-154.

250. Jenkins, K.D., Brown, D.A., and Oshida, P., "Detoxification of Metal in Sea Urchins," in *Coastal Water Research Project Biennial Report for the Years 1981-1982*, W. Bascom (ed.) (Long Beach, CA: Southern California Coastal Water Research Project, 1982), pp. 173-178.

251. Johnson, E.L., "Seventh District Court Decision on TMDLs, Memorandum to Regional Water Management Division Directors and Regional Environmental Services Division Directors" (Washington, DC: U.S. Environmental Protection Agency, Office of Water Regulations and Standards, Oct. 23, 1984).

252. Joklik, W.K., *Zinsser Microbiology* (Norwalk, CT: Appleton-Century-Crofts, 1984).

253. Jones, G.E., "The Fate of Freshwater Bacteria in the Sea," *Developments in Industrial Microbiology* 12:141-151, 1971.

254. Josephson, J., "Protecting Virginia's Waterways," *Environmental Science and Technology* 15(10): 1125-1127, October 1981.

255. JRB Associates, *Assessment of the Impacts of Industrial Discharges on Publicly Owned Treatment Works*, prepared for U.S. Environmental Protection Agency, Office of Water Enforcement (McLean, VA: Nov. 20, 1981).

256. JRB Associates, *Addendum to the Report Entitled: Assessment of the Impacts of Industrial Discharges on Publicly Owned Treatment Works*, prepared for U.S. Environmental Protection Agency, EPA Contract No. 68-01-5052, DOW 54 (McLean, VA: Feb. 25, 1983).

257. JRB Associates, *Inventory of Air Pollution Control, Industrial Wastewater Treatment and Water Treatment Sludges*, prepared for U.S. Environmental Protection Agency, EPA Contract No. 68-01-6348 (McLean, VA: Dec. 19, 1983).

258. JRB Associates, *Pretreatment Compliance Assessment of Electroplating Facilities Operated by Large Corporations* (McLean, VA: August 1984).

259. Kamlet, K.S., "Dredged-Material Ocean Dumping: Perspectives on Legal and Environmental Impacts," in *Wastes in the Ocean*, vol. 2, D.R. Kester, et al. (eds.) (New York: John Wiley & Sons, 1983), pp. 29-70.

260. Kay, S.H., *Potential for Biomagnification of Contaminants Within Marine and Freshwater Food Webs*, Technical Report D-84-7 (Vicksburg, MS: U.S. Army Engineer Waterways Experiment Station, 1984).

261. Kester, D.R., Ketchum, B.H., et al., "Have the Questions Concerning Dredged-Material Disposal Been Answered?" in *Wastes in the Ocean*, vol. 2, D.R. Kester, et al. (eds.) (New York: John Wiley & Sons, 1983), pp. 275-287.

262. Keystone Ocean Project, *The Role of the Oceans in Hazardous Waste Management, Appendices* (Keystone, CO: The Keystone Center, 1986).

263. Keystone Ocean Project, *A Decisionmaking Process for Evaluating the Use of the Oceans in Hazardous Waste Management* (Keystone, CO: The Keystone Center 1987).

264. Keystone Siting Process Group, *The Keystone Siting Process Handbook—A New Approach to Siting Hazardous Waste Management Facilities* (Austin, TX: Texas Department of Health, January 1984).

265. Khalid, R.A., et al., "Toxic Contaminant Uptake in Dredged Sediment Marshes," in *Dredging and Dredged Material Disposal (Conference Proceedings)* (New York: American Society of Civil Engineers, November 1984), pp. 500-508.

266. Kimball, K.D., and Levin, S.A., "Limitations of Laboratory Bioassays: The Need for Ecosystem-Level Testing," *BioScience* 35(3):165-171, March 1985.

267. Kimbrough, R., et al., "Induction of Liver Tumors in Sherman Strain Female Rats by Polychlorinated Biphenyl Aroclor 1260," *Journal of the National Cancer Institute* 55:1453-1459, 1975.

268. King, K.A., and Cromartie, E., "Mercury, Cadmium, Lead, and Selenium in Three Waterbird Species Nesting in Galveston Bay, Texas, USA," *Colonial Waterbirds* 9:90-94, 1986.

269. Kizlauskas, A.G., and Homer, D., "RCRA EP Toxicity Test Applied to Dredged Material" in *Dredging and Dredged Material Disposal (Conference Proceedings)* (New York: American Society of Civil Engineers, November 1984), pp. 361-370.

270. Klaassen, C.D., and Doull, J., "Evaluation of Safety: Toxicologic Evaluation," in *Toxicology: The Basic Science of Poisons*, J. Doull, et al. (eds.) (New York: Macmillan Publishing Co., Inc., 1980), pp. 11-27.

271. Klauda, R.J., Peck, T.H., and Rice, G. K., "Accumulation of Polychlorinated Biphenyls in Atlantic Tomcod (*Microgadus Tomcod*) Collected From the Hudson River Estuary, New York," *Bulletin of Environmental Contamination and Toxicology* 27(6):829-835, December 1981.

272. Klein, L., et al., "Sources of Metals in New York City's Wastewater" (New York: N.Y. Water Pollution Control Association, 1974).

273. Klesch, W. L., "Need for Long-Term Planning for Dredged Material Containment Areas," in *Dredging and Dredged Material Disposal (Conference Proceedings)* (New York: American Society of Civil Engineers, November 1984), pp. 674-683.

274. Kluwe, W., et al., "Carcinogenicity Testing of Phthalate Esters and Related Compounds by the National Toxicology Testing Program and the National Cancer Institute," *Environmental Health Perspectives* 45:129-133, 1982.

275. Kneip, T., "Public Health Risk of Toxic Substances," in *Ocean Disposal of Municipal Wastewater: Impacts on the Coastal Environment*, vol. 2, E.P. Myers (ed.), MITSG 83-33 (Cambridge, MA: MIT Sea Grant College Program, 1983), pp. 579-610.

276. Knowles, S.C., and Fabel, D.J., *Water Quality Assessment, FY 1984-1985* (Columbia, SC: South Carolina Department of Health and Environmental Control, 1986).

277. Knudson, T.J., "Struggling L.I. Baymen Face a New Threat: Development," *New York Times*, pp. B1 and B17, April 1986.

278. Knudson, T.J., "With Striped Bass Ban, A Way of Life is Fading," *New York Times*, pp. 1,2,9, May 1986.

279. Koenigsberger, M.D., "3M," paper presented at Governor's Conference on Pollution Prevention Pays, Nashville, TN, March 1986.

280. Koh, R.C.Y., "Delivery Systems and Initial Dilution," in *Ocean Disposal of Municipal Wastewater: Impacts on the Coastal Environment*, vol. 1, E.P. Myers (ed.), MITSG 83-33 (Cambridge, MA: MIT Sea Grant College Program, 1983), pp. 129-175.

281. Kolbye, A., and Carr, C., "The Evaluation of the Carcinogenicity of Environmental Substances," *Regulatory Toxicology & Pharmacology* 4:350-354, 1984.

282. Kraft, M.E., and Kraut, R., "The Impact of Citizen Participation on Hazardous Waste Policy Implementation: The Case of Clermont County, Ohio," *Policy Studies Journal* 14(1):52-61, 1985.

283. Krier, J.E., "Ocean Discharge of Municipal Wastes: Legal and Institutional Aspects," in *Ocean Disposal of Municipal Wastewater: Impacts on the Coastal Environment*, vol. 2, E.P. Myers (ed.), MITSG 83-33 (Cambridge, MA: MIT Sea Grant College Program, 1983), pp. 659-705.

284. Krom, M.D., "An Evaluation of the Concept of Assimilative Capacity as Applied to Marine Waters," *Ambio* 15(4):208-214, 1986.

285. Ktsanes, V.K., Anderson, A.C., and Diem, J.E., *Health Effects of Swimming in Lake Pontchartrain at New Orleans*, EPA-600/1-81-027 (Washington, DC: March 1981).

286. Kullenberg, G.E., "Physical Oceanography Studies Related to Waste Disposal in the Sea," in *Wastes in the Ocean*, vol. 1, I.W. Duedall, et al. (eds.) (New York: John Wiley & Sons, 1983), pp. 87-101.

287. Kuratsune, M., Masuda, Y., and Nagayama, J., "Some of the Recent Findings Concerning Yusho," *Proceedings of the National Conference on PCBs*, EPA-560/6-75-004 (Washington, DC: 1976), pp. 14-29.

288. Kurgan, S.J., Balestrino, J.M., and Daley, J.R., *Coal Combustion By-Products Utilization Manual, Vol. I: Evaluating the Utilization Option* (Palo Alto, CA: Electric Power Research Institute, 1984).

289. Kyzer, B.I., "Management of Dredged Material Disposal Areas," in *Dredging and Dredged Material Disposal (Conference Proceedings)* (New York: American Society of Civil Engineers, November 1984), pp. 803-810.

290. Ladner, C., and Franks, J.S., *A Contingency Guide to the Protection of Mississippi Coastal Environments From Spilled Oil, Protection Priorities and Related Environmental Information* (Long Beach, MS: Mississippi Department of Wildlife Conservation, April 1984).

291. Lahey, W., "Ocean Dumping of Sewage Sludge: The Tide Turns From Protection to Management," *Harvard Environmental Law Review* 6(2):395-431, 1982.

292. Lahey, W., and Connor, M., "The Case for Ocean Waste Disposal," *Technology Review* 86(6):60-68, August/September 1983.

293. Landolt, M.L., et al., *Potential Toxicant Exposure Among Consumers of Recreationally Caught Fish From Urban Embayments of Puget Sound*, NOAA Technical Memorandum NOS OMA 23 (Rockville, MD: National Oceanic and Atmospheric Administration, November 1985).

294. LA/OMA Project, *Proposed Sludge Management Program for the Los Angeles/Orange County Metropolitan Area, Final Facilities Plan/Program and Summary of Final EIS/EIR* (San Francisco, CA: U.S. Environmental Protection Agency Region IX, October 1980).

295. Larsson, P., "Contaminated Sediments of Lakes and Oceans Act as Sources of Chlorinated Hydrocarbons for Release to Water and Atmosphere," *Nature* 317:347-349, 1985.

296. Lazor, R.L., Calhoun, C.C., and Patin, T.R., "The Corps' Environmental Effects of Dredging Program," in *Dredging and Dredged Material Disposal (Conference Proceedings)* (New York: American Society of Civil Engineers, November 1984), pp. 150-156.

297. LeChevallier, M.W., et al., "Changes in Virulence of Waterborne Enteropathogens With Chlorine Injury," *Applied and Environmental Microbiology* 50:412-419, 1985.

298. Ledo, A., et al., "Effect of Depuration Systems on the Reduction of Bacteriological Indicators in Cultured Mussels (*Mytilus edulis Linnaeus*)," *Journal of Shellfish Research* 1:59-64, 1983.

299. Lee, C.R., Peddicord, R.K., et al., *Application of the Resource Conservation and Recovery Act of 1976 to Dredged Material*, Internal Working Document D-86-1 (Vicksburg, MS: U.S. Army Engineer Waterways Experiment Station, 1986).

300. Lee, C.R., and Skogerboe, J.G., "Surface Runoff Water Quality From Contaminated Dredged Material Placed Upland," in *Dredging and Dredged Material Disposal (Conference Proceedings)* (New York: American Society of Civil Engineers, November 1984), pp. 509-513.

301. Lee, C.R., Smart, R.M., et al., *Prediction of

Heavy Metal Uptake by Marsh Plants Based on Chemical Extraction of Heavy Metals From Dredged Material, Technical Report D-78-6 (Vicksburg, MS: U.S. Army Engineer Waterways Experiment Station, 1978).

302. Lee, V., and Robadue, D.D., Jr., *Upper Narragansett Bay, An Urban Estuary in Transition, Preliminary Report*, Marine Technical Report 79 (Kingston, RI: University of Rhode Island Coastal Resources Center, 1980).

303. Lentz, S.A., "Plastics in the Marine Environment: A Call For International Action," paper presented at 6th International Ocean Disposal Symposium, Monterey Bay, CA, Apr. 21-25, 1986.

304. Leonard, H.J., "Confronting Industrial Pollution in Rapidly Industrializing Countries: Myths, Pitfalls, and Opportunities," *Ecology Law Quarterly* 12(4):779-816, 1985.

305. Leschine, T.M., "Economic Charges as Incentives for Pollution Control," in *Puget Sound Notes* (Seattle, WA: U.S. Environmental Protection Agency Region X, September 1986), pp. 3-6.

306. Leschine, T.M., and Broadus, J.M., "Economic and Operational Considerations of Offshore Disposal of Sewage Sludge," in *Wastes in the Ocean*, vol. 5, D.R. Kester, et al. (eds.) (New York: John Wiley & Sons, 1985), pp. 287-315.

307. Livingston, R.J., "Trophic Response of Fishes to Habitat Variability in Coastal Seagrass Systems," *Ecology* 65(4):1258-1275, August 1984.

308. Livingston, R.J., *The Ecology of the Apalachicola System: An Estuarine Profile*, FWS/OBS-82/05 (Washington, DC: U.S. Fish and Wildlife Service, September 1984).

309. Livingston, R.J., *Historic Trends of Human Impacts on Seagrass Meadows in Florida* (St. Petersburg, FL: Florida Department of Natural Resources, Bureau of Marine Research, manuscript dated Aug. 1, 1986).

310. Loehr, L.C., and Collias, E.E., "A Case Study of the Abuse of Science in Sewage Planning," *Marine Pollution Bulletin* 15(12):439-443, 1984.

311. Lomnitz, E., Bruins, R., and Fradkin, L., "Screening Chemicals in Municipal Sludges," *BioCycle* 26(7):52-54, October 1985.

312. London Dumping Convention, *Review of the Scientific and Technical Considerations Relevant to the Proposal for the Amendment of the Annexes to the London Dumping Convention Related to the Dumping of Radioactive Wastes*, report prepared by the Expanded Scientific Panel of Experts for June 3-7, 1985, Meeting (Agenda Item 2), LDC/PRAD.1/2 (London: Apr. 12, 1985).

313. Long, E.R., and Chapman, P.M., "A Sediment Quality Triad: Measures of Sediment Contamination, Toxicity and Infaunal Community Composition in Puget Sound," *Marine Pollution Bulletin* 16(10):405-415, 1985.

314. Lopik, J. et al., *Report of South Atlantic and Gulf Region Conference on Marine Pollution Problems* (Washington, DC: National Oceanic and Atmospheric Administration, 1980).

315. Louisiana Department of Environmental Quality, Office of Water Resources, Water Pollution Control Division, *Louisiana Water Quality Inventory Report* (Baton Rouge, LA: 1984).

316. Lowrance, W., *Of Acceptable Risk: Science and the Determination of Safety* (Los Altos, CA: William Kaufmann, 1976).

317. Loxham, M., et al., "Safe Landfilling of Contaminated Dredged Material," in *Dredging and Dredged Material Disposal (Conference Proceedings)* (New York: American Society of Civil Engineers, November 1984), pp. 625-633.

318. Lue-Hing, C., Zenz, D.R., and Prakasam, T.B.S., "Treatment Technologies and Effluent Quality of the Future, 2000+," in *Public Waste Management and the Ocean Choice*, K.D. Stolzenback, et al. (eds.), MITSG 85-36 (Cambridge, MA: MIT Sea Grant College Program, April 1986), pp. 27-72.

319. Lunz, J.D., Clarke, D.G., and Fredette, T.J., "Seasonal Restrictions on Bucket Dredging Operations," in *Dredging and Dredged Material Disposal (Conference Proceedings)* (New York: American Society of Civil Engineers, November 1984), pp. 371-383.

320. Lunz, J.D., Diaz, R.J., and Cole, R.A., *Upland and Wetland Habitat Development With Dredged Material: Ecological Considerations*, Technical Report DS-78-15 (Vicksburg, MS: U.S. Army Engineer Waterways Experiment Station, 1978).

321. Luoma, S.N., and Cloern, J.E., "The Impact of Waste-Water Discharge on Biological Communities in San Francisco Bay," in *San Francisco Bay: Use and Protection*, W.J. Kockelman, et al. (eds.) (Lawrence, KS: Allen Press, 1982), pp. 137-160.

322. Lynn, F.M., "The Interplay of Science and Values in Assessing Environmental Risk," *Science, Technology, Human Values* 11(2):40-50, Spring 1986.

323. Lytle, T.F., and Lytle, J.S., *Pollutant Transport in Mississippi Sound* (Ocean Springs, MS: Mississippi-Alabama Sea Grant Consortium, 1985).

324. Macknis, J., "Chesapeake Bay Nonpoint Source Pollution," in *Perspectives on Nonpoint Source Pollution*, EPA 440/5-85/-001 (Washington, DC: U.S. Environmental Protection Agency, Office of Water Regulations and Standards, 1985), pp. 165-171.

325. MacLeod, W., et al., *Analysis of Residual Chlo-*

rinated Hydrocarbons and Aromatic Hydrocarbons and Related Compounds in Selected Sources, Sinks, and Biota of New York Bight, NOAA Technical Memorandum, OMPA-6 (Boulder, CO: National Oceanic and Atmospheric Administration, 1981).

326. Malins, D.C., et al., "Toxic Chemicals in Marine Sediment and Biota From Mukilteo, Washington: Relationships With Hepatic Neoplasms and Other Hepatic Lesions in English Sole (Parophrys vetulus)," Journal of the National Cancer Institute 74(2):487-494, February 1985.

327. Management Advisory Group to the EPA Construction Grants Program, "Report to EPA: Municipal Compliance With the National Pollutant Discharge Elimination System" (Washington, DC: June 1986).

328. Manheim, B., "International Regulation of Plastic Pollution of the Marine Environment" (Washington, DC: manuscript dated Dec. 19, 1985).

329. Manheim, B., Testimony Before the Subcommittee on Coast Guard and Navigation, House Committee on Merchant Marine and Fisheries, U.S. Congress, Hearing on "Plastic Pollution in the Marine Environment," 99th Cong., 2d sess., Serial No. 99-47 (Washington, DC: U.S. Government Printing Office, 1986).

330. Maragos, J.E., Evans, C., and Holthus, P., "Reef Corals in Kaneohe Bay Six Years Before and After Termination of Sewage Discharges (Oahu Hawaiian Archipelago)," HAWAU-R-85-017 (UNIHI-SG-CP-85-09) (Honolulu, HI: Hawaii Sea Grant, 1985).

331. Marine Technology Society, et al., Monitoring Strategies Symposium, vol. 3 of Oceans '86 Conference Record (Washington, DC: Sept. 23-25, 1986).

332. Marine Technology Society, et al., Organotin Symposium, vol. 4 of Oceans '86 Conference Record (Washington, DC: Sept. 23-25, 1986).

333. Marion, K.R., and Settine, R.L., "Organic Pollutant Levels in Bivalves of Mobile Bay, Annual Summary" (Birmingham, AL: University of Alabama at Birmingham, 1984).

334. Martin, S.O., "Last Chance for Chesapeake Bay," Planning Magazine 52:12-19, June 1986.

335. Maryland Department of Health and Mental Hygiene, Office of Environmental Programs, Maryland Water Quality Inventory (Baltimore, MD: Apr. 15, 1986).

336. Massachusetts Department of Environmental Quality Engineering, Division of Water Pollution Control, Commonwealth of Massachusetts Summary of Water Quality, 1986; Appendix III—Basin/Segment Information (Westborough, MA: 1986).

337. Massachusetts Executive Office of Environmental Affairs, Department of Fisheries, Wildlife & Environmental Law Enforcement, Massachusetts Marine Fisheries, Assessment at Mid-Decade (Boston, MA: November 1985).

338. Massachusetts Institute of Technology Sea Grant Program, "Sewage Disposal and the Ocean: The Sea Grant Role," MITSG Report No. 85-6 (Cambridge, MA: MIT Sea Grant College Program, 1985).

339. Massachusetts Office of Coastal Zone Management, PCB Pollution in the New Bedford, Massachusetts Area: A Status Report (Boston, MA: 1982).

340. Matta, M.B., Mearns, A.J., and Buchman, M.F., Trends in DDT and PCBs in U.S. West Coast Fish and Invertebrates (Seattle, WA: National Oceanic and Atmospheric Administration, National Ocean Service, March 1986).

341. Matthews, T.D., and Marcus, J.M., "The Need for Biological Resource Data in Addition to Contaminant Concentration Data," in Oceans '86 Conference Record, vol. 3, 86CH2363-0 (Washington, DC: Marine Technology Society and IEEE Ocean Engineering Society, Sept. 23-25, 1986), pp. 999-1003.

342. Matthiesen, P., Men's Lives: The Surfmen and Baymen of the South Fork (New York: Random House, 1986).

343. Mayer, G.F. (ed.), Ecological Stress and the New York Bight (Columbia, SC: Estuarine Research Federation, 1982).

344. McConnell, K.E., and Industrial Economics, Inc., The Damages to Recreational Activities From PCBs in the New Bedford Harbor, report prepared for National Oceanic and Atmospheric Administration, Ocean Assessment Division, RA7-0400 (Rockville, MD: undated).

345. McConnell, K.E., and Morrison, B.G., Assessment of Economic Damages to the Natural Resources of New Bedford Harbor: Damages to the Commercial Lobster Fishery, report prepared for National Oceanic and Atmospheric Administration, Ocean Assessment Division, RA7-0200 (Rockville, MD: June 4, 1986).

346. McCormick, J.M., et al., "Partial Recovery of Newark Bay, NJ, Following Pollution Abatement," Marine Pollution Bulletin 14(5):188-197, 1983.

347. McGovern, T., State of Maine 1986 Water Quality Assessment (Augusta, ME: Department of Environmental Protection, 1986).

348. Meade, N.F., and Leeworthy, V., Public Expend-

itures on Outdoor Recreation in the Coastal Areas of the USA (Washington, DC: National Oceanic and Atmospheric Administration, March 1986).

349. Meade, R.H., "Landward Transport of Bottom Sediments in Estuaries of the Atlantic Coastal Zone," *Journal of Sedimentary Petrology* 39(1): 222-234, 1969.

350. Mearns, A.J., "Ecological Effects of Ocean Sewage Outfalls: Observations and Lessons," *Oceanus* 24(1):45-54, 1981.

351. Mearns, A.J., *Biological Effects Studies at Various Levels Along the U.S. Pacific Coast*, paper presented at NATO Advanced Research Workshop on Scientific Basis for the Role of the Oceans as a Waste Disposal Option, Vilamoura, Portugal, Apr. 24-30, 1985.

352. Mearns, A.J., et al., "Effects of Nutrients and Carbon Loadings on Communities and Ecosystems," in *Ecological Stress and the New York Bight: Science and Management*, G.F. Mayer (ed.) (Columbia, SC: Estuarine Research Federation, 1982), pp. 53-66.

353. Mearns, A.J., et al., *The Historical Trend Assessment Program, PCBs and Chlorinated Pesticide Contamination in U.S. Fish and Shellfish: An Assessment Report* (Seattle WA: National Oceanic and Atmospheric Administration, Ocean Assessments Division, November 1986).

354. Meistrell, J.C., and Montagne, D.E., "Waste Disposal in Southern California and Its Effects on the Rocky Subtidal Habitat," in *The Effects of Waste Disposal on Kelp Communities*, W. Bascom (ed.) (La Jolla, CA: University of California Institute of Marine Resources, 1983), pp. 84-102.

355. Mendelsohn, R. *Assessment of Economic Damages: Analysis of Residential Property Values in the New Bedford Area*, report prepared for National Oceanic and Atmospheric Administration, Ocean Assessments Division (Rockville, MD: June 2, 1986).

356. Meyer, E.L., "History of Sewerage Facilities Serving the City of Tijuana, Baja California, Mexico," staff report (San Diego, CA: California Regional Water Quality Control Board, Sept. 16, 1983).

357. Meyers, P.A., "Processes Contributing to the Concentration of Polychlorinated Biphenyls at the Sea-Surface Microlayer," paper presented at the Sea-Surface Microlayer Workshop, Arlie, VA, Dec. 18, 1985.

358. Minnesota Waste Management Board, *Hazardous Waste Management Plan—Revised Draft* (St. Paul, MN: February 1984).

359. Mississippi Department of Natural Resources, *1984 Water Quality Report to Congress* (Jackson, MS: June 1984).

360. Mitchell, R.C., "Public Opinion and Environmental Politics in the 1970s and 1980s," in *Environmental Policy in the 1980s: Reagan's New Agenda*, N.J. Vig and M.E. Kraft (eds.) (Washington, DC: Congressional Quarterly, 1984), pp. 51-74.

361. Mitchell, R.C., and R. Carson, "Property Rights, Protest, and the Siting of Hazardous Waste Facilities," *American Economic Review* 76(2):285-290, May 1986.

362. Monaco, M.E., et al., *Strategic Assessments of the Nation's Estuaries: Activities of NOAA's Ocean Assessments Division* (Rockville, MD: National Oceanic and Atmospheric Administration, 1986).

363. Montgomery, R.L., "Engineering Aspects of Capping Dredged Material," Proceedings of the 16th Dredging Seminar (College Station, TX: Texas A&M University, 1985).

364. Moon, R.E., "Point Source Discharge in the Tampa Bay Area," in *Proceedings Tampa Bay Area Scientific Information Symposium*, S.F. Treat, et al. (eds.) (New York: Bellwether, 1985), pp. 551-562.

365. Moore, J., et al., "Comparative Toxicity of Three Halogenated Dibenzofurans in Guinea Pigs, Mice and Rhesus Monkeys," *Annals of the New York Academy of Sciences* 320:151-163, 1979.

366. Morita, R.Y., "Starvation and Miniaturization of Heterotrophs, With Special Emphasis on Maintenance of the Starved Viable State," in *Bacteria in Their Natural Environments*, M. Fletcher and G.D. Floodgate (eds.) (Orlando, FL: Academic Press, Inc., 1985), pp. 111-130.

367. Morse, D.L., Guzewich, J.J., et al., "Widespread Outbreaks of Clam- and Oyster-Associated Gastroenteritis," *New England Journal of Medicine* 314:678-681, Mar. 13, 1986.

368. Motes, M.L., Jr., "Effect of Chlorinated Wash Water on *Vibrio cholerae* in Oyster Meats," *Journal of Food Science* 47:1028-1029, 1982.

369. Mrak, E., *Report of the Secretary's Commission on Pesticides and Their Relationship to Environmental Health* (Washington, DC: U.S. Department of Health, Education, and Welfare, 1969).

370. Mueller, J.A., et al., "Contaminants in the New York Bight," *Water Pollution Control Federation Journal* 48(10):2309-2326, 1976.

371. Muir, W.D., "History of Ocean Disposal in the Mid-Atlantic Bight," in *Wastes in the Ocean*, vol. 1, I.W. Duedall, et al. (eds.) (New York: John Wiley & Sons, 1983), pp. 273-291.

372. Murai, Y., and Juroiwa, Y., "Peripheral Neuropathy in Chlorobiphenyl Poisoning," *Neurology* 21:1173-1176, 1971.

373. Murchelano, R.A., "Epizootic Carcinoma in the Winter Flounder, *Pseudopleuronectes americanus*," *Science* 228:587-589, May 3, 1985.

374. Murphy, L.S., and Hoar, P.R., "Effect of Pollutants on Marine Phytoplankton at 106-Mile Site," in *Assessment Report on the Effects of Waste Dumping in 106-Mile Ocean Waste Disposal Site: Dumpsite Evaluation Report 81-1* (Boulder, CO: National Oceanic and Atmospheric Administration, 1981), pp. 309-319.

375. Myers, E.P. (ed.), *Ocean Disposal of Municipal Wastewater: Impacts on the Coastal Environment*, 2 vols., MITSG 83-33 (Cambridge, MA: MIT Sea Grant College Program, 1983).

376. Myers, M.S., and Rhodes, L.D., "Morphologic Similarities and Parallels in Geographic Distribution of Suspected Toxicopathic Liver Lesions in Rock Sole (*Lepidopsetta bilineata*), Starry Flounder (*Platichthys stellatus*), Pacific Staghorn Culpin (*Leptocottus armatus*), and Dover Sole (*Microstomus pacificus*) as Compared to English Sole (*Parophrys vetulus*) From Urban and Non-Urban Embayments in Puget Sound, Washington," paper presented at Toxic Chemicals and Aquatic Life Symposium, sponsored by National Marine Fisheries Service, Seattle, WA, Sept. 16-18, 1986.

377. National Advisory Committee on Oceans and Atmosphere, *The Role of the Ocean in a Waste Management Strategy, A Special Report to the President and the Congress* (Washington, DC: U.S. Government Printing Office, 1981).

378. National Advisory Committee on Oceans and Atmosphere, *Nuclear Waste Management and the Use of the Sea, A Special Report to the President and the Congress* (Washington, DC: April 1984).

379. National Research Council, Committee on Salmonella, *An Evaluation of the Salmonella Problem* (Washington, DC: 1969).

380. National Research Council, Commission on Natural Resources, *Assessing Potential Pollutants: A Report of the Study Panel on Assessing Potential Ocean Pollutants to the Ocean Affairs Board* (Washington, DC: 1975).

381. National Research Council, *Drinking Water and Health* (Washington, DC: National Academy of Sciences, 1977).

382. National Research Council, Commission on Natural Resources, *Multimedium Management of Municipal Sludge* (Washington, DC: National Academy of Sciences, 1978).

383. National Research Council, *Lead in the Human Environment* (Washington, DC: National Academy Press, 1980).

384. National Research Council, Marine Board, *Drilling Discharges in the Marine Environment* (Washington, DC: National Academy Press, 1983).

385. National Research Council, Board on Ocean Science and Policy, *Disposal of Industrial and Domestic Wastes, Land and Sea Alternatives* (Washington, DC: National Academy Press, 1984).

386. National Research Council, Commission on Life Sciences, *Toxicity Testing: Strategies to Determine Needs and Priorities* (Washington, DC: National Academy Press, 1984).

387. National Research Council, Marine Board, *Ocean Disposal Systems for Sewage Sludge and Effluent* (Washington, DC: National Academy Press, 1984).

388. National Research Council, Marine Board, *Dredging Coastal Ports: An Assessment of the Issues* (Washington, DC: National Academy Press, 1985).

389. National Research Council of the United States and The Royal Society of Canada, *The Great Lakes Water Quality Agreement: An Evolving Instrument for Ecosystem Management* (Washington, DC: National Academy Press, 1985).

390. Neal, R.W., *Evaluation of Submerged Discharge of Dredged Material Slurry During Pipeline Dredge Operations*, Technical Report D-78-44 (Vicksburg, MS: U.S. Army Engineer Waterways Experiment Station, 1978).

391. New Jersey Public Interest Research Group, *Enforcement Under the Federal Water Pollution Control Act by the U.S. Environmental Protection Agency Region II, 1975-1980* (Trenton, NJ: 1981).

392. New York State Assembly, *SPDES—A System in Stress* (Albany, NY: Committee on Environmental Conservation, Dec. 15, 1982).

393. New York State Department of Environmental Conservation, "Regulatory Impact Statement, Required Under the State Administrative Procedures Act for the Amend of Numeral Six NYCRR Parts 10, 11, and 43" (New York: July 15, 1986).

394. New York State Department of Environmental Conservation, *New York State Water Quality, 1986* (New York: 1986).

395. Nichols, F.H., et al., "The Modification of an Estuary," *Science* 231:567-573, 1986.

396. Noga, E.J., and Dykstra, M.J., "Oomycete Fungi Associated With Ulcerative Mycosis in Menhaden, *Brovoortia tyrannus* (Latrobe)," *Journal of Fish Diseases* 9:47-53, 1986.

397. North Carolina Department of Natural Resources

and Community Development, Division of Environmental Management, *Chowan Albermarle Action Plan*, Report No 82-02 (Raleigh, NC: 1982).

398. North Carolina Department of Natural Resources and Community Development, Division of Environmental Management, *The Impact of Pulp Mill Effluent on the Chowan River Herring Fishery*, Report No. 83-08 (Raleigh, NC: 1983).

399. North Carolina Department of Natural Resources and Community Development, Division of Environmental Management, *Assessment of Surface Water Quality in North Carolina*, Report No. 85-01 (Raleigh, NC: 1985).

400. Norton, M.G., and Champ, M.A., "The Influence of Site-Specific Characteristics of the Effects of Sewage Sludge Dumping," paper presented at 4th International Ocean Disposal Symposium, Plymouth, England, April 1983.

401. Norton, M.G., "Colliery-Waste and Fly-Ash Dumping Off the Northeastern Coast of England," in *Wastes in the Ocean*, vol. 4, I.W. Duedall, et al. (eds.) (New York: John Wiley & Sons, 1985), pp. 423-448.

402. Norton, M.G., "Disposal of Sewage Sludge—International Outlook," undated manuscript.

403. Norton, V., Smith, T., and Strand, I. (eds.), *Stripers: The Economic Value of the Atlantic Coast Commercial and Recreational Striped Bass Fisheries*, MDU-T-84-001 (College Park, MD: Maryland Sea Grant, 1984).

404. Nriagu, J. (ed.), *The Biogeochemistry of Mercury in the Environment* (New York: Elsevier, 1979).

405. Nriagu, J. (ed.), *Cadmium in the Environment* (New York: Wiley-Interscience, 1981).

406. O'Brien, J.F., "PCB-Striped Bass," Interoffice Memo to J.M. Cronan and J. Stolgitis (Providence, RI: State of Rhode Island and Providence Plantations, Division of Fish and Wildlife, dated Mar. 17, 1986).

407. O'Connor, J.M., and Huggett, R.J., "Aquatic Pollution Problems, North Atlantic Coast, Including Chesapeake Bay," paper presented at Toxic Chemicals and Aquatic Life Symposium, sponsored by National Marine Fisheries Service, Seattle, WA, Sept. 16-18, 1986.

408. O'Connor, J.M., Klotz, J.B., and Kneip, T.J., "Sources, Sinks, and Distribution of Organic Contaminants in the New York Bight Ecosystem," in *Ecological Stress and the New York Bight: Science and Management*, G.F. Mayer (ed.) (Columbia, SC: Estuarine Research Federation, 1982), pp. 631-654.

409. O'Connor, J.M., and Kneip, T.J., *Human Health Effects of Waste Constituents*, contract prepared for U.S. Congress, Office of Technology Assessment (Tuxedo Park, NY: New York University Medical Center, 1986).

410. O'Connor, J.M., and Rachlin, J.W., "Perspectives on Metals in New York Bight Organisms: Factors Controlling Accumulation and Body Burdens," in *Ecological Stress and the New York Bight: Science and Management*, G.F. Mayer (ed.) (Columbia, SC: Estuarine Research Federation, 1982), pp. 655-673.

411. O'Connor, J.M., Wiedow, A., et al., "Cadmium in Blue Crabs From the Hudson River Estuary: Preliminary Health Risk Assessment," *Estuaries* 6:320, 1983.

412. O'Connor, J.S., and Dewling, R.T., "Indices of Marine Degradation, Their Utility," paper presented at 8th National Symposium on Statistics, Law, and Environment, Washington, DC, National Academy of Sciences, Oct. 15-16, 1984.

413. O'Connor, J.S., Murchelano, R.A., and Ziskowski, J.J., *Index of Pollutant-Induced Fish and Shellfish Disease*, NOAA Special Report (Washington, DC: National Oceanic and Atmospheric Administration, 1987).

414. O'Connor, J.S., Pugh, W.L., Wolfe, D.A., and Dewling, R.T., "Protection of Natural Resources Through Environmental Indices," *Sea Technology* 27(9):31-33, September 1986.

415. O'Connor, J.S., and Swanson, R.L., "Unreasonable Degradation of the Marine Environment—What Is It," in *Oceans 82 Conference Proceedings* (Washington, DC: Marine Technology Society and Institute of Electrical and Electronics Engineers, Sept. 20-22, 1982), pp. 1125-1132.

416. O'Connor, T.P., et al., "Projected Consequences of Dumping Sewage Sludge at a Deep Ocean Site Near New York Bight," *Canadian Journal of Fisheries and Aquatic Sciences* 40 (Suppl. 2):228-241, 1981.

417. O'Connor, T.P., et al., "Scales of Biological Effects," in *Oceanic Processes in Waste Disposal, Volume 2, Physical and Chemical Processes*, T.P. O'Connor, et al. (eds.) (Malabar, FL: Krieger Press, in press).

418. Odum E.P., *Fundamentals of Ecology* (Philadelphia: W.B. Saunders Co., 1971).

419. Officer, C.B., et al., "Chesapeake Anoxia: Origin, Development, and Significance," *Science* 223: 22-27, 1984.

420. O'Halloran, R.L., "Ocean Dumping: Progress Toward A Rational Policy of Dredged Waste Disposal," *Environmental Law* 12:745-772, spring 1982.

421. Ohlendorf, H.M., et al., "Organochlorines and Mercury in Eggs of California Coastal Terns and Herons," undated manuscript, circa 1985.

422. Ohlendorf, H.M., et al., "Reproduction and Or-

ganochlorine Contaminants in Terns at San Diego Bay," *Colonial Waterbirds* 8(1):42-53, 1985.

423. Ohlendorf, H.M., et al., "Tissue Distribution of Trace Elements and DDE in Brown Pelicans," *Bulletin of Environmental Contamination and Toxicology* 35:183-192, 1985.

424. Ohlendorf, H.M.. et al., "Selenium and Heavy Metals in San Francisco Bay Diving Ducks," *Journal of Wildlife Management* 50(1):64-71, 1986.

425. Oregon Department of Environmental Quality, *Oregon 1986 Water Quality Program Assessment and Program Plan for Fiscal Year 1987* (Salem, OR: 1986).

426. Orr, M.H., and Baxter L., "Dispersion of Particles After Disposal of Industrial and Sewage Wastes," in *Wastes in the Ocean*, vol. 1, I.W. Duedall, et al. (eds.) (New York: John Wiley & Sons, 1983), pp. 117-137.

427. Orth, R.J., and Moore, K.A., "Chesapeake Bay: An Unprecedented Decline in Submerged Aquatic Vegetation," *Science* 222:51-53, 1983.

428. Overstreet, R.M., "Aquatic Pollution Problems, Southeastern U.S. Coasts: Histopathological Indicators," *Aquatic Toxicology*, in press, 1987.

429. Page, A.L., et al. (eds.), *Utilization of Municipal Wastewater and Sludge on Land, Proceedings of Workshop* (Riverside, CA: University of California, 1983).

430. Pain, S., "Are British Shellfish Safe to Eat?" *New Scientist* 111:29-33, Aug. 28, 1986.

431. Palermo, M.R., "Design of Confined Disposal Areas for Retention of Contaminants," in *Dredging and Dredged Material Disposal (Conference Proceedings)* (New York: American Society of Civil Engineers, November 1984), pp. 858-865.

432. Palermo, M.R., Montgomery, R.L., and Poindexter, M.E., *Guidelines for Designing, Operating, and Managing Dredged Material Containment Areas*, Technical Report DS-78-10 (Vicksburg, MS: U.S. Army Engineer Waterways Experiment Station, 1978).

433. Parker, J.H., et al., "Coal-Waste Blocks for Artificial-Reef Establishment: A Large-Scale Experiment," in *Wastes in the Ocean*, vol. 4, I.W. Duedall, et al. (eds.) (New York: John Wiley & Sons, 1985), pp. 537-556.

434. Payton, B., "Ocean Dumping in the New York Bight," *Environment* 27:26-32, November 1985.

435. Payton, B.M., "Blight in the Bight: Sewage and Water Don't Mix," *Oceans* 18(3): 63-67, May 1985.

436. Peddicord, R.K., "Aquatic Bioassays in Dredged Material Management," in *Dredging and Dredged Material Disposal (Conference Proceedings)* (New York: American Society of Civil Engineers, November 1984), pp. 347-360.

437. Peddicord, R.K., and Hansen, J.C., "Technical Implementation of the Regulations Governing Ocean Disposal of Dredged Material," in *Wastes in the Ocean*, vol. 2, D.R. Kester, et al. (eds.) (New York: John Wiley & Sons, 1983), pp. 71-88.

438. Peddicord, R.K., Lee, C.R., et al., "General Decisionmaking Framework for Management of Dredged Material: Example Application to Commencement Bay, Washington," Miscellaneous Paper (Vicksburg, MS: U.S. Army Engineer Waterways Experiment Station, 1986).

439. Pequegnat, W.E., *An Assessment of the Potential Impact of Dredged Material Disposal in 'the Open Ocean*, Technical Report D-78-2 (Vicksburg, MS: U.S. Army Engineer Waterways Experiment Station, 1978).

440. Pequegnat, W.E., "Some Aspects of Deep Ocean Disposal of Dredged Material," in *Wastes in the Ocean*, vol. 2, D.R. Kester, et al. (eds.) (New York: John Wiley & Sons, 1983), pp. 229-252.

441. Pequegnat, W.E., *Disposal of Dredged Material, Case Studies*, contract report prepared for U.S. Congress, Office of Technology Assessment (College Station, TX: Jan. 6, 1986).

442. Pequegnat, W.E., *An Overview of the Scientific and Technical Aspects of Dredged Material Disposal in the Marine Environment*, contract report prepared for U.S. Congress, Office of Technology Assessment (College Station, TX: Jan. 18, 1986).

443. Perret, W., and Condrey, R.E., "Living Resources: Management Conflicts, Potentials," paper presented at Estuary of the Month Seminar: Louisiana's Estuaries, sponsored by NOAA Estuarine Programs Office, U.S. EPA, and U.S. Fish and Wildlife Service, Washington, DC, November 1986.

444. Pershagen, G., "The Epidemiology of Human Arsenic Exposure," in *Biological and Environmental Effects of Arsenic*, B. Fowler (ed.) (New York: Elsevier, 1983), pp. 199-232.

445. Pierce, J.J., and Cahill, L., "State Programs to Control Municipal Sludge," *Journal of Environmental Engineering* 110(1):15-26, 1984.

446. Pipkin, B.W., et al., *Laboratory Exercises in Oceanography* (San Francisco, CA: W.H. Freeman & Co., 1977).

447. Pizza, J., and O'Connor, J., "PCB Dynamics in Hudson River Striped Bass, II. Accumulation From Dietary Sources," *Aquatic Toxicology* 3:313-327, 1983.

448. Poiger, H., and Schlatter, C., "Animal Toxicology of Chlorinated Dibenzo-p-dioxins," *Chemosphere* 12:447-462, 1983.

449. Poindexter, M.E., "Long-Term Management of Confined Disposal Areas," in *Dredging and Dredged Material Disposal (Conference Proceedings)* (New York: American Society of Civil Engineers, November 1984), pp. 886-895.

450. Portney, K.E., "The Potential of the Theory of Compensation for Mitigating Public Opposition to Hazardous Waste Treatment Facility Siting: Some Evidence From Five Massachusetts Communities," *Policy Studies Journal* 14(1):81-89, 1985.

451. Portnoy, B.L., et al., "Oyster Associated Hepatitis—Failure of Shelfish Certification Programs to Prevent Outbreaks," *Journal of the American Medical Association* 233:1065, 1975.

452. Poukish, C.A., and Allison, J.T., "1985 Maryland Fish Kill Summary," draft (Annapolis, MD: Maryland Department of Health and Mental Hygiene, Office of Environmental Programs, 1986).

453. Price, C.M., "Private Enforcement of the Clean Water Act," *Natural Resources and Environment* 1(4):31-33,59, winter 1986.

454. Priede-Sedgwick, Inc., *Sarasota Bay Water Quality Study* (Tallahassee, FL: Florida Department of Environmental Regulation, Jan. 21, 1983).

455. Pruell, R.J., Hoffman, E.J., and Quinn, J.G., "Total Hydrocarbons, Polycyclic Aromatic Hydrocarbons and Synthetic Ogranic Compounds in the Hard Shell Clam, *Mercenaria mercenaria*, Purchased at Commercial Seafood Stores," *Marine Environmental Research* 11:163-181, 1984.

456. Pruter, A.T., and Alverson, D.L. (eds.), *The Columbia River Estuary and Adjacent Ocean* (Seattle, WA: University of Washington Press, 1972).

457. Public Voice for Food and Health Policy, *Great American Fish Scandal: Health Risks Unchecked* (Washington, DC: 1986).

458. Puckett, J., Dodds, R., and Crosby, L., *License to Pollute: The National Pollutant Discharge Elimination System in Washington State* (Seattle, WA: Greenpeace, May 21, 1985).

459. Puget Sound Water Quality Authority, *Annual Report* (Olympia, Washington: 1984).

460. Puget Sound Water Quality Authority, *Combined Sewer Overflow*, Issue Paper (Seattle, WA: April 1986).

461. Puget Sound Water Quality Authority, *Industrial Pretreatment*, Issue Paper (Seattle, WA: April 1986).

462. Puget Sound Water Quality Authority, *Industrial and Municipal Discharges*, Issue Paper (Seattle, WA: May 1986).

463. Puget Sound Water Quality Authority, *The State of the Sound 1986* (Seattle, WA: July 1986).

464. Puget Sound Water Quality Authority, *1987 Puget Sound Water Quality Management Plan; Environmental Impact Statement*, draft (Seattle, WA: Sept. 17, 1986).

465. Rabalais, N.N., and Boesch, D.F., "Extensive Depletion of Oxygen in Bottom Waters of the Louisiana Shelf During 1985" (draft manuscript, 1986).

466. Rabalais, N.N., Dagg, M.J., and Boesch, D.F., *Nationwide Review of Oxygen Depletion and Eutrophication in Estuarine and Coastal Waters: Gulf of Mexico*, final report to National Oceanic and Atmospheric Administration (LA: Louisiana Universities Marine Consortium, 1985).

467. Rappe, C., Buser, H., and Bosshardt, H., "Dioxins, Dibenzofurans, and Other Polyhalogenated Aromatics: Production, Use, Formation, and Destruction," *Annals of the New York Academy of Sciences* 320:1-18, 1979.

468. Ravan, J.E., Testimony Before the Subcommittee on Investigations and Oversight, House Committee on Public Works and Transportation, U.S. Congress, "Hearings on the Implementation of the Federal Clean Water Act," Mar. 8 and Sept. 19, 1984, 98th Cong., Committee Serial 98-81 (Washington, DC: U.S. Government Printing Office, 1985).

469. ReNewsletter, "Metro Begins Large-Scale Silvigrow Applications," *ReNewsletter* vol. 2, No. 3 (Seattle, WA: Metro, November 1985).

470. Reddy, C., Dorn, C., et al., "Municipal Sewage Sludge Application on Ohio Farms: Tissue Metal Residues and Infections," *Environmental Research* 38:360-376, 1985.

471. Reddy, C., and Dorn, C., "Municipal Sewage Sludge Application on Ohio Farms: Estimation of Cadmium Intake," *Environmental Research* 38:377-388, 1985.

472. Reed, A.W., *Ocean Waste Disposal Practices* (Park Ridge, NJ: Noyes Data Corp., 1975).

473. Reed, M., and Bierman, Jr., V.J., (eds.), *Proceedings of a Workshop for the Development of a Scientific Protocol for Ocean Dump Site Designation*, prepared for the U.S. Environmental Protection Agency, Office of Water Regulations and Standards, Criteria and Standards Division, held at University of Rhode Island, Kingston, RI, Feb. 14-17, 1983.

474. Reid, B., "TBT, A Chemical On Trial," special report in the Daily Press and Times-Herald newspapers (Richmond, VA: Apr. 13-20, 1986).

475. Reinisch, C.L., Charles, A.M., and Stone, A.M.,

"Epizootic Neoplasia in Soft Shell Clams Collected From New Bedford Harbor," *Hazardous Waste* 1(1):73-81, 1984.

476. Renaud, M.L., "Hypoxia in Louisiana Coastal Waters During 1983: Implications for Fisheries," *Fishery Bulletin* 84(1):19-26, January 1986.

477. Resources For the Future, "An Overview of the RFF Environmental Data Inventory: Methods, Sources and Preliminary Results, Volume 1" (Washington, DC: September 1984).

478. Resources For the Future, "Pollutant Discharges to Surface Waters in Coastal Regions," contract prepared for U.S. Congress, Office of Technology Assessment (Washington, DC: February 1986).

479. Rhan, W.R., Jr., "The Role of *Spartina alterniflora* in the Transfer of Mercury in a Salt Marsh," M.S. Thesis (Atlanta, GA: Georgia Institute of Technology, 1973).

480. Rhode Island Department of Environmental Management, Division of Water Resources, *The State of the State's Waters—Rhode Island: A Report to Congress* (Providence, RI: 1986).

481. Rhodes, M.W., and Kator, H.I., *In Situ Survival of Enteric Bacteria in Estuarine Environments*, Bulletin 140 (Blacksburg, VA: Virginia Polytechnic Institutes and State University, Virginia Water Resources Research Center, 1983).

482. Rieser, A., and Spiller, J., *Regulating Drilling Effluents on Georges Bank and the Mid-Atlantic Outer Continental Shelf: A Scientific and Legal Analysis* prepared for the States of Maine, New Hampshire, Massachusetts, and New Jersey (Augusta, ME: April 1981).

483. Riley, R.G., et al., *Organic and Inorganic Toxicants in Sediment and Marine Birds From Puget Sound*, NOAA Technical Memorandum NOS OMS 1 (Rockville, MD: National Oceanic and Atmospheric Administration, National Ocean Service, June 1983).

484. Risebrough, R.W., Reichle, P., et al., "Polychlorinated Biphenyls in the Global Ecosystem," *Nature* 220:1098-1102, 1968.

485. Risebrough, R.W., Sibley, F.C., and Kirven, M.N., "Reproductive Failure of the Brown Pelican on Anacapa Island in 1969," *American Birds* 25(1):8-9, February 1971.

486. Robinson, K., *New Jersey 1986 State Water Quality Inventory Report* (Trenton, NJ: New Jersey Department of Environmental Protection, Division of Water Resources, July 1986).

487. Rosbe, W., and Gulley, R., "The Hazardous and Solid Waste Amendments of 1984: A Dramatic Overhaul of the Way America Manages Its Hazardous Wastes," *Environmental Law Reporter* 14(12): 10458-10467, 1984.

488. Rose, C.D., Ward, T.J., de Pass, V.E., "Ecological Assessment for Coal Ash Dumped Off the Northeastern Coast of England," in *Wastes in the Ocean*, vol. 4, I.W. Duedall, et al. (eds.) (New York: John Wiley & Sons, 1985), pp. 389-422.

489. Rosenberg, M.L., et al., "Shigellosis From Swimming," *Journal of the American Medical Association* 236:1849-1852, 1976.

490. Royston, M.G., *Pollution Prevention Pays* (New York: Pergamon Press, 1979).

491. Rubinstein, N., Lorea, E., and Gregory, N., "Accumulation of PCBs, Mercury, and Cadmium by *Nereis virens*, *Mercenaria mercenaria* and *Palaemnetes pugio* From Contaminated Harbor Sediment," *Aquatic Toxicology* 3:249-260, 1983.

492. Rulifson, R.A., et al., *Anadromous Fish in the Southeastern United States and Recommendations for Development of a Management Plan* (Atlanta, GA: U.S. Fish and Wildlife Service, Fishery Resources, 1982).

493. Russell, P.P., et al., "Water and Waste Inputs to San Francisco Estuary—An Historical Perspective," in *San Francisco Bay: Use and Protection*, W.J. Kockelman, et al. (eds) (Lawrence, KS: Allen Press, 1982), pp. 127-136.

494. Safe, S., "Polychlorinated Biphenyls (PCBs) and Polybrominated Biphenyls (PBBs): Biochemistry, Toxicology and Mechanism of Action," *CRC Critical Reviews in Toxicology* 13:319-393, 1984.

495. Sass, S.L., and Murchelano, R.A., "Hepatic Tumors and Other Liver Pathology in Massachusetts Flatfish," paper presented at Toxic Chemicals and Aquatic Life Symposium, sponsored by National Marine Fisheries Service, Seattle, WA, Sept. 16-18, 1986.

496. Saucier, R.T., et al., *Executive Overview and Detailed Summary*, Technical Report DS-78-22 (Vicksburg, MS: U.S. Army Engineer Waterways Experiment Station, 1978).

497. Save the Bay, Inc., *Down the Drain: Toxic Pollution and the Status of Pretreatment in Rhode Island* (Providence, RI: 1986).

498. Sawyer, C.N., and McCarty, P.L., *Chemistry for Environmental Engineering* (New York: McGraw-Hill Book Co., 1978).

499. Schafer, H.A., "Chlorinated Hydrocarbons in Marine Mammals," in *Coastal Water Research Project Biennial Report, 1983-1984*, W. Bascom (ed.) (Long Beach, CA: Southern California Coastal Water Research Project, 1984), pp. 108-114.

500. Schubel, J.R., et al., *Field Investigations of the*

Nature, Degree and Extent of Turbidity Generated by Open-water Pipeline Disposal Operations, Technical Report D-78-30 (Vicksburg, MS: U.S. Army Engineer Waterways Experiment Station, 1978).

501. Schwartz, R.E., and Hackett, D.P., "Citizen Suits Against Private Industry Under the Clean Water Act," *Natural Resources Lawyer* 17(3):327-372, 1984.

502. Science Applications International Corp., *Assessment of Ocean Waste Disposal Case Studies,* contract report prepared for U.S. Congress, Office of Technology Assessment (McLean, VA: Oct. 3, 1985).

503. Science Applications International Corp., *Overview of Sewage Sludge and Effluent Management,* contract prepared for U.S. Congress, Office of Technology Assessment (McLean, VA: 1986).

504. SEAMOcean, Inc, "Preliminary Assessment of the Cost Implications of Multimedia Waste Management Decisions That Include an Ocean Option," final report for National Oceanic and Atmospheric Administration, Office of Oceanic and Atmospheric Research, NOAA Contract No. 40-AANR502485 (Wheaton, MD: February 1986).

505. Sea Technology, "A National Strategy for Monitoring and Research," *Sea Technology* 27(9):10-17, September 1986.

506. Segar, D.A., "Design of Monitoring Studies to Assess Waste Disposal Effects on Regional to Site Specific Scales," in *Public Waste Management and the Ocean Choice,* MITSG 85-36, K.D. Stolzenbach, et al. (eds.) (Cambridge, MA: MIT Sea Grant College Program, April 1986), pp. 189-206.

507. Segar, D.A., and Davis, P.G., *Contamination of Populated Estuaries and Adjacent Coastal Ocean —A Global View,* NOAA Technical Memorandum NOS OMA 11 (Rockville, MD: National Oceanic and Atmospheric Administration, National Ocean Service, 1984).

508. Segar, D.A., and Stamman, E., "Monitoring in Support of Estuarine Pollution Management Needs," *Oceans '86 Conference Record,* vol. 3, 86CH2363-0 (Washington, DC: Marine Technology Society and IEEE Ocean Engineering Society, Sept. 23-25, 1986), pp. 874-877.

509. Segar, D.A., Stamman, E., and Davis, P.G., "Beneficial Use of Municipal Sludge in the Ocean," *Marine Pollution Bulletin* 16(5):186-191, May 1985.

510. Sheifer, I.C., "San Francisco Bay Considered as a Recreational Resource: Indicators of Economic Value," in *Oceans '86 Conference Record,* vol.

2 (Washington, DC: Marine Technology Society and IEEE Ocean Engineering Society, Sept. 23-25, 1986), pp. 627-631.

511. Shields, D.F., Jr., and Montgomery, R.L., "Fundamentals of Capping Contaminated Dredged Material," in *Dredging and Dredged Material Disposal (Conference Proceedings)* (New York: American Society of Civil Engineers, November 1984), pp. 446-460.

512. Shigenaka, G., and Calder, J.A., *Organic Compounds and Metals in Selected Fish, Bivalve Molluscs, and Sediments of Chesapeake Bay: Preliminary Results From the National Status and Trends Program* (Rockville, MD: National Oceanic and Atmospheric Administration, National Status and Trends Program for Marine Environmental Quality, August 1986).

513. Shigenaka, G., and Leschine, T., "Puget Sound Water Quality Management: Setting the Agenda" (Seattle, WA: University of Washington Institute for Marine Studies, unpublished manuscript dated 1986).

514. Shrimp Notes Inc., *Assessment of Shrimp Industry Potentials and Conflicts, Volume Two, Report V: Potential Actions of Tariff and Quota Legislation* (New Orleans, LA: August 1983).

515. Sindermann, C.J., "Fish and Environmental Impacts," Proceedings of the Fourth Congress of European Ichthyologists, *Archiv für Fischereiwissenschaft* 35:125-160, 1984.

516. Sindermann, C.J., "Pollution-Associated Diseases of Marine Fish," in *Principal Diseases of Marine Fish and Shellfish* (New York: Academic Press, in press).

517. Sittig, M., *Handbook of Toxic and Hazardous Chemicals and Carcinogens* (Park Ridge, NJ: Noyes Press, 1985).

518. Sloan, R., et al., *PCB in Striped Bass From the Marine District of New York,* Technical Report 86-1 (BEP) (Albany, NY: New York State Department of Environmental Conservation, April 1986).

519. Sloan, R., Brown, M., Brandt, R., et al., "Hudson River PCB Relationships Between Resident Fish, Water, and Sediment," *Northeastern Environmental Science* 3:138-152, 1984.

520. Sludge Newsletter, "Slants and Trends," *Sludge Newsletter* 10(13):97, June 24, 1985.

521. Smith, S.V., "Environmental Status of Hawaiian Estuaries," in *Estuarine Pollution Control and Assessment: Proceedings of a Conference,* EPA 440/1-77-007 (Washington, DC: U.S. Environmental Protection Agency, Office of Water Planning Standards, 1977), pp. 297-306.

522. Snyder, C., et al., "Evidence for Hematotoxicity

and Tumorigenesis in Rats Exposed to 100 ppm Benzene,'' *American Journal of Industrial Medicine* 5:429-434, 1984.

523. Sobsey, M.D., Rullman, V.A., and Davis, A.L., ''Influence of Temperature, Salinity, and Availability of Food Supply on the Elimination of Hepatitis A Virus and Poliovirus Type 1 From the Eastern Oyster, *Crassostrea virginica*, Under Depuration Conditions,'' in *U.S./French Symposium on Discharge of Urban Wastes in Marine and Coastal Waters* (College Park, MD: University of Maryland Sea Grant Program, 1986).

524. Sommers, L.E., ''Chemical Composition of Sewage Sludges and Analysis of Their Potential Use as Fertilizers,'' *Journal of Environmental Quality* 6(2):225-232, 1977.

525. Sopper, W.E., Seaker, E.M., and Bastian, R.K., (eds.), *Land Reclamation and Biomass Production With Municipal Wastewater and Sludge* (University Park, PA: Pennsylvania State University Press, 1982).

526. Soule, D.F., ''Changes in a Harbor Ecosystem Following 'Improved' Waste Treatment,'' in *Marine Environmental Pollution*, vol. 2, R.A. Geyer (ed.) (Amsterdam: Elsevier, 1981), pp. 313-378.

527. Southern California Coastal Water Research Project, *The Effects of Ocean Disposal of Municipal Wastes* (El Segundo, CA: June, 1978).

528. Southern California Coastal Water Research Project, *Coastal Water Research Project Biennial Report, 1983-1984* (Long Beach, CA: 1984).

529. Spies, R.B. et al., *Pollutant Body Burdens and Reproduction in Platichthys stellatus From San Francisco Bay*, report prepared for National Oceanic and Atmospheric Administration Coastal and Estuarine Assessment Branch, UCID 19993-84 (Livermore, CA: Lawrence Livermore Laboratory, 1985).

530. Spies, R.B., et al., ''Reproductive Success, Xenobiotic Contaminants and Hepatic Mixed-Function Oxidase (MFO) Activity in *Platichthys stellatus* Populations From San Francisco Bay,'' *Marine Environmental Research* 17:117-121, 1985.

531. Spirer, J. ''The Ocean Dumping Deadline: Easing the Mandate Milestone,'' *Fordham Urban Law Journal* 11(1):1-49, 1982.

532. Squires, D.F., *The Ocean Dumping Quandary—Waste Disposal in the New York Bight* (Albany, NY: State University of New York Press, 1983).

533. Stanfield, R.L., ''Money Down the Drain?'' *National Journal*, June 2, 1984, p. 1097.

534. Stanley, D.E., *Nationwide Review of Oxygen Depletion and Eutrophication in Estuarine and Coastal Waters: Southeast Region* (Greenville, NC: Institute for Coastal and Marine Resources, 1985).

535. Stegeman, J.J., Kloepper-Sams, P.J., and Farrington, J.W., ''Monoxygenase Induction and Chlorobiphenyls in the Deep-Sea Fish *Coryphaenoides armatus*,'' *Science* 231:1287-1289, 1986.

536. Steidinger, K.A., and Haddad, K., ''Biologic and Hydrographic Aspects of Red Tides,'' *BioScience* 31(11):814-819, December 1981.

537. Steimle, F.W., et al., ''Organic and Trace Metal Levels in Ocean Quahog, Arctica Islandica Linne, From the Northwestern Atlantic,'' *Fishery Bulletin* 84(1):133-40, January 1986.

538. Stein, R., and Woods, J., ''Review of the Great Lakes Water Quality Agreement,'' *Environment* 28(6):25-27, July/August 1986.

539. Stull, J.K., et al., ''Long-Term Changes in the Benthic Community on the Coastal Shelf of Palos Verdes, Southern California,'' *Marine Biology* 91(4):539-553, 1986.

540. Suflita, J.M., et al., ''Kinetics of Microbial Dehalogenation of Haloaromatic Substrates in Methanogenic Environments,'' *Applied and Environmental Microbiology* 45(5):1466-1473, 1983.

541. Sullivan, K., Atlas, E., and Giam, C., ''Adsorption of Phthalic Acid Esters From Seawater,'' *Environmental Science and Technology* 16:428-433, 1982.

542. Summer, F., ''Sewage Treatment: When the Federal Government Pulls the Plug,'' *The Environmental Forum* 4:8-13, April 1986.

543. Sun, M., ''The Chesapeake Bay's Difficult Comeback,'' *Science* 233:715-717, Aug. 15, 1986.

544. Susani, L., *Liver Lesions in Feral Fish: A Discussion of Their Relationship to Environmental Pollutants*, NOAA Technical Memorandum NOS OMA 27 (Rockville, MD: National Oceanic and Atmospheric Administration, 1986).

545. Suszkowski, D.J., and Santoro, E.D., ''Marine Monitoring in the New York Bight,'' in *Oceans '86 Conference Record*, vol. 3 (Washington, DC: Marine Technology Society and IEEE Ocean Engineering Society, 1986), pp. 754-759.

546. Swanson, R.L., and Sindermann, C.J., *Oxygen Depletion and Associated Benthic Mortalities in New York Bight*, NOAA Professional Paper 11 (Rockville, MD: National Oceanic and Atmospheric Administration, 1976).

547. Swanson, R.L., et al., *Long Island Beach Pollution: June 1976* (Boulder, CO: National Oceanic and Atmospheric Administration, Environmental Research Laboratory, 1977).

548. Swanson, R.L., et al., ''Sewage Sludge Dumping in the New York Bight Apex: A Comparison With Other Proposed Dumpsites,'' in *Wastes in the Ocean*, vol. 6, I.W. Duedall, et al. (eds.) (New York: John Wiley & Sons, 1985), pp. 461-488.

549. Swartz, R.C., et al., "Sediment Toxicity to a Marine Infaunal Amphipod: Cadmium and Its Interaction With Sewage Sludge," *Marine Environmental Research* 18:133-153, 1985.

550. Takizawa, Y., "Epidemiology of Mercury Poisoning," in *The Biogeochemistry of Mercury in the Environment*, J. Nriagu (ed.) (New York: Elsevier, 1979), pp. 325-366.

551. Tarr, J.A., et al., "Water and Wastes: A Retrospective Assessment of Wastewater Technology in the United States, 1800-1932," *Technology and Culture* 25(2):226-263, 1984.

552. Taylor, M., "Trouble on the Pamlico," *Wildlife in North Carolina* 50(5):4-11, May 1986.

553. Temple, Barker, & Sloane, Inc., *Costs of Ocean Disposal of Municipal Sewage Sludge and Industrial Wastes*, prepared for U.S. Environmental Protection Agency, Office of Analysis and Evaluation (Lexington, MA: September 1982).

554. Texas Department of Water Resources, *The State of Texas Water Quality Inventory*, LP-59 (Austin, TX: 1984).

555. Thackston, E.L., Montgomery, R.L., and Palermo, M.R., "Settling of Dredged Material Slurries," in *Dredging and Dredged Material Disposal (Conference Proceedings)* (New York: American Society of Civil Engineers, November 1984), pp. 849-857.

556. Thom, N., and Agg, A., 1975. "The Breakdown of Synthetic Organic Compounds in Biological Processes," *Proceedings of the Royal Society of London, Series B* 189:347-357, 1975.

557. Tramontozzi, P., "Reforming Water Pollution Regulation," Formal Publication Number 69 (St. Louis, MO: Washington University Center for the Study of American Business, August 1985).

558. Tribe, L.H. "Ways Not To Think About Plastic Trees: New Foundations for Environmental Law," *Yale Law Journal* 83(7):1315-1319,1325-1327, 1329-1332,1341-1346, June 1974.

559. United Kingdom Department of the Environment, Radioactive Waste Division, *Assessment of Best Practicable Environment Options (BPEOs) for Management of Low- and Intermediate-Level Solid Radioactive Wastes* (London: Her Majesty's Stationery Office, March 1986).

560. U.S. Army Corps of Engineers, *1980 Report to Congress on Administration of Ocean Dumping Activities*, Pamphlet 82-P1 (Fort Belvoir, VA: Water Resources Support Center, May 1982).

561. U.S. Army Corps of Engineers, *Ocean Dumping Report for Calendar Year 1981*, Summary Report 82-SO2 (Fort Belvoir, VA: Water Resources Support Center, June 1982).

562. U.S. Army Corps of Engineers, *Dredged Material Disposal Alternatives: Port of New York and New Jersey*, NANP 1145-2-2 (New York: New York District, May 1983).

563. U.S. Army Corps of Engineers, *All Dredged Up and No Place to Go*, NANP 1145-2-3 (New York: New York District, July 1983).

564. U.S. Army Corps of Engineers, *Ocean Dumping Report for Calendar Year 1982*, Summary Report 83-SR1 (Fort Belvoir, VA: Water Resources Support Center, October 1983).

565. U.S. Army Corps of Engineers, *Beneficial Uses of Dredged Material*, Engineer Manual EM 1110-2-5026 (Washington, DC: Office of the Chief of Engineers, 1985).

566. U.S. Bureau of the Census, *Statistical Abstract of the United States: 1986 (106th ed.)* (Washington, DC: U.S. Government Printing Office, 1985).

567. U.S. Congress, *A Legislative History of the Clean Water Act of 1977, A Continuation of the Legislative History of the Federal Water Pollution Control Act*, 95th Cong., 2d sess., Senate Committee on Environment and Public Works, Committee Print 95-14 (Washington, DC: 1978).

568. U.S. Congress, Congressional Budget Office, *Hazardous Waste Management: Recent Changes and Policy Alternatives* (Washington, DC: U.S. Government Printing Office, May 1985).

569. U.S. Congress, Congressional Budget Office, *Efficient Investments in Wastewater Treatment Plants* (Washington, DC: U.S. Government Printing Office, June 1985).

570. U.S. Congress, Congressional Research Service, *Ocean Dumping: A Time to Reappraise?* Issue Brief No. IB81088 (Washington, DC: 1983).

571. U.S. Congress, Congressional Research Service, *Water Quality: Implementing the Clean Water Act*, update of Issue Brief IB83030 (Washington, DC: 1984).

572. U.S. Congress, Congressional Research Service, *Clean Water: Section 404 Dredge and Fill Permit Program*, Issue Brief IB83011 (Washington, DC: 1985).

573. U.S. Congress, Congressional Research Service, *Wastewater Treatment Programs: Impact of Gramm-Rudman-Hollings Act and Prospects for Federal Funding (With Appendix)*, Issue Brief IB86018 (Washington, DC: February 1986).

574. U.S. Congress, General Accounting Office, *More Effective Action by the Environmental Protection Agency Needed To Enforce Industrial Compliance With Water Pollution Control Discharge Permits* (Washington, DC: Oct. 17, 1978).

575. U.S. Congress, General Accounting Office, *Billions Could Be Saved Through Waivers for Coastal Wastewater Treatment Plants*, CED-81-68 (Washington, DC: May 1981).

576. U.S. Congress, General Accounting Office, *Wastewater Dischargers Are Not Complying With EPA Pollution Control Permits* (Washington, DC: Dec. 2, 1983).

577. U.S. Congress, House Committee on Energy and Commerce, *Health Implications of Toxic Chemical Contamination of the Santa Monica Bay*, Hearings Before the Subcommittee on Health and the Environment of the House Committee on Energy and Commerce, Serial No. 99-74 (Washington, DC: U.S. Government Printing Office, 1986).

578. U.S. Congress, House Committee on Merchant Marine and Fisheries, *Ocean Dumping Reauthorization (FY 1980) and Oversight, Ocean Dumping Deadline Oversight, Ocean Dumping Reauthorization (FY 1981) and Oversight*, Hearings Before the Subcommittee on Oceanography and the Subcommittee on Fisheries and Wildlife Conservation and the Environment of the Committee on Merchant Marine and Fisheries, 96th Cong., Serial No. 96-40 (Washington, DC: U.S. Government Printing Office, 1980).

579. U.S. Congress, House Committee on Merchant Marine and Fisheries, *Ocean Dumping Amendments Act of 1983*, 98th Cong., 1st sess., Report 98-200, Part 1 (Washington, DC: May 16, 1983).

580. U.S. Congress, House Committee on Merchant Marine and Fisheries, *Plastic Pollution in the Marine Environment*, Hearing Before the Subcommittee on Coast Guard and Navigation of the Committee on Merchant Marine and Fisheries, 99th Cong., 2d sess., Serial No. 99-47 (Washington, DC: U.S. Government Printing Office, Aug. 12, 1986).

581. U.S. Congress, House of Representatives, *Ocean Dumping Amendments Act of 1985*, 99th Cong., 1st sess., Report 99-107, Part 2 (Washington, DC: U.S. Government Printing Office, June 19, 1985).

582. U.S. Congress, Office of Technology Assessment, *Wetlands: Their Use and Regulation*, OTA-O-206 (Washington, DC: U.S. Government Printing Office, March 1984).

583. U.S. Congress, Office of Technology Assessment, *Superfund Strategy*, OTA-ITE-252 (Washington, DC: U.S. Government Printing Office, April 1985).

584. U.S. Congress, Office of Technology Assessment, "Workshop On the Disposal of Dredged Material in the Marine Environment" (Washington, DC: July 23, 1985).

585. U.S. Congress, Office of Technology Assessment, *Subseabed Disposal of High-Level Radioactive Waste*, staff paper (Washington, DC: May 1986).

586. U.S. Congress, Office of Technology Assessment, *Ocean Incineration, Its Role in Managing Hazardous Waste*, OTA-O-313 (Washington, DC: U.S. Government Printing Office, August 1986).

587. U.S. Congress, Office of Technology Assessment, *Serious Reduction of Hazardous Waste*, OTA-ITE-317 (Washington, DC: U.S. Government Printing Office, September 1986).

588. U.S. Congress, Office of Technology Assessment, *Ecological Consequences of Waste Disposal in Marine Environments*, staff paper (Springfield, VA: National Technical Information Service, 1987).

589. U.S. Congress, Senate Committee on Environment and Public Works, *Ocean Dumping*, Hearing Before the Subcommittee on Environmental Pollution of the Committee on Environment and Public Works, Senate Hearing 99-206 (Washington, DC: U.S. Government Printing Office, 1985).

590. U.S. Congress, Senate Committee on Governmental Affairs, *The Future Direction of the Chesapeake Bay Program*, Hearing Before the Senate Subcommittee on Governmental Efficiency and the District of Columbia of the Committee on Governmental Affairs, 99th Cong., 2d sess., Senate Hearing 99-950, June 24, 1986 (Washington, DC: U.S. Government Printing Office, 1986).

591. U.S. Department of Agriculture, *Utilization of Sewage Sludge Compost as a Soil Conditioner and Fertilizer for Plant Growth*, Agriculture Information Bulletin Number 464 (Washington, DC: Agriculture Research Service, 1984).

592. U.S. Department of Commerce, National Oceanic and Atmospheric Administration, National Marine Fisheries Service, Northeast Monitoring Program, *Annual NEMP Report on the Health of the Northeast Coastal Waters of the United States, 1980*, NOAA Technical Memorandum NMFS-F/NEC-10 (Washington, DC: U.S. Government Printing Office, 1981).

593. U.S. Department of Commerce, National Oceanic and Atmospheric Administration, *Gulf and Atlantic Survey for Selected Organic Pollutants in Finfish* (Washington, DC: NOAA, 1982).

594. U.S. Department of Commerce, National Oceanic and Atmospheric Administration, Ocean Assessments Division, *Assessment of Ocean Dumping North of Puerto Rico*, NOAA Technical Memorandum NOS 28 (Rockville, MD: 1983).

595. U.S. Department of Commerce, National Oceanic Atmospheric Administration, *Report to the Congress on Ocean Pollution, Monitoring and Research, October 1982 through September 1983* (Washington, DC: July 1984).

596. U.S. Department of Commerce, National Oceanic

and Atmospheric Administration, National Marine Pollution Program Office, *Federal Action Plan for Ocean Pollution, Research, Development, and Monitoring*, Draft (Rockville, MD: Dec. 7, 1984).

597. U.S. Department of Commerce, National Oceanic and Atmospheric Administration, National Marine Fisheries Service, Northeast Monitoring Program, *Annual NEMP Report on the Health of the Northeast Coastal Waters, 1982*, NOAA Technical Memorandum NMFS-F/NEC-35 (Washington, DC: U.S. Government Printing Office, January 1985).

598. U.S. Department of Commerce, National Oceanic and Atmospheric Administration, National Marine Pollution Program, *Catalog of Federal Projects, FY 1984 Update* (Rockville, MD: June 1985).

599. U.S. Department of Commerce, National Oceanic and Atmospheric Administration, National Marine Pollution Program, *Agency Program Summaries, FY 1984 Update* (Rockville, MD: July 1985).

600. U.S. Department of Commerce, National Oceanic and Atmospheric Administration, Ocean Assessments Division, *The National Coastal Pollutant Discharge Inventory* (Rockville, MD: July 1985).

601. U.S. Department of Commerce, National Oceanic and Atmospheric Administration, National Marine Pollution Program, *National Marine Pollution Program, Federal Plan For Ocean Pollution Research, Development, and Monitoring, Fiscal Years 1985-1989* (Washington, DC: September 1985).

602. U.S. Department of Commerce, National Oceanic and Atmospheric Administration, Ocean Assessments Division, *National Estuarine Inventory Data Atlas: Physical and Hydrologic Characteristics* (Rockville, MD: November 1985).

603. U.S. Department of Commerce, National Oceanic and Atmospheric Administration, and Department of Health and Human Services, Food and Drug Administration, *1985 National Shellfish Register of Classified Estuarine Waters* (Washington, DC: December 1985).

604. U.S. Department of Commerce, National Oceanic and Atmospheric Administration, Office of Sea Grant, *The National Sea Grant Program, Annual Retreat Report, FY 85* (Rockville, MD: December 1985).

605. U.S. Department of Commerce, National Oceanic and Atmospheric Administration, National Marine Fisheries Service, *Marine Recreational Fishery Statistics Survey, Atlantic and Gulf Coasts, 1983-1984*, Current Fishery Statistics No. 8326 (Washington, DC: 1985).

606. U.S. Department of Commerce, National Oceanic and Atmospheric Administration, National Marine Fisheries Service, *Marine Recreational Fishery Statistics Survey, Pacific Coast, 1983-1984*, Current Fishery Statistics No. 8325 (Washington, DC: 1985).

607. U.S. Department of Commerce, National Oceanic and Atmospheric Organization, Ocean Assessments Division, *Gulf of Mexico Coastal and Ocean Zones Strategic Assessment: Data Atlas* (Rockville, MD: 1985).

608. U.S. Department of Commerce, National Oceanic and Atmospheric Administration, Ocean Assessments Division, ''Tables From National Coastal Pollutant Discharge Inventory,'' tables prepared for U.S. Congress, Office of Technology Assessment (Rockville, MD: 1985-1986).

609. U.S. Department of Commerce, National Oceanic and Atmospheric Administration, National Marine Fisheries Service, *The Habitat Conservation Program of the National Marine Fisheries Service, 1984/1985* (Washington, DC: NOAA, January 1986).

610. U.S. Department of Commerce, National Oceanic and Atmospheric Administration, National Ocean Service, *Report to the Congress on Ocean Pollution, Monitoring, and Research, October 1984 Through September 1985* (Rockville, MD: May 1986).

611. U.S. Department of Commerce, National Oceanic and Atmospheric Administration, National Ocean Service, *Strategic Assessment Projects: Fact Sheets* (Rockville, MD: June 1986).

612. U.S. Department of Commerce, National Oceanic and Atmospheric Administration, Office of Oceanography and Marine Assessment, *A National Atlas, Health and Use of Coastal Waters, United States of America* (Rockville, MD: November 1986).

613. U.S. Department of Commerce, National Oceanic and Atmospheric Administration, National Marine Pollution Program, *National Marine Pollution Program, Summary of Federal Programs and Projects, FY 1985 Update* (Washington, DC: 1986).

614. U.S. Department of Commerce, National Oceanic and Atmospheric Administration, National Marine Fisheries Service, *Fisheries of the United States, 1985*, Current Fishery Statistics No. 8380 (Washington, DC: 1986).

615. U.S. Department of Commerce, National Oceanic and Atmospheric Administration, Ocean Assessments Division, *The National Status and Trends Program for Marine Environmental Quality, Briefing Guide* (Rockville, MD: 1986).

616. U.S. Department of Commerce, National Oceanic and Atmospheric Administration, Ocean Assessments Division, *The National Status and Trends Program for Marine Environmental Quality, FY86 Program Description* (Rockville, MD: 1986).

617. U.S. Department of Commerce, National Oceanic and Atmospheric Administration, Office of Sea Grant, *NOAA's National Sea Grant College Program* (Rockville, MD: 1986).

618. U.S. Department of Commerce, National Oceanic and Atmospheric Administration, *FY 1987 Budget Summary* (Washington, DC: 1986).

619. U.S. Department of Commerce, National Oceanic and Atmospheric Administration, Ocean Assessments Division, *National Estuarine Inventory, Living Marine Resources Component, West Coast* (Rockville, MD: 1986).

620. U.S. Department of Commerce, National Oceanic and Atmospheric Administration, Ocean Assessments Division, *The Consequences of Contaminants to Living Marine Resources and Human Health* (Rockville, MD: 1986).

621. U.S. Department of Commerce, National Oceanic and Atmospheric Administration, Ocean Assessments Division, *Progress Report and Preliminary Assessment of the Findings of the 1984 Benthic Surveillance Project* (Rockville, MD: 1987).

622. U.S. Department of Energy, Oak Ridge National Laboratory, *Integrated Data Base for 1986: Spent Fuel and Radioactive Waste Inventories, Projections, and Characteristics*, DOE/RW-0006, Rev.2 (Oak Ridge, TN: September 1986).

623. U.S. Department of the Interior, *The National Estuarine Pollution Study, Report of the Secretary of the Interior to the U.S. Congress Pursuant to Public Law 89-753, The Clean Water Restoration Act of 1966,* 91st Cong., 2d sess., Senate Document No. 91-58 (Washington, DC: U.S. Government Printing Office, 1970).

624. U.S. Department of the Interior, Fish and Wildlife Service, *The Canvasback* (Annapolis, MD: 1985).

625. U.S. Department of the Interior, Fish and Wildlife Service, *Striped Bass* (Annapolis, MD: 1986).

626. U.S. Department of the Interior, National Park Service, *1982-1983 Nationwide Recreation Survey* (Washington, DC: U.S. Government Printing Office, April 1986).

627. U.S. Department of the Interior, National Park Service, *National Park Statistical Abstract 1985* (Denver, CO: 1986).

628. U.S. Department of the Interior, Fish and Wildlife Service, and Department of Commerce, Bu-
reau of the Census, *1980 National Survey of Fishing, Hunting, and Wildlife-Associated Recreation* (Washington, DC: U.S. Government Printing Office, 1982).

629. U.S. Department of the Navy, *Final Environmental Impact Statement on the Disposal of Decommissioned, Defueled Naval Submarine Reactor Plants* (Washington, DC: May 1984).

630. U.S. Environmental Protection Agency and U.S. Army Corps of Engineers, *Ecological Evaluation of Proposed Discharge of Dredged Material into Ocean Waters* (Vicksburg, MS: U.S. Army Engineer Waterways Experiment Station, 1977).

631. U.S. Environmental Protection Agency, Office of Water Program Operations, *Report to Congress on Control of Combined Sewer Overflow in the United States*, EPA-430/9-78-006 (Washington, DC: October 1978).

632. U.S. Environmental Protection Agency, Region II, *Final Environmental Impact Statement on the Ocean Dumping of Sewage Sludge in the New York Bight* (New York: 1978).

633. U.S. Environmental Protection Agency, Center for Environmental Research Information, *Environmental Pollution Control Alternatives—Municipal Wastewater*, EPA-625/5-79-012 (Cincinnati, OH: November 1979).

634. U.S. Environmental Protection Agency, Office of Water Programs Operations, *Primer for Wastewater Treatment,* MCD-65 (Washington, DC: July 1980).

635. U.S. Environmental Protection Agency, Center for Environmental Research Information, *Process Design Manual for Land Treatment of Municipal Wastewater*, EPA 625/1-81-013 (Cincinnati, OH: October 1981).

636. U.S. Environmental Protection Agency, Region I, *Impacts of Wastewater Disinfection Practices on Coldwater Fisheries,* EPA 901-82-000 (Boston, MA: July 1982).

637. U.S. Environmental Protection Agency, Effluent Guidelines Division, *Fate of Priority Pollutants in Publicly Owned Treatment Works, Final Report, Volume I*, EPA 440/1-82/303 (Washington, DC: September 1982).

638. U.S. Environmental Protection Agency, Sludge Task Force, *Interim Report of the EPA Sludge Task Force for the Sludge Policy Committee* (Washington, DC: August 1983).

639. U.S. Environmental Protection Agency, Center for Environmental Research Information, *Process Design Manual for Land Application of Municipal Sludge*, EPA-625/1-83-016 (Cincinnati, OH: October 1983).

640. U.S. Environmental Protection Agency, Region III, Chesapeake Bay Program, *Chesapeake Bay: A Profile of Environmental Change*, PB 84119197 (Philadelphia, PA: 1983).

641. U.S. Environmental Protection Agency, Office of Research and Development, *Health Effects Criteria for Marine Recreational Waters* (Washington, DC: August 1983).

642. U.S. Environmental Protection Agency, *Annual Report to Congress on Administration of the Marine Protection, Research and Sanctuaries Act of 1972*, draft (Washington, DC: 1983).

643. U.S. Environmental Protection Agency, Sludge Task Force, *Workplan for Development of an Integrated Sludge Management Regulatory Program* (Washington, DC: February 1984).

644. U.S. Environmental Protection Agency, Office of Water Regulations and Standards, Effluent Guidelines Division, *Paragraph 4(c) Program Summary Report* (Washington, DC: March 1984).

645. U.S. Environmental Protection Agency, *Final Development Document for Effluent Limitations and Standards for the Copper Forming Point Source Category*, EPA 440/1-84/074, Table X-19 (Washington, DC: March, 1984).

646. U.S. Environmental Protection Agency, Public Affairs Office, "EPA Designates Disposal Sites 106 Miles Off Atlantic Coast," *Environmental News*, Apr. 26, 1984.

647. U.S. Environmental Protection Agency, Office of Water, *Combined Sewer Overflow Toxic Pollutant Study*, EPA 440/1-84/304 (Washington, DC: April 1984).

648. U.S. Environmental Protection Agency, Office of Water Regulations and Standards, *Report to Congress, January 1981-December 1983, On Administration of the Marine Protection, Research, and Sanctuaries Act of 1972, As Amended (P.L. 92-532) and Implementing the International London Dumping Convention* (Washington, DC: June 1984).

649. U.S. Environmental Protection Agency, Office of Water Program Operations, *Report on the Implementation of Section 301(h)*, EPA 430/9-84-007 (Washington, DC: August 1984).

650. U.S. Environmental Protection Agency, Intra-Agency Sludge Task Force, *Use and Disposal of Municipal Sludge*, EPA 625/10-84-003 (Washington, DC: September 1984).

651. U.S. Environmental Protection Agency, "Additional Information to Supplement the GAO Report on NPDES Noncompliance in the Lower Mississippi River Basin," in *Implementation of the Federal Clean Water Act (EPA Enforcement of the National Pollution Discharge Elimination System Permit Program)*, Hearings Before the Subcommittee on Investigations and Oversight of the Committee on Public Works and Transportation, House of Representatives, 98th Cong., 2d sess. (Washington, DC: Mar. 7-8, Sept. 19, 1984).

652. U.S. Environmental Protection Agency, *Risk Assessment and Management: Framework for Decision Making*, EPA 600/19-85-002 (Washington, DC: December 1984).

653. U.S. Environmental Protection Agency, Office of Water Enforcement and Permits, *Pretreatment Implementation Review Task Force: Final Report to the Adminstrator* (Washington, DC: Jan. 30, 1985).

654. U.S. Environmental Protection Agency, Office of Municipal Pollution Control, *Assessment of Needed Publicly Owned Wastewater Treatment Facilities in the United States*, EPA 430/9-84-011 (Washington, DC: February 1985).

655. U.S. Environmental Protection Agency, Office of Policy, Planning, and Evaluation, *Action Needed to Support State-EPA Implementation of the National Municipal Policy, Final Report on a Study by the Program Evaluation Division, Office of Policy, Planning, and Evaluation* (Washington, DC: June 1985).

656. U.S. Environmental Protection Agency, Office of Water Regulations and Standards, *Summary of Environmental Profiles and Hazard Indices for Constituents of Municipal Sludge: Methods and Results* (Washington, DC: July 1985).

657. U.S. Enviromental Protection Agency, Office of Policy Analysis, *The Effluent Charge System in the Federal Republic of Germany*, EPA 230-07-85-011 (Washington, DC: July 1985).

658. U.S. Environmental Protection Agency, Office of Water Regulations and Standards, *National Water Quality Inventory, 1984 Report to Congress*, EPA 440/4-85-029 (Washington, DC: August 1985).

659. U.S. Environmental Protection Agency, Office of Water, *Technical Support Document for Water Quality-Based Toxics Control* (Washington, DC: September 1985).

660. U.S. Environmental Protection Agency, Center for Environmental Research Information, *Municipal Wastewater Sludge Combustion Technology*, EPA/625-4-85/015 (Cincinnati, OH: September 1985).

661. U.S. Environmental Protection Agency, Office of Research and Development, *Handbook—Estimating Sludge Management Costs*, EPA/625/6-85/010 (Washington, DC: October 1985).

662. U.S. Environmental Protection Agency, Office of

Water Regulations and Standards and Office of Research and Development, *Pathogen Risk Assessment Feasibility Study* (Washington, DC: November 1985).

663. U.S. Environmental Protection Agency, Great Lakes National Program Office, *Five-Year Program Strategy for Great Lakes National Program Office, 1986-1990* (Chicago, IL: 1985).

664. U.S. Environmental Protection Agency, Office of Research and Development and Office of Water Regulations and Standards, *Bacteriological Ambient Water Quality Criteria for Marine and Freshwater Recreational Waters*, EPA-440/5-84-002 (Washington, DC: January 1986).

665. U.S. Environmental Protection Agency, Office of Water Regulations and Standards, Monitoring and Data Support Division, *Summary of Effluent Characteristics and Guidelines for Selected Industrial Point Source Categories: Industry Status Sheets* (Washington, DC: Feb. 28, 1986).

666. U.S. Environmental Protection Agency, Office of Water Regulations and Standards, *Report to Congress on the Discharge of Wastes to Publicly Owned Treatment Works*, EPA 530/SW-86-004 (Washington, DC: February 1986).

667. U.S. Environmental Protection Agency, "Hearing Officer's Report on the Tentative Determination to Issue the Incineration-At-Sea Research Permit HQ-85-001" (Washington, DC: May 1, 1986).

668. U.S. Environmental Protection Agency, Office of Water Regulations and Standards, *Quality Criteria for Water, 1986*, EPA 440/5-86-001 (and Update No. 1, Sept. 2, 1986) (Washington, DC: May 1986).

669. U.S. Environmental Protection Agency, Office of Water, "State Breakout of NMP Majors Construction Required, Status at End 3rd Quarter FY1986" (Washington, DC: data as of July 1, 1986).

670. U.S. Environmental Protection Agency, Office of Marine and Estuarine Protection, "Near-Coastal Waters Strategic Options Paper" (Washington, DC: Aug. 12, 1986).

671. U.S. Environmental Protection Agency, Office of Enforcement and Compliance Monitoring, *1986 Update to Clean Water Act Civil Penalty Analysis* (Washington, DC: Dec. 12, 1986).

672. U.S. Environmental Protection Agency, "Summary of EPA Enforcement Activity for 1980-1986," press release (Washington, DC: Dec. 16, 1986).

673. U.S. Environmental Protection Agency, Office of Water, *The Enforcement Management System—National Pollutant Discharge Elimination System* (Washington, DC: 1986).

674. U.S. Environmental Protection Agency Region V, Illinois Environmental Protection Agency, Indiana Department of Environmental Management, Michigan Department of Natural Resources, and Wisconsin Department of Natural Resources, *Lake Michigan Toxic Pollutant Control/Reduction Strategy* (Chicago, IL: July 1986).

675. U.S. Environmental Protection Agency, Science Advisory Board, *Review of EPA Water Quality Based Approach Research Program*, SAB-EC-87-011 (Washington, DC: Dec. 11, 1986).

676. U.S. Environmental Protection Agency, Office of Municipal Pollution Control, *1986 Needs Survey Report to Congress, Assessment of Needed Publicly Owned Wastewater Treatment Facilities in the United States*, EPA 430/9-87-001 (Washington, DC: February 1987).

677. U.S. Food and Drug Administration, *Evaluation Report: Phthalate Esters in Foods Survey* (Washington, DC: 1974).

678. U.S. Food and Drug Administration, Industry Programs Branch, Center for Food Safety and Applied Nutrition, *Compliance Policy Guides Manual* (Washington, DC: October 1986).

679. University of Rhode Island Sea Grant Marine Advisory Service, "Sea Grant Grandchild," *Marine Resources Information* 14D:1-2, July-August 1985.

680. Utility Solid Wastes Activities Group and the National Rural Electric Cooperative Association, *Report and Technical Studies on the Disposal and Utilization of Fossil Fuel Combustion By-Products*, prepared for the U.S. Environmental Protection Agency (Washington, DC: Edison Electric Institute, 1982).

681. Virginia Water Control Board, *Water Quality Inventory, 305(b) Report, Virginia*, vol. 2, chs. 4 through 12, Information Bulletin 558 (Richmond, VA: 1984).

682. Walker, A., et al., "The Toxicology and Pharmacodynamics of Dieldrin (HEOD): Two-Year Oral Exposures of Rats and Dogs," *Toxicology and Applied Pharmacology* 15:345-373, 1969.

683. Walsh, J., "Delaware Bay on the Rebound," *Science* 223:1375, Mar. 30, 1984.

684. Ward, B., and Harris, C., "The 1984 Hazardous and Solid Waste Amendments: A Bold Experiment in Hazardous Waste Management," *Journal of the Air Pollution Control Association* 35(3): 254-258, March 1985.

685. Water and Wastewater Equipment Manufacturers Association, Inc., Testimony Before the Subcommittee on Water Resources, House Public Works and Transportation Committee, "Hearings on Possible Amendments to the Federal Water Pol-

lution Control Act,'' 99th Cong., Committee Serial 99-9 (Washington, DC: U.S. Government Printing Office, 1985).

686. Watling, L., et al., ''Evaluation of Sludge Dumping Off Delaware Bay,'' *Marine Pollution Bulletin* 5:39-42, 1974.

687. Weaver, G., ''PCB Contamination In and Around New Bedford, Mass.,'' *Environmental Science and Technology* 18(1):22A-27A, 1984.

688. Welsh, B.L., *Long Island Sound, The Pelagic Ecosystem*, paper presented at Estuary of the Month Seminar, sponsored by National Oceanic and Atmospheric Administration Estuarine Programs Office and U.S. Environmental Protection Agency, Washington, DC, May 10, 1985.

689. Wessinger, H.J., Testimony Before the Subcommittee on Investigations and Oversight, House Committee on Public Works and Transportation, U.S. Congress, ''Hearings on the Implementation of the Federal Clean Water Act,'' Sept. 19, 1984, 98th Cong., Committee Serial 98-81 (Washington, DC: U.S. Government Printing Office, 1985).

690. Westat, Inc., *National Survey of Hazardous Waste Generators and Treatment, Storage and Disposal Facilities Regulated under RCRA in 1981*, contract prepared for U.S. Environmental Protection Agency, Office of Solid Waste (Rockville, MD: Apr. 20, 1984).

691. Whipple, J.A., *The Impact of Estuarine Degradation and Chronic Pollution on Populations of Anadromous Striped Bass [Morone saxatilis] in the San Francisco Bay Delta, California, A Summary for Managers and Regulators*, Administrative Report T-84-01 (Tiburon, CA: Tiburon Fisheries Laboratory, 1984).

692. White, H.H. (ed), *Concepts in Marine Pollution Measurements* (College Park, MD: University of Maryland Sea Grant, 1984).

693. White, H.H., and Champ, M.A., ''The Great Bioassay Hoax, and Alternatives,'' in *Hazardous and Industrial Solid Waste Testing: Second Symposium*, R.A. Conway and W.P. Gulledge (eds.), ASTM STP 805 (Philadelphia: American Society for Testing and Materials, 1983), pp. 288-312.

694. Whitledge, T.E., *Nationwide Review of Oxygen Depletion and Eutrophication in Estuarine and Coastal Waters: Northeast Region*, BNL 36580 (Upton, NY: Brookhaven National Laboratory, Department Applied Science, March 1985).

695. Whitledge, T.E., *Nationwide Review of Oxygen Depletion and Eutrophication in Estuarine and Coastal Waters*, BNL 37144 (Upton, NY: Brookhaven National Laboratory, Department Applied Science, September 1985).

696. Wiedow, M., *Distribution and Binding of Cadmium in the Blue Crab (Callinectes sapidus): Implications in Human Health*, Ph.D. Thesis (New York: New York University, 1981).

697. Wiedow, M., Kneip, T., and Garte, S., ''Cadmium Binding Proteins From Blue Crabs, *Callinectes sapidus*, Environmentally Exposed to Cadmium,'' *Environmental Resesarch* 28:164-170, 1982.

698. Wiemeyer, S.N., Swineford, D.M., and Spitzer, P.R., ''Organochlorine Residues in New Jersey Osprey Eggs,'' *Bulletin of Environmental Contamination and Toxicology* 19:56-63, 1978.

699. Wilkins, L.P., ''The Implementation of Water Pollution Control Measures—Section 208 of the Water Pollution Control Act Amendments,'' *Land and Water Law Review* 15(2):479-502, 1980.

700. Wilson, R., Lieb, S., et al., ''Non-O Group 1 *Vibrio cholerae* Gastroenteritis Associated With Eating Raw Oysters,'' *American Journal of Epidemiology* 114:293, 1981.

701. Windom, H.L., ''Environmental Aspects of Dredging in the Coastal Zone,'' *CRC Critical Reviews in Environal Control* 7:91-109, March 1976.

702. Windom, H., and Kendall, D., ''Accumulation and Biotransformation of Mercury in Coastal and Marine Biota,'' in *The Biogeochemistry of Mercury in the Environment*, J. Nriagu (ed.) (New York: Elsevier, 1979), pp. 303-324.

703. Windsor, J.G., Jr., *Nationwide Review of Oxygen Depletion and Eutrophication in Estuarine and Coastal Waters: Florida Region*, report to Brookhaven National Laboratory and the National Oceanic and Atmospheric Administration (Melbourne, FL: Florida Institute of Technology, 1985).

704. Wisconsin Department of Natural Resources, *An Analysis of Wisconsin's Policy on the Disinfection of Wastewater* (Madison, WI: Bureau of Water Quality, July 1977).

705. Wise, J.P., ''Fisheries Statistics—Boring Stuff Until You Need Them for Monitoring,'' in *Oceans '86 Conference Record*, vol. 3, 86CH2363-0 (Washington, DC: Marine Technology Society and IEEE Ocean Engineering Society, 1986), pp. 904-907.

706. Wood, J.D., Jr. (ed) *Proceedings 1985 National Outdoor Recreation Trends Symposium II, Volume II-Concurrent Sessions* (Atlanta, GA: National Park Service Science Publications Office, 1985).

707. Woolson, E., ''Emissions, Cycling and Effects of Arsenic in Soil Ecosystems,'' in *Biological and Environmental Effects of Arsenic*, B. Fowler (ed.) (New York: Elsevier, 1983), pp. 51-140.

708. Wright, T.D., *Aquatic Dredged Material Disposal*

Impacts, Technical Report DS-78-1 (Vicksburg, MS: U.S. Army Corps of Engineers Waterways Experiment Station, 1978).

709. Young, R.A., et al., "Dispersal Pathways for Particle-Associated Pollutants," *Science* 229:431-435, 1985.

710. Zafiriou, D.C., et al., "Photochemistry of Natural Waters," *Environmental Science and Technology* 18(12):358-371, 1984.

711. Zdanowicz, V.S., Gadbois, D.F., and Newman, M.W., "Levels of Organic Contaminants in Sediments and Fish Tissues and Prevalences of Pathological Disorders in Winter Flounder From Estu-

aries of the Northeast United States, 1984," in *Oceans '86 Conference Record*, vol. 2, 86CH2363-0 (Washington, DC: Marine Technology Society and IEEE Ocean Engineering Society, 1986), pp. 578-585.

712. Zeppetello, M.A., "National and International Regulation of Ocean Dumping: The Mandate to Terminate Marine Disposal of Contaminated Sewage Sludge," *Ecology Law Quarterly* 12:619, 1985.

713. Ziskowski, J.J. et al., *Disease in Economically Important Fish Stocks in the Northwest Atlantic*, draft (Highlands, NJ: Sandy Hook Laboratory, 1986).

Index

internal regulation by organisms: 94
pathways to and impacts on humans: 129-133
Oslo Convention: 149
Outer Continental Shelf Lands Act: 258,260
oxygen-demanding substances: 90
biochemical oxygen demand: 6,62,90,211,213
hypoxia and anoxia: 16,17,20,22,90,100,101,107,115,
116,118,119,223

Pacific coast
California and Hawaii: 13,61,62,64,66,72,109,114,
115,199
northern: 62,64,109,113,114
pathogens: 91,134,263
bacteria: 133-140
contamination of shellfish: *see* shellfish
depuration: 139,140
disinfection: *see* disinfection
genetic engineering: 13,222
monitoring: 60,91,137
parasites: 134
pathways to and impacts on humans: 16,135-138
shortcomings of current standards: 138
viability on land: 224
viability in marine waters: 91,135,138,139
viruses: 133-140
penalties: *see* NPDES
Permit Compliance System: *see* NPDES
permits: *see* NPDES; Marine Protection, Research, and
Sanctuaries Act
pharmaceutical waste: *see* industrial wastes
philosophical perspectives: 39,40,44-46
balancing factors: 45,143
comparing CWA v. MPRSA: 46,47,143,177
global view: 45
plastics: 76,77
point sources: *see* industrial dischargers, municipal dischargers
pollutant control programs: *see* NPDES, National
Pretreatment Program
pollutants (also *see* individual pollutants)
conventional: 6,59,63,152,179
expected reductions: *see* industrial dischargers
hazardous: *see* RCRA
inputs, quality of information: 57,58
linking with impacts: 16,90-94,99
nonconventional: 6,59,63,152,179
pathways to humans: *see* human health impacts
persistent v. labile: 123,127
priority pollutants, list: 6,7,58,152,179,186-189,194,
206,212
regulated v. unregulated: 7,8,179
relative contribution by major sources: 15,62-64,193
toxic: 6,63,152,177,179,212,224 (also *see* metals, organic chemicals)
unregulated pollutants: 26,28,58,156,188,193,194,227
polychlorinated biphenyls (PCBs): 8,15,20,39,43,44,65,90,
93,94,96,103,110,111,113,115,116,119,123,124,
129-132,136,210,215,225,242,245

polycyclic aromatic hydrocarbons (PAHs): 60,94,102,103,
123,129-131,133,136
Port and Tanker Safety Act: 145
productivity of marine waters: 34,81,82,84,86
public concerns and issues: 3,39,40,49-53
equity: 40,49,50
liability: 49,51
NIMBY: 50
public participation: 40,50,51,159,164
risk acceptability: 9,40,50,51,53
siting: 49,50
Public Health Advisories: 131
publicly owned treatment works (POTWs): *see* municipal
sewage treatment
Puget Sound: 10,19,32,70,86,102,103,112-114,129,131,
136,154,163,194,196,224
Puget Sound Water Quality Authority: 19,29,155,157,
160,162-164,169,170,195,252
pycnoclines: 82

radioactive waste
high-level radioactive waste: 3,74
low-level radioactive waste: 6,68-70,145
land-based disposal: 68
marine disposal: 68,69,74
moratorium: 33,68,69,74
State compacts: 70
RCRA (Resource Conservation and Recovery Act): 10,
28,47,48,146,147,204,210,212,227,242,248,260,262,
264
Domestic Sewage Exemption: 28,147,204,210,212,227
extraction procedure (EP) toxicity test: 242
Hazardous and Solid Wastes Amendments: 147
small quantity generators: 147,204
toxic characteristic leachate procedure (TCLP): 242
Resources for the Future: 16,57,58
reversal and recovery: 16,17,95
accommodative capacity: 95,96
assimilative capacity: 95

Safe Drinking Water Act: 145
San Francisco Bay: 4,19,20,70,85,108,112,115,125,131,
136,154,156,163
San Francisco Bay Regional Water Quality Control
Board: 20,160,163,169,170
Santa Monica Bay: 110,131
Science Advisory Board: 30,228
screening
pollutants: 59
waterbodies: 32
seafood processing wastes: *see* industrial wastes
Sea Grant Program: 166
sediment quality criteria: 31,242,248
sediments
contamination with toxic material: 17,26,103,104
physical modification: 16,103
sedimentation, accumulation of particles: *see* fate of
pollutants
sewage effluent (also *see* municipal sewage treatment)